THE CAVENDISH LABORATORY
1874–1974

The old and the new Cavendish Laboratories

THE CAVENDISH LABORATORY 1874-1974

J. G. CROWTHER

SCIENCE HISTORY PUBLICATIONS

NEW YORK

First published in the United States by
Science History Publications
a division of
Neale Watson Academic Publications, Inc.
156 Fifth Avenue, New York 10010

Library of Congress Cataloging in Publication Data

Crowther, James Gerald, 1899-

The Cavendish Laboratory, 1874-1974.

1. Cambridge. University. Cavendish Laboratory. I. Title.
QC51.G72C353   530'.07'2   74-13623
ISBN 0-88202-029-3

'... the opinion seems to have got abroad, that in a few years all the great physical constants will have been approximately estimated, and that the only occupation which will then be left to men of science will be to carry on these measurements to another place of decimals... But we have no right to think thus of the unsearchable riches of creation, or of the untried fertility of those fresh minds into which these riches will be poured.'

James Clerk Maxwell

Introductory Lecture as first Cavendish Professor of Experimental Physics, 25 October 1871

Contents

List of Illustrations

Nobel Laureates

Nobel Laureates who carried out fundamental researches, or started on their line of research, in the Cavendish Laboratory; together with the year in which they received their award.

Rayleigh 1904
J. J. Thomson 1906
E. Rutherford 1908
W. H. Bragg 1915
W. L. Bragg 1915
C. G. Barkla 1917
F. W. Aston 1922
C. T. R. Wilson 1927
O. W. Richardson 1928
J. Chadwick 1935
G. P. Thomson 1937

E. V. Appleton 1947
P. M. S. Blackett 1948
C. F. Powell 1950
J. D. Cockcroft 1951
E. T. S. Walton 1951
M. F. Perutz 1962
J. C. Kendrew 1962
F. H. C. Crick 1962
J. D. Watson 1962
M. H. F. Wilkins 1962
B. W. Josephson 1973

Preface

This book, the aim of which is to present the Cavendish Laboratory as a feature in British civilisation and world science, is intended for the general as well as the scientific reader.

It contains an account of how the Laboratory came into existence, the Cavendish Professors and the main contributions which they and their pupils have made to science. Twenty-two Nobel Prize Awards have been made to Cavendish men who accomplished or started their fundamental researches, or lines of research, in ᵥthe Laboratory. Besides these, other Nobel prizemen acquired the Cavendish outlook and technique in the Laboratory during their student days, which influenced them in making discoveries of great importance elsewhere. No other physics laboratory in the world has a comparable record. The identification of some of the factors that have contributed to the Laboratory's achievement has been attempted.

The original Laboratory was opened on 16 June 1874. The new Laboratory, erected on an extensive site a little more than a mile from the centre of Cambridge, consists of a series of buildings, the first three of which have been named respectively after Mott, Bragg and Rutherford.

At the date of the centenary of the opening of the Laboratory on 16 June 1974, a large part of the new buildings was already in use.

The new Laboratory is of industrial proportions. The Mott Building alone has 187 rooms, and there are elaborate services and scientific equipment. On a normal working day in term time about 700 persons are active in research, teaching and ancillary duties in the Cavendish Laboratory organisation.

The evolution of the Laboratory from a small building in the heart of Cambridge costing £10000 and staffed by a professor and a handful of colleagues, into a complex of industrial style costing more than £3 million, built on the outskirts of Cambridge and staffed with six professors, more than sixty other university officers and more than two hundred research workers, is outlined.

The change in character of the Laboratory, in response to University needs, the individual genius of professors, and the influence of social factors on the subjects and direction of research, has been kept in view.

The book is concluded with some reflections on the nature and significance of the Laboratory's achievement.

It is hoped that this glance at the first century of the Laboratory's life may reinforce the determination of the British people, as well as of the University of Cambridge and the physicists, to cherish this institution as one of the chief British contributions to science and civilisation.

J. G. Crowther
May 1974

Acknowledgments

Professor Brian Pippard and his colleagues in the Cavendish Laboratory have given courteous and generous help with information in the preparation of this book. It has been most gratefully appreciated.

I am deeply indebted to Cavendish scientists and others who have talked to me about their experiences and thoughts concerning the Laboratory, specially for this book. They are:

Professor P. W. Anderson, Mr W. H. Andrews, Mr P. S. Bixley, Lord Blackett, Sir James Chadwick, Dr V. E. Cosslett, Dr F. H. C. Crick, Mr J. Deakin, Professor P. I. Dee, Mr E. H. K. Dibden, Professor N. Feather, Professor O. R. Frisch, Professor A. Hewish, Professor P. B. Hirsch, Dr J. C. Kendrew, Dr W. B. Lewis, Sir Harrie Massey, Sir Nevill Mott, Dr A. D. I. Nicol, Mr J. R. Payne, Dr M. F. Perutz, Professor A. B. Pippard, Mr J. A. Ratcliffe, Sir Martin Ryle, Dr D. Shoenberg, Sir Geoffrey Taylor, Dr W. H. Taylor, Sir George Thomson, Professor D. H. Wilkinson, Dr Nora Wooster and Dr W. A. Wooster.

Alice, Lady Bragg and Mr Stephen Bragg very kindly gave me permission to read papers of the late Sir Lawrence Bragg, concerning the Cavendish Laboratory.

I also owe very much to the many Cavendish men who have talked to me during visits to the Laboratory during the period 1917–73, that is, during more than half of the Laboratory's first hundred years. Personal acquaintance with five of the first seven Cavendish Professors has been a privilege and a great help.

The responsibility for all opinions expressed in the book is mine.

The main books and papers that have been drawn upon are given in the references at the end of the book.

I am most grateful to the Cavendish Laboratory for permission to reproduce illustrations. I am particularly indebted to Mr W. H. Andrews for assistance in the choice and preparation of these illustrations.

Dr. M. F. Perutz kindly supplied prints of the haemoglobin, myoglobin and DNA molecules.

The print of the Seventh Duke of Devonshire was supplied by the Radio Times Hulton Picture Library. The floor plans of the original Cavendish Laboratory are taken from *Nature*. The portraits of Professor Pippard and Sir Martin Ryle are by Edward Leigh, who also provided the prints of the two groups of 1973; of Sir Lawrence Bragg by Bassano; and Sir Nevill Mott by Godfrey Argent. The two plans of the New Cavendish are based on drawings by Robert Matthew, Johnson-Marshall and Partners. The picture of the 5-km radio telescope is by British Insulated Callender's Cables.

I am indebted to Messrs Edward Arnold for permission to quote from *John William Strutt: Third Baron Rayleigh* by Robert John Strutt, Fourth Baron Rayleigh; to the Cambridge University Press to quote from *The Life of J. J. Thomson* by Rayleigh and *Rutherford* by A. S. Eve; to Sir Mark Oliphant and the Elsevier Publishing Company to quote from *Rutherford; Recollections of the Cambridge Days* by Sir Mark Oliphant; and to Messrs Longman for permission to quote from *The Cavendish Laboratory 1871–1910*.

Introduction

The creation of the Cavendish Laboratory arose out of a complex situation, in which there were several major factors. Among these were the existing science tradition in Cambridge, the academic politics involved in a change of its direction and the ideas and personalities of two of the chief people involved: James Clerk Maxwell and the Seventh Duke of Devonshire.

One of the first questions that arises is: why was there no University physics laboratory? It seems all the more strange because Cambridge had enjoyed the inspiration of Isaac Newton's discoveries and prestige for more than a hundred years. The answer lies deep in the history of Cambridge, in the study of science and the way in which it developed in response to social interests.

The University was founded in the Middle Ages to train men for the Church and the administration of the State. It existed for centuries before the development of science in Britain began to acquire any specifically British features. The change began with the discovery of America, which transferred the centre of the Western World from the Mediterranean to the Atlantic Ocean. This placed Britain, hitherto on the fringe of Western civilisation, at its centre. She took advantage of her position, manifested by achievements in government, exploration, naval warfare, literature and science.

The rapid expansion of sea power depended on navigation. This technique was improved from many different directions by invention and research. It provided the subject for the first characteristically British development of physical science. A new scientific instrument industry arose, especially in London, in which skilled craftsmen, or 'practitioners' as they were called, made astronomical instruments for observations at sea, magnetic compasses, rulers for measuring maps and other devices. Some of the practitioners experimented with these instruments in order to improve them, and to make new observations of natural phenomena.

A parallel theoretical and mathematical development started with the improvement of methods of calculating, and the invention of logarithms. It led to the appointment of the first professor of mathematics in England, Henry Briggs, in 1596. This was at Gresham College, which had been founded specifically to teach the geometry and astronomy that navigators and traders now required, and the law and arts needed to make them competent and cultivated citizens.

The aims of the Gresham professors were brilliantly expressed by Christopher Wren, when at the age of twenty-five in 1657 he was appointed Professor of Astronomy. In his inaugural lecture he explained why astronomy, and in particular *mathematical* astronomy, was so important.

The City of London was an unsurpassed centre of wealth, power, mechanical arts and trade. The goods of the world were brought to the City, and the City's goods went out to the ends of the earth. 'And now since navigation brings with it both wealth and splendour, politeness and learning, what greater happiness can I wish to the Londoners? Than that they may continually deserve to be deemed as formerly the great navigators of the world; that they always may be, what the Tyrians first, and then the Rhodeans were called, '*The Masters of the Sea*'; and that London may be an Alexandria, the established residence of mathematical arts.'

Wren indicated that the primary scientific problem of the age was the discovery of an exact theory of the universe. This would provide, among other things, a precise technical basis for world navigation. He said that the solution was immanent in Kepler's laws, and he forecast that their mathematical explanation was 'justly to be expected from men of our own nation at this day living, and known to this auditory.'

Newton was then fourteen years old. He was more absorbed at that time in learning Latin and fighting his school-fellows than in probing the secrets of the universe. Wren was probably thinking of his own school-fellow and distant relative, Robert Hooke.

Newton's contribution to mathematics, astronomy and physics was intellectually an extension of the Gresham school of Wren, Hooke and Barrow. He conferred his own immense reputation on Cambridge science, but he did not fundamentally modify the organisation of science at Cambridge. The University remained primarily an institution for educating prospective clergymen. Newton himself had originally gone there for that purpose. Newton's works were duly esteemed and taught, but the University was not reorganised or reconstructed to adapt it to extend the science he had so greatly advanced. In effect, the ancient medieval and monastic university added the *Principia* to the list of sacred works, and for the next one hundred and fifty years taught the *Principia* rather as the theologians taught the Bible. Newtonian science tended to become scholastic and static.

Newton's views on how experimental science should be prosecuted in Cambridge may be illustrated by the deed with which Thomas Plume's chair of mathematics was founded in 1704. It was drawn up with the advice of Newton among others. The Plumian professor was to provide a residence, an observatory and an assistant, as well as the necessary instruments. He was to carry out observations bearing on solar, lunar and planetary theories. He was to give courses at his residence on astronomy, optics, trigonometry, statics, hydrostatics, magnetics and pneumatics. His pupils were to pay for the experiments, and provide the professor with an adequate reward.

The range of subjects agreed closely with those pursued by Newton. It seems clear that Newton regarded experimental research as a personal matter, to be conducted by the scientist in the manner of a master craftsman. He worked in his own rooms, and added to his living by selling instruction in experimental physics to the appropriate purchasers: the students.

There was no suggestion of the University providing rooms and equipment, that is, a laboratory, for experimental physics. There appears to be no evidence that Newton ever conceived the idea of a university laboratory of experimental physics. His conception of how experimental physics should be pursued was pre-industrial.

Cambridge could not find in Newton an inspiration for the creation of a University laboratory of experimental physics. This did not arise until deep social changes altered both the direction and the method of physical research and teaching.

The Industrial Revolution focused attention on the properties of materials and their transformation, and on heat and power. A demand from outside the universities arose for a new kind of science, elucidating the scientific principles underlying these phenomena.

During the eighteenth century the creation of machine and power industry posed a new series of questions to nature, leading to the substantiation of the atomic theory of matter, and through the study of power to the concept of energy and thermodynamics.

By the middle of the nineteenth century these sciences of the new industrial age had displaced astronomy, the science of the exploratory and mercantile age, from its previous position. The shaping of scientific research and education to meet the new scientific needs did not develop quickly. The appropriate ideas were conceived one by one and published, but a century passed before they were articulated into theoretical and practical systems of experimental science.

Besides making the properties and transformation of matter the primary subject of physical science, industrial society also provided the most appropriate social form for their investigation and teaching. Before the development of the power-driven factory, physics laboratories were essentially intellectual versions of the craftsman's workshop. The Royal Institution was the classic example, where Davy and Faraday, who, like workmen, knew no mathematics beyond elementary arithmetic and geometry, made discoveries by personal efforts and manipulation. Faraday, who was the son of a blacksmith, did not have one technical assistant or pupil in his life.

The idea of organised research and teaching of experimental physics was taken over from the factory system, where transformations of matter were carried out in an organised manner for productive purposes, and the technique was taught by the foreman to the numerous operatives.

The need for organised laboratories of experimental physics, in which the subjects were taught, and numbers of pupils could pursue guided researches, appeared at about the same time in widely separated countries, showing that it arose in response to similar social demands.

Liebig's celebrated laboratory at Giessen, founded in 1826, started systematic teaching and research in chemistry. A physics laboratory was founded in Berlin in 1863, and R. B. Clifton (who was a Cambridge man) started teaching experimental physics in a single room at Oxford in 1867. The Clarendon Laboratory at Oxford was begun in the following year, and the first classes given in it in 1870. Balfour Stewart introduced laboratory teaching at Manchester in 1870.

At University College London, G. Carey Foster started an experimental physics class in one room. He 'was convinced that there could be no sound teaching of physics apart from practical work; students must have personal acquaintance with phenomena before they can profitably reason about them'. In 1877 W. Grylls Adams at King's College London, began systematic instruction in practical physics, also in a single room. Queen Victoria presented the College with the scientific apparatus from King George III's museum, forming a nucleus for a collection of teaching apparatus.

J. C. Adams visited Germany and France to collect ideas on what Cambridge should do. He found practical physics classes only in Jamin's laboratory at the Sorbonne, where 'students were already engaged in the determination of Physical Constants'.

The increasing demand for scientists stimulated the teaching of classes rather than individuals in experimental physics. When only a few were needed practical training in the apprentice manner — as assistants to a professor as master — was sufficient. This was the system that had developed in William Thomson's laboratory at Glasgow.

Classes of students could not be taught in this manner, so a new system of instruction had to be devised. Laboratories that were not part of the professor's personal property and equipment were required. As they belonged to the university they were public organisations and provided instruction for the academic community as a whole.

By the middle of the nineteenth century it was becoming recognised that Cambridge methods of teaching physical science were out of date, and were not properly educating young men for the scientific requirements of the contemporary world.

The reform of Cambridge science to fit it better for the new industrial society was a difficult task. It was hindered by events such as the Napoleonic wars, which isolated Britain and increased British prejudice against any new mathematics and science coming from France. The most serious obstacle was not the petrified condition of the Newtonian inheritance, but the social strength of the old university system, which made it very difficult to introduce new ways of dealing with science. Cambridge was still governed by much the same statutes as Queen Elizabeth had given it in 1570. Historically, the college system was older than the university itself, which had come into existence to provide regulations and facilities for the students of pre-existing colleges. These students had been primarily candidates for the church. Even as late as 1830 about a third of the undergraduates at Trinity College, which at that time constituted about a quarter of the whole University, were studying for holy orders.

The students felt that their main loyalty was to the college with which they were so intimately identified. When they left Cambridge they remembered their college before the University; they bequeathed gifts to their old college in their wills. In feudal centuries the land was the chief source of wealth, so the colleges gradually inherited considerable estates. Few men thought of bequeathing land to the University, so, relative to it, the colleges became wealthy. If any new project was to be carried out, the colleges rather than the University were looked to for the money required.

The rise of the mercantile and trading society which began in the sixteenth century undermined the feudal principle, but did not destroy the ancient universities; it changed and strengthened them. Henry VIII founded Trinity College at Cambridge, and Wolsey Christ Church at Oxford. The old feudal estates became more profitable by the application of the new business attitude. This produced a revolution in agriculture. In the eighteenth century City merchants and overseas traders bought country estates and set up as gentlemen, but they tended to manage their new property in land with their customary business acumen. Through the improvement of agricultural technique the incomes of the Cambridge colleges increased.

The profitability of the land after the agricultural revolution of the eighteenth century was one of the largest sources, if not the largest, of the capital that financed the industrial revolution. Improving landlords were one of the features of the mercantile age. Their agricultural concerns directed their attention to natural history. One of the most conspicuous was Joseph Banks. He was born in 1743, and came into his father's estate in 1764. It was then worth £6000 a year; when he died in 1820 it was worth £30000 a year. When he entered Oxford in 1760 he already had a considerable knowledge of botany. He joined Cook's first voyage round the world, taking scientific equipment and nine assistants, at a cost of £10000 to himself. Rutherford once pointed out that if it had not been for Cook's discovery of New Zealand, his own ancestors would not have emigrated there, and he would never have come into existence. Banks was elected President of the Royal Society in 1778, a position to which Rutherford was to be elected in 1925.

By the first half of the eighteenth century the ancient University of Cambridge was badly decayed. In Elizabeth I's time the number of undergraduates had exceeded 1300, but it had now fallen to less than four hundred. The same kind of decay had occurred at Oxford. Adam Smith entered that university in 1740, and resided there continuously for six years. He was shocked by the professors and dons, who did almost no research and scamped their teaching. It strengthened his opinion that professors should be paid only for those lectures that their students found useful.

The conflict between Adam Smith and Oxford illustrated the contrast between the decaying atmosphere of a dying age, and the spirit of the rising industrial society.

The success of the agricultural revolution had the effect of making the colleges rich, and able to pay their fellows handsomely for tutoring their undergraduates. The university remained relatively poor; it could not pay its professors adequately. They regarded their chair as a symbol of intellectual prestige, and did not look upon their teaching responsibilities seriously. The colleges took the teaching into their own hands, and strove to keep it away from the professors. The university had little funds, and in general the colleges did not see why they should create facilities for others than themselves. The professorial lectures were usually above the heads of the undergraduates, and college tutors advised their men to ignore them.

In the Middle Ages degrees were awarded on disputations, and examinations had a minor part. By the beginning of the nineteenth century the disputation had given way to the examination, which became more and more mathematical. This was because mathematics lent itself most conveniently to examinations; answers were either right or wrong, and gave examiners the minimum of trouble. Tutors were strongly in favour of it. Besides this, the newtonian tradition had strengthened the prestige of mathematics.

Thus, in a university where the largest category of students were prospective clergymen, mathematics became the chief examination subject. Even the advocates of Latin and Greek were fighting for their place, and the natural sciences were scarcely thought of as a degree subject.

A few men of knowledge and foresight were aware of the situation, but the talented students of the time, Babbage, Herschel and Peacock, led the revolt. They were followed by Airy and Whewell, who were acquainted with the new French mathematical science. They began to teach it, and work it into the university system. Within mathematics they were mildly progressive. Whewell, for example, introduced the modern word 'scientist', and as the advisor to Faraday on terminology coined terms such as 'ion' and 'cathode', which are now standard. But in the relations of mathematics to experimental science, he was reactionary in spite of his admiration for Bacon.

Both of them were skilled academic politicians. Whewell, the son of a carpenter, with the figure of a pugilist and a domineering personality, did not scruple to solicit for the mastership of Trinity College. As for Airy, the future Astronomer Royal, his son wrote that 'the great secret of his long and successful official career was that he was a good servant and thoroughly understood his position. He never set himself in opposition to his masters, the Admiralty.'

Men of this type were able to make some way through the complexities of the Cambridge university politics of the period, but they did not have the conceptions or qualities to effect radical changes.

Whewell solicited the Prince Consort to become Chancellor of the University. He thought he might be able to control the University through him. The Prince was then twenty-seven. He had received mathematical instruction from Quetelet, and had attended the modern University of Bonn. He wanted to do something useful, but was precluded from taking any part in politics. He had asked Sir Robert Peel for advice, and Peel had suggested in 1841 that he should devote himself to the apparently neutral but highly esteemed subjects of art and science. Peel strongly supported the proposal that he should accept the chancellorship of Cambridge.

The Prince took his duty to Cambridge seriously. He asked for details of the university system, one account of which was supplied by Whewell. With his modern education he was mystified by the system, which on the face of it seemed irrational. Whewell was still contending that 'mathematical knowledge is entitled to *paramount* consideration, because it is conversant with indisputable truths — that such departments of science as Chemistry are not proper subjects of academical instruction. . .'.

Chemistry was at that time almost entirely an experimental and laboratory science, so in effect Whewell was saying that courses in experimental science were not suitable media for education.

The Prince referred Whewell's and other papers to Peel for his opinion. Peel was the son of a textile magnate who had been in the forefront of the industrial revolution, and for whom chemistry was one of the most important of the sciences. Peel replied that chemistry should not be ignored 'because there is controversy respecting important facts and principles, and constant accession of information from new discoveries — and danger that students may lose their reverence for Professors, when they discover that the Professors cannot maintain doctrines as indisputable as mathematical or arithmetical truths.'

8

Whewell had contended that a *century should pass* before new discoveries in science were admitted into the course of academic instruction. Peel said that this 'exceeds in absurdity anything which the bitterest enemy of University Education would have imputed to its advocates... Are the students of Cambridge to hear nothing of electricity?' asked the statesman.

Whewell's attitude reflected the strength of the opposition to reform. This was illustrated by Clerk Maxwell's experience with the famous mathematical coach Isaac Todhunter, who edited Whewell's works. He once asked Todhunter whether he would like to see an experimental demonstration of conical refraction. 'No', replied Todhunter, 'I have been teaching it all my life, and I do not want to have my ideas upset.'

The Prince as Chancellor, with Peel's backing, stood his ground. He encouraged the more progressive reformers. He noted with satisfaction in his diary in 1848 that his plan for the reform of Cambridge studies had been carried by a large majority.

The University seemed to be reforming itself from the inside, but the Whig Prime Minister, Lord John Russell, suspected that the conservatives in the University would frustrate the reforms by delaying tactics. He appointed a Royal Commission to inquire into the State, Discipline, Studies, and Revenues of the University of Cambridge. Its members were John Graham, Bishop of Chester, G. Peacock, J. F. W. Herschel, John Romilly and Adam Sedgwick. Its Secretary was W. H. Bateson (father of the geneticist William Bateson).

It reported in 1852, and confirmed what the Prince Consort had regarded as his reforms of 1848. The Commissioners said that time, that 'great innovator', and 'the operation of social causes little within her control' had caused the University to be 'left out of her true position, and become imperfectly adapted to the present wants of the country, so as to stand in need of external help to bring about some useful reforms'.

The Commissioners stressed the need for teaching engineering and modern languages. They proposed that a second chair of chemistry should be founded, and the salaries of scientific and mathematical professors should be improved. The Lucasian professor received only £157 a year, and the professor of chemistry only £241. In contrast, Lady Margaret's Professor of Divinity received £1854.

The terms of the professorships varied very widely, according to the ideas of the founders, and the times in which they were founded. With such varied systems it was difficult to secure professors of uniformly high standard. The university did not appear to be appointing its professors according to any definite principle.

The Commissioners were not able, however, to carry the case entirely against Whewell and the conservatives, who vehemently opposed the introduction of examinations in heat and electricity, on the ground that these were immature sciences. They recommended

that certain Mathematico-physical theories, which had obtained a temporary and questionable footing in the Examination, and which were felt to be in a considerable state of obscurity, involving great mathematical difficulties, and rather marking the frontier of science, than coming as yet fully within its ascertained range (those, namely, of Electricity, Magnetism and Heat), should not be admitted as subjects of examination.

However, Mr Ellis had told them that Thomson's theory of electrical images, and Abel's theorem of elliptic integrals were sufficiently clear to be suitable for examination questions.

One of the most important witnesses was William Hopkins, the great teacher who in twenty years coached 175 Wranglers, including Clerk Maxwell, and 17 First Wranglers. He said that the existing mathematics examination had contributed to the students' exact knowledge of detail, but had obscured 'the perception of the logical connection of one part of a subject with another'. It promoted dexterity with symbols, but neglected fundamental ideas. In particular, 'In Electricity and Magnetism, there can hardly be said to have arisen any axiomatic principles whatsoever'.

The Commissioners emphasised the need for lectures illustrated by experiments, and for instruments of the kind 'which are daily in use in the hands of working men of science', for carrying out experiments. Stokes told them that 'the want of such a collection is strangely inconsistent with the high character which Cambridge maintains for the study of the Exact Sciences, and astonishes foreigners who visit the University'.

The Commissioners reported that there should be a 'complete and thoroughly equipped laboratory' for chemistry. Besides the professor any member of the University who wished to study chemistry should be free to work in it. The absence of such a laboratory was in startling contrast with those that already existed in London at the Royal College of Chemistry, the University and King's Colleges, and the Royal Institution.

The Commission also thought that experimental science might help to solve the student problem of their day. The University was threatened by a dangerous class of gentleman-undergraduates who were exposed to the temptation of finding 'mean and frivolous ways in which wealth may be squandered, and leisure abused'. The Commission recommended that the University should 'teach them handling of meteorological and magnetic instruments, and cameras'. If this were done, some of them might 'become Humboldts'.

The Commission expressed the opinion that if their recommendations were adopted, there did not appear to be any reason 'why Cambridge should not become as great a School of physical and experimental as it is already of mathematical and classical instruction'.

Through the Commission the State now added its voice to that of the Prince Consort and his advisers, calling for the development of the natural and experimental sciences at Cambridge.

Such teaching as was being done, was carried out in buildings erected in the Botanic Garden, which had been founded near the centre of Cambridge in 1762. The first was put up in 1786, and enlarged in 1833. These lecture rooms, though ill-arranged and inconvenient, served for the teaching of anatomy, physics, botany, chemistry and applied mechanics, or such part of them as the Jacksonian professor chose to deal with.

After the Natural Sciences Tripos was established in 1851 the number of science students increased, and more accommodation was required for their instruction. Meanwhile, the Botanic Garden became surrounded by city buildings, and less and less suited to its purpose. Its collections were accordingly moved to another site. This was begun in 1846 and completed by 1852. Some were in favour of selling the old site for use as a market place or other business, since it was in a very convenient and financially valuable situation for such purposes.

Fortunately, the majority were against this proposal. The University secured full rights over the site, and requested the committee, which had dealt with the negotiations over the rights, to advise on what should be done in regard to the construction of new museums and lecture rooms. A scheme was proposed, but its estimated cost was more than could be raised. Nothing more was done until 1860, when the proposal was reconsidered; the first attention was given to the question of finance.

This time sufficient was found for a start. The professor of chemistry, G. D. Liveing, laid the foundation stone in 1863. By 1865 accommodation was provided for zoology, comparative anatomy, chemistry, mineralogy and botany.

In 1861 the Prince Consort had died at the early age of forty-two. This had deprived the University of his progressive influence. Fortunately, it secured as his successor the Duke of Devonshire, who was almost as aristocratic and still more qualified in science and practical business.

Thus the struggle for reform proceeded at the top. But what were the students doing, down below? This may be illustrated by examining what was happening to the most significant student of the period: James Clerk Maxwell. He entered Cambridge in 1850, and lived and worked through the intensest period of the struggle for the reform of Cambridge science.

James Clerk Maxwell was born in Edinburgh in 1831. He entered Cambridge in 1850, when he was nineteen years of age and had already published original research. His upbringing as well as his genius was to have a deep influence on the creation of the Cavendish Laboratory. Other qualities besides sheer intellectual ability were required in performing this task, and a number of these were fostered by his inheritance and the circumstances of his early life. He was descended from the Clerk family, which had been known for three centuries. It had had political connections with Mary Queen of Scots, and one of its members made a fortune by commercial operations in Paris. He bought the estate of Penicuik near Edinburgh on which C. T. R. Wilson was to be born in 1869. His son married a granddaughter of William Drummond of Hawthornden, and his grandson became an eminent lawyer, and one of the Commissioners of the Union of England and Scotland.

The Commissioner's sons were sent to study under Boerhaave at Leyden. One of the sons married Agnes Maxwell, heiress of the estate of Middlebie in south-west Scotland. Their daughter Dorothea married a first cousin, George Clerk, who afterwards adopted the surname of Maxwell. This George Clerk Maxwell also studied at Leyden. He became interested in geology and prospecting and friendly with James Hutton, the founder of modern geology. He became a trustee for the improvement of fisheries and manufactures, and his brother John Clerk wrote a book on *Naval Tactics*, from which, he claimed, Rodney had learned the tactics he applied successfully at the naval battle off Dominique in 1782.

The Clerk Maxwell family was distinctly inbred, which helps to account for the concentration of genius that appeared in James Clerk Maxwell. He himself strongly believed that human qualities were very much determined by heredity.

George Clerk Maxwell's younger son James became a naval captain in the East India Company. His younger son John inherited Middlebie. The estate had been encumbered by his father's losses in mining and manufacturing speculations, and a large part had been sold to liquidate debts.

John Clerk Maxwell was a laird, but on the remains of his estate there was not even a house. At first he lived in Edinburgh with a small but secure income. He attended the University, and engaged desultorily in the law; but his interests were inspired by the industrial revolution in Scotland, and fostered by the atmosphere created by Joseph Black, James Hutton, and other founders of the Scottish development in science.

John Clerk Maxwell lived with his mother in Edinburgh until she died in 1824, when he was thirty-six. Shortly afterwards he married Frances Cay from Northumberland. He then began to act seriously on the improvement of his estate. He built a modest but solid mansion on ground rising from a stream. This became known as Glenlair. He attended to every detail of the building and grounds. He planned the outbuildings, and made the working drawings for the masons himself. John Clerk Maxwell's constructional interests and modes of operation were reflected in James Clerk Maxwell's approach to the problem of designing the Cavendish Laboratory. James made preliminary sketches for the Cavendish in the manner of his father.

Mrs John Clerk Maxwell was the daughter of a judge. She had a strong and resolute nature. She died of cancer at the age of forty-seven, when her son was eight years old. James Clerk Maxwell was to die of the same disease at almost the same age.

James's exceptional curiosity was already noted at the age of three. One of his favorite phrases was: 'What's the go of it?' When dissatisfied with the explanation he would ask: 'But what's the *particular* go of it?' Among his toys was a phenakistoscope, a primitive form of motion-picture apparatus invented by Faraday. He used the principle years later in his invention of the colour top. He added lenses to the phenakistoscope, bringing it nearer to the cinematograph, and he applied it to make motion pictures of the collision of smoke rings, the first application of motion pictures to scientific research.

As a boy he played with the children of his father's employees,

13

referring to them as 'vassals'. In the most amiable way he grew up with the outlook of a member of the governing class.

After somewhat crude instruction by a raw tutor, James was sent at the age of ten to the recently established Edinburgh Academy. He arrived in eccentric clothes which had been designed by his father. The boys did not like his peculiar garb, nor his broad Galloway dialect. James came home minus various parts of his suit, and no one could get out of him exactly what had happened, but his tendency to hesitations in speech and obliquity in expression became worse. He was nicknamed 'Dafty'. For several years he showed little interest in school; his mind and heart were still in the country life at Glenlair. When his father came to Edinburgh he took him to see such things as 'electro-magnetic machines'.

Before he began the study of geometry at school he constructed models of a tetrahedron and a dodecahedron, and 'two other hedrons', the names of which he didn't know. He adapted himself to the school atmosphere with Spartan determination, and after he had mastered the situation his intellectual development was rapid. At the age of fourteen he won the medal for mathematics. His father was pleased, and took James to meetings of the Royal Society of Edinburgh, and the Society of Arts. At the latter James heard D. R. Hay speak on how the Greeks drew the ovals for their 'eggs-and-dart' architectural friezes. James began to think about the problem, and discovered a new way of drawing ovals. His father drew the attention of Hay and of J. D. Forbes, the distinguished professor of natural philosophy at Edinburgh University, to it. Forbes was impressed, and looked up the literature. He found that the previous contributors to the subject included Descartes, Newton and Huyghens!

James, still fourteen, wore short trousers, so it was not thought proper that he should personally address the Royal Society of Edinburgh; Forbes read his paper for him. He allowed James to experiment in his laboratory, and James read his well-known paper on the physics of glaciers. This interested him in the properties of jellies and elastic substances. James's preoccupation with research retarded his progress in school, but in his last year he was first in mathematics and English, and high in Latin. Maxwell valued his classical education, and held that mastering the exact meaning of words, and expressing it, was one of the best ways of training the mind. Becoming interested in Newton's rings and the polarisation of light, his uncle took him to see Nicol, the inventor of the famous prism.

14

At the age of sixteen James entered Edinburgh University, where he worked for three years, almost without supervision. His interest in optics led him to study the application of double refraction to the investigation of the strains in elastic materials. He painted water-colours of the patterns obtained, and sent them to Nicol. When he was nineteen he published a long paper on the mathematical analysis of the bending of beams and the twisting of cylinders, with confirmation of the results by experiments with polarised light and distorted trans-parent jellies, founding this now classical method, especially valuable in aircraft and other kinds of advanced engineering design.

Forbes called on Maxwell's father, and advised that he should be sent to Cambridge. Peterhouse was then the fashionable mathematical college, which was winning the prizes. William Thomson had passed through with an immense reputation, and Maxwell's school-fellow P. G. Tait had recently gone there.

Maxwell entered Peterhouse in 1850. His unusual manners and dialect did not allow him to adjust easily to the clear-cut conventional undergraduate life, and he found the strongly routined first-year ele-mentary teaching boring. The Peterhouse tutor advised him that he had not much prospect of gaining one of the few Peterhouse fellow-ships, for which the competition was very fierce, and suggested he should go to another college.

Maxwell went to see the tutor at Trinity, who asked why they should admit him. He produced a bundle of reprints of his original papers, and said that perhaps these were sufficient evidence that he was not unfit to enter the college.

At Trinity Maxwell soon found able and sympathetic friends. He grew less odd, and became happier. In due course he entered the training for the mathematical Tripos examination. This was still conducted like a horse-race, for the recommendation of the Royal Commission of 1852 had scarcely begun to have an effect. He was elected a member of the *Apostles*, a club restricted to twelve of the leading undergraduates.

He trained for the examination under William Hopkins, who had prepared Stokes and William Thomson. Hopkins was shocked by the state of Maxwell's knowledge. Its variety and depth were extraor-dinary, but as a whole, it appeared to be chaotic. It may well be that Hopkins's influence had a great deal to do with developing Maxwell's sense of systematisation, which became such a masterly feature of his work on electromagnetic theory. Hopkins said that Maxwell was by

15

far the most remarkable of all his famous pupils. He was almost incapable of thinking incorrectly on physical subjects, though his command of formal mathematics was defective.

Maxwell's friends realised that he had a very uncommon personality. How he appeared to fellow students was described by Montagu Butler in a sermon ten days after Maxwell's death in 1879. He said: 'When I came up to Trinity twenty-eight years ago, James Clerk Maxwell was just beginning his second year. His position among us — I speak in the presence of many who remember that time — was unique. He was the one acknowledged man of genius among the undergraduates ....'. Those who knew Maxwell less well thought him charming and amusing, but did not appreciate his intellectual weight.

Maxwell took the Tripos examination in 1854. By a sheer effort of concentration he gained second place. E. J. Routh, who was first, became the most famous of all the Cambridge mathematical tutors. Maxwell's father was pleased, telling him that he had done better than generally had been expected.

Maxwell's reform of Cambridge teaching in physical science and his creation of the intellectual policy of the Cavendish Laboratory could not have been achieved if he had not first mastered the existing scientific and social milieu in Cambridge. Through his wide intellectual interests and social sympathies he secured the friendship and understanding of able men who were not scientists, and who supported him in his recommendations for science.

After he had graduated in 1854 he looked round for a field of research. He wrote to William Thomson, who had already published important individual discoveries in electrical theory, asking whether he would mind entering his field. Thomson generously encouraged him, and Maxwell wrote to his father that Thomson was 'very glad that I should poach on his electrical preserves'.

Maxwell read Faraday's researches very carefully. He perceived that Faraday used concepts that were capable of mathematical description, though he never applied anything more than elementary mathematics to them. Maxwell chose Faraday's conception of lines of force as the subject of his first electrical paper, which was published in 1855–6. He started by examining the philosophical state of the subject, and then each of the chief contemporary conceptions of electrical phenomena. He remarked that while mathematical accounts of some individual electrostatic, current and electromagnetic phenomena had been given, no general theory connecting all these phenomena had yet been found.

No one except Maxwell was able to develop the ideas in this first paper. Mathematical physicists were the prisoners of action-at-a-distance theories, which were supported by the enormous prestige of newtonian gravitational theory. Airy declared that he could hardly imagine how anyone could hesitate for a moment between the simplicity and precision of action-at-a-distance, and the vagueness of varying lines of force.

After his first electrical paper Maxwell decided to compete for the Adams Prize. The subject happened to be one of the most difficult unsolved problems in the purest newtonian gravitational dynamics: the nature and stability of the rings of the planet Saturn. He proved that they must consist of clouds of revolving satellites. It demonstrated his mastery of conventional action-at-a-distance theory. It earned the approval of Airy, who declared it was one of the most remarkable applications of mathematics to physics he had ever seen.

Maxwell was now accepted as a master of traditional physical theory. His strange ideas on electricity could no longer be dismissed as ingenious but unsound speculations. He was beginning to establish a strong tactical position for promoting post-newtonian science in Cambridge.

He heard that the chair of natural philosophy at Marischal College, Aberdeen, was vacant, and decided to apply for it. His father remarked that if the post proved disappointing it could be given up, and in any case it occupied only half the year. So much for chairs in the opinion of the Scottish laird!

Maxwell's father died shortly afterwards, and Maxwell decided to accept the chair. After all, it was nearer his little estate in Scotland, which he had now inherited. He wrote to a friend that 'the transition state from a man into a Don must come at last'.

It is said that Maxwell's investigation of the properties of the clouds of small satellite particles led him into his researches on the statistical theory of gases. He read his first important contribution to this theory at the Aberdeen meeting of the British Association in 1859.

Maxwell was very active at Aberdeen. Besides pursuing his researches, he married the daughter of the Principal of Marischal College. Then in 1860 Marischal College and King's College were amalgamated. There was room for only one chair of natural philosophy. This was given to David Thomson, the professor at King's. He was a very good teacher, senior to Maxwell, and a nephew of Faraday. The extinction of a chair occupied by Maxwell must seem a strange event in the perspective of history.

Maxwell was glad to have a good excuse for retiring to Glenlair. He was passed over in favour of Tait for the Edinburgh chair. It was explained that while Maxwell's talent was already acknowledged, there was 'a deficiency' in his powers of oral exposition. Later in the same year, 1860, the chair at King's College London became vacant and Maxwell was appointed. His five years there were extremely productive. During them he developed the statistical theory of gases, and discovered the equations of the electromagnetic field.

During his strenuous years at King's Maxwell supervised the experimental determination of electrical units for the British Association. The results became the basis for the international system of electrical units. Maxwell interested his pupils in the problems of standardisation and measurement of units.

Maxwell suffered serious illnesses during this period. He decided again to retire to Glenlair. While in retirement there, he worked on the manuscript of his *Treatise on Electricity and Magnetism*. A special postbox was put up at the bottom of his garden to receive the proofs as they arrived from the Clarendon Press.

He kept in touch with Cambridge as examiner in the Mathematical Tripos in 1866, 1867, 1869 and 1870. He took the leading tactical part in the reform of this examination by extending the introduction of problems involving the new theories of electricity and heat. Through this crucial academic operation, Maxwell promoted the adaptation of Cambridge physical science to the needs of contemporary society.

While Clerk Maxwell was at the head of those tactically introducing the new physics as material for the Mathematical Tripos examination, another major figure was assisting the reform of Cambridge from a different perspective. This was the Chancellor, or official head of the University, William Cavendish, Seventh Duke of Devonshire (see figure 1). He succeeded the Prince Consort as Chancellor in 1861, and held that office until his death in 1891. He was born in 1808, a descendant of the fourth Duke, who had married Charlotte Elizabeth Boyle: besides belonging to the Cavendishes, who had produced Henry Cavendish, he also belonged to the Boyles, who had produced Robert Boyle; he was therefore related to two of the most distinguished physical scientists in British history. Like Clerk Maxwell, he came of inbred families of high ability. He was educated at Eton, and then at Trinity College, Cambridge. He was quiet, reserved, and studious. In 1829 he graduated as Second Wrangler, First Smith's Prizeman,

1 *The Seventh Duke of Devonshire*

and Eighth Classic; the man who had beaten him in the mathematical examination was beaten by him in the contest for the prize.

His achievement was significant from several points of view. It was highly unusual for a prospectively very wealthy landed aristocrat to excel in examinations, especially in mathematics. Charles Babbage, then Lucasian professor, was one of his mathematical examiners. His marks were so high that Babbage gave his papers to a second examiner for an independent assessment, in case there should be any accusation of favouritism. The second examiner's marks substantially confirmed Babbage's.

Not only had William Cavendish done very well: the form of the examination was comparatively new. It was one of the earliest Cambridge degree examinations in which printed questions were put to candidates; traditionally they had been put verbally. Thus, William Cavendish's examination was in itself part of the gradual process of the reform of Cambridge mathematics teaching.

He followed the Whig political tradition of his family. When he was young he was on their left. The Cambridge University Whigs and Liberals persuaded him to stand, with Palmerston, as their candidate in the General Election of 1829. Babbage campaigned energetically on his behalf. Cavendish and Palmerston won. Cavendish had no taste for speaking, but he strongly supported Parliamentary reform. He argued that unless it was carried out the country would become unstable. If there was danger of revolution it would not come from reform, but from the lack of it. The aristocracy would lose some of their power, but they would be induced to rely more on their talents as the source of their legitimate influence. For Cavendish, this was a very convenient philosophy, for he was intellectually very well endowed. His programme was too much for the Cambridge University electors, who rejected both him and Palmerston in the General Election of 1831.

Three years later his grandfather died. He inherited the earldom of Burlington, and thenceforth sat in the House of Lords. His cousin, the Sixth Duke of Devonshire died in 1858, whereupon he inherited the dukedom. He was now aged fifty. He found the Cavendish estates in poor shape. He applied the same capable management to them as he had done to his own Burlington estates, which included the Boyle properties in Ireland. He promoted real-estate projects, converting three villages in Sussex into the new seaside resort of Eastbourne, and built railways in Ireland. His eldest son, the Marquis of Hartington, became a Liberal prime minister; his second son, Lord Frederick Cavendish, then Chief Secretary for Ireland, was shot in Phoenix Park in Dublin in 1882. The Duke, who had supported Gladstone's Irish policy, became strongly opposed to Home Rule, and a leader of the new Unionist movement.

When he inherited large estates he applied good management and

20

scientific agricultural methods. In 1839 he became a founder of the
Royal Agricultural Society, of which he served as President in 1869.
He bred shorthorn cattle, and persuaded Cambridge University to
introduce the study of agricultural science.

The most important of the Duke's industrial ventures was his
exploitation of the rich deposits of high-grade haematite iron ore discovered on his estates in north Lancashire. The works built to make
steel from the iron ore became one of the largest Bessemer plants
in Europe. A railway was built to transport ore and products to
Barrow, where a new port was created. The Duke could remember a
time when Barrow was a fishing village of a hundred inhabitants; when
he died in 1891 it was a port, iron and armaments city of 57 000
people. He and his partners even had visions of developing Barrow to
compete as a port with Liverpool.

The Duke left a fortune of £1 790 870. He had become one of the
greatest British industrial magnates of the nineteenth century. After
the death of the Duke of Wellington the name of 'The Iron Duke'
was, on other grounds, applied to him.

Besides being among the foremost figures of the new industrial
age, he promoted the application of science to industry. In 1869,
when the Iron and Steel Institute was founded, he became its first
President. In his first presidential address he referred to the growing
recognition of the importance of those branches of science of interest
to the iron-master. He mentioned that proposals were under consideration for the more systematic study of experimental science. Several
branches of it had already been of great value in improving the processes in their furnaces and iron works. In all probability, they were
destined to have a still greater influence. Spectrum analysis promised
important improvements in the operation of the Bessemer process.
Among other topics to which he referred was Whitworth's process
for high-pressure steel casting, and the social influence of iron,
through its use in the construction of railways, guns and agricultural
machinery.

No one was in a better socially strategic position for changing the
direction of science from that traditional in the age in which he was
born, to the one required in the new industrial age in which he
matured and reached the forefront.

It is not surprising that he of all men was to be the founder of the
Cavendish Laboratory. No one could have been more socially and
scientifically appropriate. His part in the founding has always been

21

highly esteemed, but historically its importance and significance are even greater than has generally been recognised.

Though the Duke did so much for science, he regretted his early application to learning. He believed that it had made him unfit to exercise the political influence that his position and wealth brought to him. Because of this, he refused to allow his son to have the same kind of education that he had received. He seems not to have foreseen that his part in the creation of the Cavendish Laboratory might have a greater influence on history than he could have expected to have had, even as a prime minister.

1

Founding the Professorship and Laboratory

After more than twenty years of agitation and deliberation the University began to take positive steps for establishing the systematic teaching and study of experimental physics. A syndicate, or committee, was appointed on 25 November 1868 to report on the matter. Their report was delivered on 27 February 1869. In it, they observed that:

The importance of cultivating a knowledge of the great branches of Experimental Physics in the University was prominently urged by the Royal Commissioners appointed in 1850, to inquire into the State, Discipline, Studies and Revenues of the University and Colleges of Cambridge.

On page 118 of their Report, after commending the manner in which the subject of Physical Optics is studied in the University, and stating that 'there is perhaps no public institution in which it is better represented, or prosecuted with more zeal and success in the way of original research', they go on to say that 'no reason can be assigned why other great branches of Natural Science should not become equally objects of attention, or why Cambridge should not become a great school of physical and experimental as it is already of mathematical and classical instruction'. And again, 'In a University so thoroughly imbued with mathematical spirit, physical study might be expected to assume within its precincts its highest and severest tone, be studied under more abstract forms, with more continual reference to mathematical laws, and therefore with better hope of bringing them one by one under the domain of mathematical investigation than elsewhere'.

In the Scheme of Examination for Honours in the Mathematical Tripos, approved by Grace of the Senate on the 2nd June, 1868, Heat, Electricity and Magnetism, if not introduced for the first time, had a much greater degree of importance assigned to them than at any previous period, and these subjects will hence demand a corresponding amount of attention from the Candidates for Mathematical Honours.

The Syndicate have limited their attention almost entirely to the question of providing public instruction in Heat, Electricity and Magnetism. They recognise the importance and advantage of tutorial instruction in these subjects in the several Colleges, but they are also alive to the great impulse given to students of this kind, and to the large amount of additional training which many Students may receive, through the instruction of a public Professor, and by knowledge gained in a well-appointed Laboratory.

In accordance with these views, and at an early period in their deliberations, they requested the Professors of the University, who are engaged in teaching Mathematical and Physical Science, to confer together upon the present means of teaching Experimental Physics, especially Heat, Electricity and Magnetism, and to inform them how far the increased requirements of the University in this respect could be met by them.

The Professors, so consulted, favoured the Syndicate with a Report on the subject, which the Syndicate now beg leave to lay before the Senate. It points out how the requirements of the University might be 'partially met'; but the Professors state distinctly that they 'do not think that they are able to meet the want of an extensive course of' Lectures on Physics treated as such, and in great measure experimentally.

As Experimental Physics may be fairly considered to come within the province of one or more of the above-mentioned Professors, the Syndicate have considered whether now or at a future time some arrangement might not be made to secure the effective teaching of this branch of science without having resort to the services of an additional Professor. They are however of opinion that such an arrangement cannot be made at the present time and that the exigencies of the case may be best met by founding a new Professorship which shall terminate with the tenure of office of the Professor first elected. The services of a man of the highest attainments in Science, devoting his life to public teaching as such Professor, and engaged in original research, would be of incalculable benefit to the University in this department.

It is evident that the reporters had a very clear notion of what they were doing, and foresaw future needs with great depth. One of the causes of the achievement of the Cavendish Laboratory may well be the high mental effort that went into the initial conception of the professorship and laboratory. It was the result of years of reflection by very able men.

The Syndicate find that the Rules which regulate the Jacksonian Professorship of Natural Philosophy are 'fanciful and obsolete', and that they require revision. The Rules also of the Plumian Professorship of Astronomy and Experimental Philosophy seem to require modification, as the range of subjects prescribed is much too large to be assigned to any one Professor in the present state of Science. If new Rules were made by competent authority for the government of these two Professorships it is possible that, on the occasion of a vacancy in one of them, the proposed new Professorship might be merged in this, though the desirableness of such an arrangement must depend upon the experience of the working of the new system.

The founding of a Professorship would be incomplete unless means were also supplied to render the Professor's teaching practical, and assistance given to him, both in the Laboratory and Lecture-room. The need of providing instruments, Apparatus and Laboratories is obvious, and it seems not less necessary to obtain some additional assistance in giving personal instruction to students in the Laboratory. On this point the Royal Commissioners state [Report, p. 119] that 'generally speaking much more than material aid is requisite for conveying to Students a practical training in the handling of scientific apparatus, and the mode of conducting experiments to a satisfactory conclusion. Personal instruction is essential; and the Professor, who ought to be at least partly occupied in original research, and whose attention must at all events be sufficiently occupied in preparing himself to impart and in actually imparting in the most luminous manner the scientific principles of his subject, is not likely to have much time at his disposal for the instruction of tyros in the use of their tools'.

The Syndicate are of opinion that due provision cannot be made for teaching the subjects herein considered without the appointment of a Demonstrator; and they further think that a collection of Instruments and Apparatus, and suitable Class rooms and Laboratories must be also provided.

The purchase of Apparatus is a matter which cannot well be deferred; and the Syndicate think that in the event of the establishment of a Professor of Experimental Physics, the Senate must be prepared to expend a sum of say £1000 on this object.

The rooms which appear to them to be necessary are one or more Apparatus rooms, Laboratories and a Class room for students, a Laboratory and a private room for the Professor, and a Lecture room, in juxtaposition with these, suited for public lectures to a large class.

The Syndicate are of opinion that this accommodation cannot be provided in the rooms now unoccupied in the new Museum and Lecture Rooms, but that additional buildings specially designed for this particular branch of science will be required. A temporary arrangement might be made, which, though imperfect, would allow the work of instruction to be begun, and would be the means of giving time to the University to make permanent and satisfactory provision as soon as circumstances will permit. The probable cost of erecting buildings for this department may be estimated at not less than £5000.

The Syndicate submit the following Scheme, under four separate heads, for the purpose of providing the means of instruction in Experimental Physics to the Students of the University:

I   A Professorship of Experimental Physics

1   There shall be established in the University a Professorship to be called the Professorship of Experimental Physics, to terminate with the tenure of office of the Professor first elected.

2   It shall be the principal duty of the Professor to teach and illustrate the laws of Heat, Electricity and Magnetism, to apply himself to the advancement of the knowledge of such subjects and to promote their study in the University.

3   The Professor shall be chosen and appointed by those persons whose names are on the Electoral Roll of the University.

4   The Stipend of the Professor shall be £500 a year, payable so long as the person appointed under these regulations shall continue to hold the Professorship.

5   The Professorship shall be governed by the regulations of the Statute for Sir Thomas Adams' Professorship of Arabic and certain other Professorships in common, and the Professor shall comply with all the previous provisions of the said Statute.

6   He shall be required to give eighteen Lectures at least in every Term, and his Scheme of Lectures shall be subject to the approval of the Board of Mathematical Studies, until the University by Grace of the Senate shall alter any part of this arrangement.

7   The fee for attendance upon the Lectures in any one Term, given in accordance with the sixth Regulation, shall be one guinea, but no further fee shall be charged after the payment of two such fees. In case of any *extra* course of Lectures, the fee for attendance shall be at the discretion of the Professor.

8   The Professorship shall be *ipso facto* vacant if the Person holding it be elected into any other Professorship in the University.

II   A Demonstrator of Experimental Physics

A Demonstrator of Experimental Physics shall be appointed, whose principal duty shall be to give personal instruction in the laboratories. He shall be under the general direction of the Professor, and his stipend shall be £100 per annum.

The appointment of such Demonstrator shall be made by the Professor of Experimental Physics, with the consent of the Vice-Chancellor; and any person so appointed shall be removable in like manner by the Professor, with the consent of the Vice-Chancellor. The appointment of the Demonstrator shall terminate with the tenure of office of the Professor.

III   A Museum and Lecture Room Attendant

The Professor shall be authorized to employ a person to attend in the Lecture Room and Laboratories, to wait on the Professor and Demonstrator, to clean, fetch and remove Apparatus, and to keep in readiness for use the various Instruments and Apparatus which may be placed in his charge. His stipend shall be £60 per annum, and an allowance shall be made to him for lodgings until rooms can be provided for him. He shall be under the direction of the Professor and removable by him.

IV   Instruments and Apparatus

An extra sum of £1000 shall be appropriated for the purpose of procuring a first stock of philosophical Instruments and Apparatus; and a further sum of £300 shall be appropriated for the purpose of fitting up the Apparatus rooms with proper cases and furniture

It will be seen that the Scheme proposed by the Syndicate involves the following items of expenditure:

1   A sum of £1300 or thereabout for Apparatus, Cases and Furniture.

2   An estimated sum of £5000 for a new building, which shall contain Lecture room, Laboratories, Class rooms and Apparatus rooms.

3   An annual sum of £660 for the stipends of a Professor of Experimental Physics, a Demonstrator of Experimental Physics, and an Attendant.

The Syndicate are of the opinion that the question of providing the necessary funds should be considered by a special Syndicate of Finance, and they recommend the appointment of such a Syndicate without delay.

The Report was signed by: E. Atkinson, Vice-Chancellor, H. W. Cookson, James Cartmell, G. M. Humphrey, G. G. Stokes, G. D. Liveing, W. M. Campion, S. G. Phear, William Walton, J. L. Hammond, E. J. Routh and T. P. Hudson.

The Plumian, Jacksonian, Lucasian, Sadlerian, and Lowndean Professors, and the Professor of Chemistry reported to the Syndicate that

the increased requirements of the University in relation to the teaching of Experimental Physics arise in relation:

1 To certain new subjects introduced as optional branches of study into the Mathematical Tripos;

2 To the Natural Sciences Tripos;

3 To the optional branch of Mechanism and Applied Science, and to that of Chemistry for the Ordinary Degree; to which may be added the introduction of Heat and Electricity into the first Examination for the M. B. degree.

For the first object, it is important that means should be afforded to enabling Students to apprehend clearly the physical principles of the subjects, so far at least as they have been reduced to a precise mathematical form. For this purpose a short course of experimental lectures would be very useful, at any rate in default of a more extensive physical course. As regards the more difficult mathematical parts of the subjects, books seem to be of more importance than lectures, which for this purpose are to be regarded rather as supplementary.

For the second object it is clear that the subjects should be treated so as to make the physical study the principal object, mathematics of an easy kind being freely used.

For the third object the subjects should be treated experimentally, not avoiding very elementary mathematics.

For the second and third objects, longer courses of experimental lectures would be advantageous.

We think it of great importance that lectures should be given in which these subjects should be treated *as branches of physics*, rather inductively than deductively. Lectures of a more mathematical and deductive character might perhaps be given in addition, and independently.

Among the Professors applied to, the Jacksonian Professor, and the Professor of Chemistry are concerned with branches of experimental physics. The Plumian Professor may to a certain extent be joined with these, as the Professorship is for Experimental Philosophy as well as Astronomy, though it had hitherto always been regarded as one of the mathematical professorships.

The Jacksonian Professor already lectures on Mechanism; and it can hardly be expected that he should now take up a course of lectures on subjects to which he has not paid special attention.

Modern Chemistry is such an extensive subject that it fully suffices to occupy a man's whole time. Prof. Liveing actually gives a course of lectures on Heat, which he finds a considerable interruption of his more proper work, and of which he would be glad to be relieved.

Professor Challis would be unwilling to consider his professorship otherwise than as mathematical, but is willing to give a course of lectures on Terrestrial Magnetism.

The Lucasian Professor might be willing to give a short course of lectures on Electricity in addition to his present course, but if so would wish to confine himself to those parts which admit of being treated mathematically.

The Lowndean Professor is mathematical by the terms of his appointment; and as a great part of his time must be devoted to the superintendence of the Observatory, he can hardly be expected to lecture except on subjects which are more or less connected with Astronomy; for example, the Figure of the Earth, or the Tides.

The Sadlerian Professor by the very terms of his professorship is required to devote his attention to Pure Mathematics.

Although the requirements of the University might be partially thus met, we do not think that we are able to meet the wants of an extensive course of lectures on physics treated as such, and in great measure experimentally.

This Report, of 3 December 1868, was signed by: J. Challis, R. Willis, G. G. Stokes, A. Cayley, J. C. Adams and G. D. Liveing.

A Syndicate was then appointed, on 13 May 1869, 'to consider the means of raising the necessary funds for establishing a Professor and Demonstrator of Experimental Physics, and for providing buildings and apparatus required for that department of Science, and further to consider other wants of the University, and the sources from which those wants may be supplied'.

The Syndicate produced an Amended Report on 3 May 1870. They said that they had proceeded to consider carefully 'the various wants of the University, and to enumerate the more obvious and pressing ones, previously to proposing some practicable scheme to provide them'. In doing so, they assigned the first place to the objects that were brought to the notice of the Senate by the Physical Science Syndicate.

The other objects were the providing stipends for a Demonstrator of Chemistry and a Teacher of Paleontology; stipends for two Teachers of Modern Languages; provision for the care and repair of Physical Apparatus; the erection of a large Examination Hall; and the increasing of the stipends of the Professors of Latin, Moral Philosophy, Chemistry, Anatomy, Zoology, Botany, Mineralogy, Political Economy, and of the Lowndean Professor of Astronomy to £500 per annum each.

The Syndicate estimated the amount which would be required for these purposes, exclusive of the cost of an Examination Hall, and of a capital sum of £3300 for buildings and apparatus, required for the department of Experimental Physics, at £3160 per annum.

They considered carefully the various means by which the necessary funds for the above purposes might be raised, and they decided upon addressing a Communication to the several Colleges of the University to enquire whether they would be willing, under proper safeguards for the due appropriation of any monies which might be entrusted to the University, to make contributions from their corporate funds for the above-mentioned objects.

The answers of the several Colleges, except that of King's, which has not yet been received, have been fully considered by the Syndicate. They indicated such a want of concurrence in any proposal to raise contributions from the Corporate Funds of Colleges, by any kind of direct taxation, that the Syndicate felt obliged to abandon the notion of obtaining the necessary funds from this source and accordingly to limit the number of objects which they should recommend the Senate to accomplish. They confined their attention therefore to the means of raising sufficient funds only for carrying out the recommendations of the Physical Science Syndicate in their Report dated February 27 1869. These were to provide the stipends of a Professor of Experimental Physics, of a Demonstrator, and an Attendant, requiring altogether a sum of £660 per annum. Also to provide a capital sum of £5000 for a new Building, and £1300 for apparatus.

The Syndicate are of opinion that these sums may be raised from the ordinary sources of revenue of the University, and that a small addition to the amount of the annual Capitation Tax will suffice for the purpose. They think also that there are circumstances connected with the fixing the amount of the Capitation Tax by the Grace of May 31, 1866, which themselves justify some increase.

The Syndicate therefore recommend, as a temporary measure, for the approval of the Senate:

I

That, for the purpose of making such an addition to the income of the Chest as will suffice to meet the charge for a Professor and Demonstrator of Experimental Physics, as recommended in the Report to the Physical Science Syndicate dated 27 Feb. 1869, the amount of the capitation-tax (17 shillings per annum) paid to the Chest by every member of the University be increased to nineteen shillings per annum, such sum of nineteen shillings to be paid to the Vice-Chancellor by four equal quarterly payments on the usual quarter-days, the first of which shall become due at Christmas 1870.

That this increase be discontinued as soon as the income of the Chest shall have been adequately increased by means of Contributions from the Colleges.

II

That in order to provide a capital sum of £5000 or thereabout, the income of the General Building Fund and of the Museums Building Fund be made available, and that a sum not exceeding £1500 be appropriated from the Government Stock belonging to the Chest for the purchase of the Apparatus, Cases and Furniture, recommended by the Physical Science Syndicate in their

Report dated 27th Feb. 1869; but that no expenditure shall be incurred for Buildings, Apparatus, Cases, and Furniture until plans and estimates have been submitted to the Senate for its approval.

The Report, which was not wholly unanimous, was signed by: E. Atkinson, H. W. Cookson, J. Power, W. M. Campion, G. G. Stokes, Henry Latham, John Lamb, T. Brocklebank, Wm. Bennett Pike, J. Clough Williams Ellis, Thomas Hewitt, G. F. Cobb, F. Pattrick and E. T. S. Carr.

The Senate finally decided on 9 February 1871:
That there be established in the University a Professorship of Experimental Physics, and that this Professorship shall be subject to the following Regulations:

Regulations for the Professorship of Experimental Physics

1   The Professorship shall be called the Professorship of Experimental Physics, and shall terminate with the tenure of office of the Professor first elected unless the University by Grace of the Senate shall decide that the Professorship should be continued.
2   It shall be the principal duty of the Professor to teach and illustrate the laws of Heat, Electricity and Magnetism, to apply himself to the advancement of the knowledge of such subjects and to promote their study in the University.
3   The Professor shall be chosen and appointed by those persons whose names are on the Electoral Roll of the University.
4   The Stipend of the Professor shall be £500 a year, payable out of the University Chest, so long as the person appointed under these regulations shall continue to hold the Professorship.
5   The Professor shall be governed by the regulations of the Statute for Sir T. Adams' Professorship of Arabic and certain other professorships in common, and the Professor shall comply with all the provisions of the said Statute.
6   He shall be required to reside within the precincts of the University for eighteen weeks during term time in every Academical year.
7   He shall be required to give one course of Lectures each of two terms at least, and to give not fewer than forty lectures, in every Academical year. The Scheme of Lectures shall be subject to the approval of the Board of Mathematical Studies.
8   The fee for attendance upon the Lectures of any one Term, given in accordance with the seventh Regulation, shall be one guinea, but no further fee shall be charged after the payment of two such fees. In cases of any courses of lectures beyond those prescribed in the seventh Regulation, the fee for attendance shall be fixed by the Professor with the sanction of the Vice-Chancellor.
9   The Professorship shall be *ipso facto* vacant if the Person holding it be elected into any other Professorship in the University.

However, the problem of finance continued to be difficult. The proposal to increase the capitation tax was rejected, and by the summer of 1870 no solution had yet been found.

Then, at the beginning of the autumn term in that year, the Vice-Chancellor of the University received a letter from the Chancellor, the Duke of Devonshire.

He wrote:

Holker Hall,
Grange,
Lancashire

October 10, 1870

My dear Mr. Vice-Chancellor,
   I have the honour to address you for the purpose of making an offer to the University, which, if you see no objection, I shall be much obliged to you to submit in such a manner as you may think fit for the consideration of the Council and the University.

I find in the Report dated February 29, 1869, of the Physical Science Syndicate, recommending the establishment of a Professor and Demonstrator of Experimental Physics, that the building and apparatus required for this department of Science are estimated to cost £6300. I am desirous to assist the University in carrying this recommendation into effect, and shall accordingly be prepared to provide the funds required for the building and apparatus, so soon as the University shall have in other respects completed its arrangements for teaching Experimental Physics, and shall have approved the plan of the building.

         I remain (&c)
         Devonshire.

The Duke, who wrote in his own hand, made a slight error in the date of the Report, which was in fact 27 February 1869

Compared with the amounts previously donated for science this was a very large sum; it set a new order of expenditure on university science. In the context of 1870 it was just as revolutionary as the expenditure on university science in 1970 in comparison with the scale a generation earlier. The new kind of laboratory, owned by the University, involved a large capital investment, like a factory. It belonged to the new industrial age: a manufactory for discoveries. It was different in principle from the earlier personal laboratory of the professor, which was often his private property, and was more like the private workshop of a master-craftsman.

After the Duke's gift the colleges became more helpful financially, and the proposals for new professorships, which had been recommended for some time, could now be implemented. Accordingly, on 28 November 1870, the Council of the University Senate proposed the foundation of a Professorship of Experimental Physics. Their proposal was confirmed on 9 February 1871.

Meanwhile, there had been much informal discussion on who the first Professor should be. In 1870 Sir William Thomson had reached the height of his fame. He was generally regarded as the most eminent British physicist, and the obvious first candidate. It appeared, however, that he did not wish to leave Glasgow. Besides having a department at the University organised exactly to suit his personal convenience, he had an interest in a flourishing Glasgow firm which manufactured his seventy patented and highly profitable inventions, especially for submarine cables and marine navigation.

Besides his industrial interests from which he gained a fortune of £161 923, he had a mansion at the coastal resort of Largs, and a large yacht. Thomson's teaching of physics in his own laboratory, after he had been appointed professor in 1846 at the age of twenty-two, had been one of the inspirations of the Cambridge development. He had, however, conceived this teaching in terms of the master-and-apprentice system. His preoccupation with his own interests and concerns prevented him from becoming the creator of systematic practical physics in Britain.

It is said that after William Thomson had refused to stand, Helmholtz was approached. Understandably, he did not wish to leave Berlin.

Maxwell (see figure 2), at the age of thirty-nine, was living in retirement at Glenlair. When it became certain that Thomson and Helmholtz would not stand for the professorship, Maxwell was pressed by many to become a candidate. There was doubt as to whether he could be persuaded to emerge from the pleasures of country life in Kirkudbright.

Among those who appealed to him was J. W. Strutt, the future Lord Rayleigh. He wrote to him from Cambridge, on 14 February 1871, that everyone was talking about the new professorship, and hoping that he would come. It seemed that Thomson had definitely declined.

There is no one here in the least fit for the post. What is wanted by most who know anything about it is not so much a lecturer as a mathematician who has actual experience in experimenting, and who might direct the energies of the younger Fellows and bachelors into a proper channel. There must be many who would be willing to work under a competent man, and who, while learning themselves, would materially assist him. . . I hope you may be induced to come; if not, I don't know who it is to be.

32

2 *Clerk Maxwell*

On the same day E. W. Blore, later Vice-Master of Trinity wrote, saying that 'Many residents of influence are desirous that you should occupy the post, hoping that in your hands this University would hold a leading place in this department'. Since it had been ascertained that Thomson would not accept, he need not fear the possibility of coming into the field against him, should he consent to stand.

On the following day, 15 February, Maxwell wrote from Glenlair to Blore

Though I feel much interest in the proposed Chair of Experimental Physics, I had no intention of applying for it when I got your letter, and I have none now, unless I come to see that I can do some good by it... I am sorry Sir W. Thomson has declined to stand. He has had practical experience in teaching experimental work, and his experimental corps have turned out very good work. I have no experience of this kind, and I have seen very little of the somewhat similar arrangements of a class of real practical chemistry. The class of Physical Investigations, which might be undertaken with the help of men of Cambridge education, and which would be creditable to the University, demand, in general, a considerable amount of dull labour which may or may not be attractive to the pupils.

A few days later Maxwell decided to stand. Stokes, who had been one of his most urgent persuaders, wrote to him on 23 February that he was glad he had so decided. On the following day, 24 February, Maxwell's candidature was announced, and on 8 March he was elected the first Professor of Experimental Physics. It was not at all widely understood at the time that the Cavendish had secured an even greater scientist than William Thomson or Helmholtz. Maxwell had not yet published his *Treatise on Electricity and Magnetism*.

He delivered his inaugural lecture on 25 October 1871, and his first regular courses in Liveing's chemistry lecture room. Liveing himself had included some instruction in heat, which was necessary for the students attending his own courses.

Maxwell was not sure how well his Introductory, or Inaugural, Lecture would be received. It was, in fact, to prove one of his most inspired efforts. Instead of delivering it in the Senate House, which was the usual place for such lectures, he was vague about the time and place, and gave it in an obscure lecture room. Horace Lamb, who was present, said there was an audience of only about twenty, nearly all recent mathematical Tripos graduates or students.

Later in the term, when he started his first course of professorial lectures, numerous members of the university, including the leading scientists Adams, Cayley and Stokes, came to the first lecture of the course, under the impression that it was his inaugural lecture. Maxwell, with a twinkle in his eyes, spent most of the lecture explaining the elementary principles of temperature measurement. Fixing his gaze on the puzzled Adams, Cayley and Stokes, he carefully explained to them the arithmetical relations between the Fahrenheit and Centigrade scales. It was suspected afterwards that Maxwell had engineered the whole incident!

2

Creating the Cavendish Tradition

Maxwell started his Introductory or Inaugural Lecture with a description of the way in which physical science had been pursued at Cambridge. He pointed out that the University, according to its tradition and law of evolution, while maintaining 'the strictest continuity between the successive phases of its history,... adapts itself with more or less promptness to the requirements of the times'. It had accordingly instituted a course of experimental physics.

This, while requiring the maintenance of 'those powers of attention and analysis', which had long been cultivated in the University, now demanded the exercise of 'our senses in observation, and our hands in manipulation'. Pen, ink and paper would no longer be sufficient, and more room would be required than that provided by a seat or desk, and a wider area than a blackboard. Owing to the munificence of the Duke of Devonshire, the material facilities for the full development of their future experiments 'will be upon a scale which has not hitherto been surpassed'.

Maxwell pointed out both the historical character and the scale of the change, and in this lecture he referred to the new laboratory, which was to be built, as 'the Devonshire Physical Laboratory'. He appears not to have regarded it as a memorial to Henry Cavendish, though he considered the editing of Henry Cavendish's unpublished papers on electricity as one of his most important duties. He devoted an intense effort to this work in his last years. The edition was completed and published in 1878. He repeated Cavendish's experiments, and made many others to elucidate how he had done them. It is the finest contribution to the history of science in the English language, and the only one in which a physicist of the first rank has made a fundamental historical study of the work of another.

Some have regretted that Maxwell devoted so much effort to this work, and have wished that he had restricted himself to purely conventional physical research. As in other ways, Maxwell's feeling for the significance of the history of science was characteristically more far-seeing.

Maxwell now proceeded to discuss by what means the University of Cambridge, regarded as a 'living body', might 'appropriate and vitalize this new organ', the 'outward shell' of which would soon rise before them. This should be done before they started on any detailed special scientific study. The course of study at Cambridge had long included natural philosophy as well as pure mathematics. To convey a sound knowledge of physics, and a correct grasp of dynamical principles had been regarded as one of its highest functions. This had been achieved to such an extent that it was now difficult to enter into the ideas of even such philosophers as Descartes, who had done their work before Newton had 'announced the true laws of the motion of bodies'. Indeed

the cultivation and diffusion of sound dynamical ideas has already effected a great change in the language and thoughts even of those who make no pretensions to science, and we are daily receiving fresh proofs that the popularization of scientific doctrines is producing as great an alteration in the mental state of society as the material applications of science are effecting in its outward life. Such indeed is the respect paid to science, that the most absurd opinions may become current, provided they are expressed in language, the sound of which recalls some well-known scientific phrase. If society is thus prepared to receive all kinds of scientific doctrines, it is our part to provide for the diffusion and cultivation, not only of true scientific principles, but of a spirit of sound criticism, founded on an examination of the evidences on which statements apparently scientific depend.

As fundamental to the Cavendish tradition, Maxwell placed the social duty to attend to the public's concern with science and to see that the scientific ideas communicated to it were sound.

One of the reasons for cultivating experimental physics was that the experimentalist, by 'the keenness of his eye, the quickness of his ear, the delicacy of his touch, and the adroitness of his fingers' would 'ensure the association of the doctrines of science with those elementary sensations which form the obscure background of all our conscious thoughts'.

In a course of experimental physics, either the physics or the experiments might be the leading feature. Experiments may be employed to illustrate a physical principle, or to exemplify a particular experimental method. 'In the order of time, we should begin, in the

Lecture Room, with a course of lectures on some branch of Physics aided by experiments of illustration, and conclude, in the Laboratory, with a course of experiments of research'.

The aim of the first kind of experiment was to throw light on a scientific idea so that the student could grasp it. This kind of experiment is so arranged that attention is concentrated on the idea in question, disentangled from other obscuring phenomena, 'as it is when it occurs in the ordinary course of nature'. To exhibit such illustrative experiments, encourage others to make them and develop the ideas on which they throw light, would be an important part of their duty.

The simpler the materials of an illustrative experiment, and the more familiar they were to the student, the more likely he was to acquire the idea intended. 'The educational value of such experiments is often inversely proportional to the complexity of the apparatus. The student who uses home-made apparatus, which is always going wrong, often learns more than one who has the use of carefully adjusted instruments, to which he is apt to trust, and which he dares not take to pieces.'

The use of self-made apparatus became a marked feature of the Cavendish tradition.

Maxwell held that it was very necessary that those who were trying to learn the facts of physical science from books, should be enabled by a few illustrative experiments to recognise these facts when they encountered them 'out of doors'. By looking at the facts of nature out of doors with a scientific eye C. T. R. Wilson was later able to produce marvellous results.

Science appeared in a very different aspect when we found out that it was not only in lecture-rooms, with pictures projected on a screen, that physical phenomena could be seen, but that illustrations of 'the highest doctrines of science' may be found 'in games and gymnastics, in travelling by land and water, in storms of the air and of the sea, and wherever there is matter in motion'.

The great achievements of the Cavendish in particle physics owe something to the tradition of ball games. Knowing from experience how a spinning ball behaves has been of assistance in interpreting physical phenomena from the time of Newton, who compared the swerve of his light particles to that of a spinning tennis ball, to the interpretation of cloud-chamber photographs of atomic tracks. The tradition of ball games, which both J. J. Thomson and Rutherford enjoyed, may well have been a factor in the discovery of the electron and the atomic nucleus.

'In experimental researches, strictly so called, the ultimate object is to measure something which we have already seen — to obtain a numerical estimate of some magnitude.' But, Maxwell continues

This characteristic of modern experiments — that they consist principally of measurements — is so prominent, that the opinion seems to have got abroad, that in a few years all the great physical constants will have been approximately estimated, and that the only occupation which will then be left to men of science will be to carry on these measurements to another place of decimals.

If this is really the state of things to which we are approaching, our Laboratory may perhaps become celebrated as a place of conscientious labour and consummate skill, but it will be out of place in the University, and ought rather to be classed with the other great workshops of our country, where equal ability is directed to more useful ends.

Maxwell then joined Bacon and Shakespeare in uttering a warning against human intellectual presumption:

But we have no right to think thus of the unsearchable riches of creation, or of the untried fertility of those fresh minds into which these riches will be poured. It may possibly be true that in some of those fields of discovery which lie open to such rough observations as can be made without artificial methods, the great explorers of former times have appropriated most of what is valuable, and that the gleanings which remain are sought after, rather for their abstruseness than for their intrinsic worth. But the history of science shows that even during that phase of her progress in which she devotes herself to improving the accuracy of the numerical measurement of quantities with which she has long been familiar, she is preparing the materials for the subjugation of new regions, which would have remained unknown if she had been contented with the rough methods of her early pioneers.

In the light of this principle, Rayleigh pursued the careful measurement of gases in order to check the accepted values. Discrepancies in the density of nitrogen obtained from different sources led him to the entirely unexpected and fundamental discovery of the noble gases.

Maxwell advocated that the Laboratory should participate in 'Experiments in concert', according to the principle proposed by Bacon. He quoted as an example the Magnetic Union, promoted by Alexander von Humboldt, worked out and directed by Gauss and Weber, and equipped with instruments made by Leyser. Through this international organisation, simultaneous observations were made on terrestrial magnetism in many different countries.

The increase in the accuracy and completeness of observations so obtained 'opened up fields of research which were hardly suspected to exist by those whose observation of the magnetic needle had been conducted in a more primitive manner'.

Any detailed account of the disturbances in the earth's magnetism would have to be reserved for a later stage in the Laboratory course. It had been learned that 'the whole character of the earth, as a great magnet, is being slowly modified', and that 'the interior of the earth is subject to the influences of the heavenly bodies'.

The system of automatic instruments recorded phenomena of 'the never-resting heart of the earth', and registered 'its pulsations and its flutterings, as well as of that slow but mighty working which warns us that we must not suppose that the inner history of our planet is ended'. A century later the great advances in terrestrial and cosmical magnetism confirmed yet again Maxwell's foresight.

Besides pointing out the potentiality of this research, Maxwell emphasised its effects on the development of measurement. The new methods of measuring forces were successfully applied by Weber to the numerical determination of all the phenomena of electricity. The electric telegraph (devised by Gauss for communicating the data of magnetic observations over distances), 'by conferring a commercial value on exact numerical measurements, contributed largely to the advancement as well as to the diffusion of scientific knowledge'.

Gauss had delivered scientists 'from that absurd method of estimating forces by a variable standard', which was still used in the Cambridge textbooks. Though dynamical equations were stated directly, they were 'usually explained there by assuming, in addition to the variable standard of force, a variable, and therefore illegal, standard of mass'.

Such were some of the scientific results that followed from 'bringing together mathematical power, experimental sagacity, and manipulative skill, to direct and assist the labours of a body of zealous observers. If therefore we desire, for our own advantage and for the honour of our University, that the Devonshire Laboratory should be successful, we must endeavour to maintain it in living union with the other organs and faculties of our learned body'.

He proceeded first to consider their relation to those mathematical studies that had flourished so long in Cambridge, dealing with the subject-matter of physics. They differed from 'experimental studies only in the mode in which they are presented to the mind'.

There was no more powerful way of introducing knowledge to the mind than by examining it from different aspects, and combining the results. It was 'natural to expect that the knowledge of physical science obtained by the combined use of mathematical and experimental research will be of a more solid, available, and enduring kind than that possessed by the mere mathematician or the mere experimenter'.

But what would be the effect on the University if men pursuing that course of reading that had produced so many distinguished mathematicians should now also have to work at experiments? Would they not break down under the strain? 'The Physical Laboratory, we are told, may perhaps be useful to those who are going out in Natural Sciences, and who do not take Mathematics, but to attempt to combine both kinds of study during the time of residence at the University is more than one mind can bear.'

It was not until theory was brought into contact with the practical, that the full effect of what Faraday called 'mental inertia' was experienced. This was not only the difficulty of recognising among concrete objects the abstract relations learned from books, but 'the distracting pain of wrenching the mind away from the symbols to the objects, and from the objects back to the symbols. This however is the price we have to pay for new ideas.'

When, however, the scientific faculty became more developed by the repetition of this exercise, the detection of scientific principles in nature, and the direction of practice by theory were no longer irksome. Indeed, it became a pleasure, so that 'at last even our careless thoughts begin to run in a scientific channel'.

It was true that mental energy is limited, and some students try to do more than is good for them. 'But the question about the introduction of experimental study is not entirely one of quantity. It is to a great extent a question of distribution of energy. Some distributions of energy, we know, are more useful than others, because they are more available for those purposes which we desire to accomplish.'

In the case of study a great part of the fatigue often arose from those mental efforts 'spent in recalling our wandering thoughts'. Here Maxwell was speaking especially for himself, for his mind teemed with ideas. If the disturbing force of mental distraction could be removed, concentration would be greater. 'A man whose soul is in his work always makes more progress than one whose aim is something not immediately connected with his occupation.'

There may be some mathematicians who cultivate mathematics purely for its own sake. 'Most men, however, think that the chief use of mathematics is found in the interpretation of nature.' A man who studies a piece of mathematics in order to understand a natural phenomenon, or to calculate the best arrangement for an experiment, 'is likely to meet with far less distraction of mind than if his sole aim had been to sharpen his mind for the successful practice of the Law, or to obtain a higher place in the Mathematical Tripos'.

He had known men who, when they were at school, could never see the good of mathematics, but later in life, when they had become eminent scientific engineers and found it useful, proceeded to master branches of abstract mathematics.

After discussing the advantage of practical science to the University, Maxwell asked what help the University might give to science, 'when men well trained in mathematics and enjoying the advantages of a well-appointed Laboratory, shall unite their efforts to carry out some experimental research which no solitary worker could attempt'.

Their principal experimental work at first would probably be in the illustration of particular branches of science. To this should be added the study of scientific methods, the same method being illustrated by applications in different branches of science. A course of experimental study could be imagined, which was based on a classification of methods, rather than of objects of investigation. But a better procedure would be a combination of both, in which care was taken 'not to dissociate the method from the scientific research to which it is applied, and to which it owes its value'. In this way he gave warning against concentration on technique for its own sake.

When elaborate experiments have been set up, and experimenters trained in the various parts of the necessary technique, advantage should be taken of the situation, not only to perform the experiment for which the set-up was originally made, but to apply the set-up, before it was dismantled, to the investigation of perhaps entirely different classes of physical phenomena.

This suggestion contained a hint of the modern method, where many experiments, proposed by scientists from different institutions, are collected, and then put through in a run in a central laboratory, in apparatus too complicated and expensive to be repeatedly set up and taken down.

Maxwell thought that the principal work of the Laboratory should be the learning and comparison of scientific methods, and the estimation of their value. It would be a result worthy of the University 'if, by the free and full discussion of the relative value of different scientific procedures, we succeed in forming a school of scientific criticism'.

In this Maxwell touched on one of the deepest qualities that distinguishes a great school of research and teaching, the creation of the finest kind of intellectual atmosphere. He endeavoured to endow the Laboratory with this intellectual quality from the start, and so encourage his successors to sustain and cultivate it.

He then discussed the question whether they might not be attributing to science too important a place in liberal education. Fortunately, there was no question that the University should be a place of liberal education, rather than preparing young men for particular professions. 'Though some of us may, I hope, see reason to make the pursuit of science the main business of our lives,' it must be one of their most constant aims 'to maintain a living connection between our work and the other liberal studies of Cambridge, whether literary, philological, historical or philosophical'.

A 'narrow professional spirit' may grow among scientists just as among men who practise other special businesses. But a university was the very place where this tendency to become 'as it were, granulated into small worlds, which are all the more worldly for their very smallness', should be overcome. It was not so long since any man who devoted himself to a science such as geometry was looked upon as a misanthrope, who had abandoned all human interests, and become insensible alike to the attractions of pleasure and the claims of duty. 'In the present day, men of science are not looked upon with the same awe or the same suspicion. They are supposed to be in league with the material spirit of the age, and to form a kind of advanced Radical party among men of learning'.

He was not there to defend literary and historical studies. He admitted that the proper study of mankind is man.

But is the student of science to be withdrawn from the study of man, or cut off from every noble feeling, so long as he lives in intellectual fellowship with men who have devoted their lives to the study of truth, and the results of whose enquiries have impressed themselves on the ordinary speech and way of thinking of men who never heard their names? Or is the student of history and of man to omit from his consideration the history of the origin and diffusion of those ideas which have produced so great a difference between one age of the world and another?

Besides conceiving the social relations of science as a proper concern for the members of the new Laboratory, Maxwell foretold the kind of situation in history that he himself was to assume. He gave the complete theory of radio waves before their existence was experimentally demonstrated, and laid the foundation of the future system

42

of radiocommunication, which has so impressed itself on the speech and thinking of men who have never heard his name.

It is true that the history of science is very different from the science of history.

We are not studying or attempting to study the working of those blind forces which, we are told, are operating on crowds of obscure people, shaking principalities and powers, and compelling reasonable men to bring events to pass in an order laid down by philosophers.

We recognize the men whose names are found in the history of science as men like ourselves. Not all of their investigations were successful. Some of the ablest men failed to find the key of knowledge, and the reputation of others has only given a firmer footing to the errors into which they fell.

The history of the development of ideas, whether normal or abnormal, is of all subjects that in which we, as thinking men, take the deepest interest. But when the action of the mind passed out of the intellectual stage, in which truth and error were alternatives, into the more violently emotional states of anger and passion, malice and envy, fury and madness, the student of science, though he is obliged to recognise the powerful influence of these wild forces 'is perhaps in some measure disqualified from pursuing the study of this part of human nature'.

It was impossible to enter into

full sympathy with these lower phases of our nature without losing some of that antipathy to them which is our surest safeguard against a reversion to a meaner type, and we gladly return to the company of those illustrious men who by aspiring to noble ends, whether intellectual or practical, have risen above the region of storms into a clearer atmosphere, where there is no misrepresentation of opinion, nor ambiguity of expression, but where one mind comes into closest contact with another at the point where both approach nearest to the truth.

Maxwell concluded his lecture by announcing that his first course of professorial lectures would be on heat. He said that, as the 'facilities for experimental work are not yet fully developed', he proposed to discuss the relations between the different branches of the science, rather than the details of experimental methods. He would begin with thermometry and calorimetry, and the relations between temperature and quantity of heat. Then he would discuss thermodynamics, in which the relations between the thermal and the dynamical properties of bodies would be explored.

'The principles of Thermodynamics throw great light on all the phenomena of nature, and it is probable that many valuable applications of those principles are yet to be made.' However, he would point out the limits of the science, and show that many problems, especially those involving the conservation of energy, 'are not capable of solution by the principles of Thermodynamics alone, but that in order to understand them, we are obliged to form some more definite theory of the constitution of bodies'.

Two theories had struggled for victory since the earliest ages of speculation: the theory of the plenum, and that of atoms and void. The theory of the plenum was associated with the doctrine of mathematical continuity, and its mathematical methods were those of differential calculus.

The theory of atoms and void leads us to attach more importance to the doctrines of integral numbers and definite proportions; but, in applying dynamical principles to the motion of immense numbers of atoms, the limitation of our faculties forces us to abandon the attempt to express the exact history of each atom, and to be content with estimating the average condition of a group of atoms large enough to be visible. This method of dealing with groups of atoms, which I may call the statistical method, and which in the present state of our knowledge is the only available method of studying the properties of real bodies, involves an abandonment of strict dynamical principles, and an adoption of the mathematical methods belonging to the theory of probability. It is probable that important results will be obtained by the application of this method, which is as yet little known and is not familiar to our minds. If the actual history of Science had been different, and if the scientific doctrines most familiar to us had been those which must be expressed in this way, it is possible that we might have considered the existence of a certain kind of contingency a self-evident truth, and treated the doctrine of philosophical necessity as a mere sophism.

Maxwell's genius was peering along the path that was to lead to quantum theory and the uncertainty principle. He said that during his first course of lectures he hoped to expound some of the evidence for the existence of molecules which, as presented to the imagination, are very different from anything with which experience has hitherto made us acquainted. He dwelt particularly on the philosophical significance of the identity of molecules. There was not only one molecule of a kind, but 'innumerable other molecules, whose constants are not approximately, but absolutely identical with those of the first molecule, and this whether they are found in the earth, in the sun, or in the fixed stars'.

He could not conjecture by what process of evolution the philosophers of the future would attempt to account for this identity in the properties of such a multitude of bodies. Was it possible that our scientific speculations have really penetrated beneath the visible appearance of things, which seem to be subject to generation and corruption, and reached the entrance of that world of order and perfection, which continues this day as it was created, perfect in number and measure and weight?

We may be mistaken. No one has as yet seen or handled an individual molecule, and our molecular hypothesis may, in its turn, be supplanted by some new theory of the constitution of matter, but the idea of the existence of unnumbered individual things, all alike and all unchangeable, is one which cannot enter the human mind and remain without fruit.

But what if these molecules, indestructible as they are, turn out to be not substances themselves, but mere affections of some other substances?

Maxwell felt that the absolute uniformity of the atomicity of matter had a particularly profound significance. He was convinced that it was inexplicable in terms of existing scientific ideas, and, indeed, that it was supernatural. He quoted J. F. W. Herschel's observation that atoms of any substance had the marks of being *manufactured* articles: they were all exactly alike. Maxwell regarded this as evidence that they had been manufactured by Somebody, namely God.

Rosenfeld has remarked that 'the importance of the uniformity of atomic structure for a coherent account of large scale regularities was properly emphasized by Maxwell, when he compared molecules with 'manufactured articles'. Incidentally

he carried the metaphor so far as to assert that the existence of these molecules, all 'manufactured' on the same pattern, pointed to the existence of a Manufacturer:— a delightful example of the indirect, but very substantial impact on physical thought of the new social outlook dominated by the development of mass production methods; natural philosophy no longer culminates into the apotheosis of the craftsman-potter or clockmaker—but in that of the self-satisfied factory owner.

From the social point of view, Herschel and Maxwell were using the characteristic theme of their own epoch of the new industrial society, the mass manufacture of identical products, to explain a feature of nature.

Maxwell said that 'no theory of evolution can be found to account for the similarity of molecules, for evolution necessarily implies change, and the molecule is incapable of growth or decay, of generation or destruction... None of the processes of Nature, since the time when Nature began, has produced the slightest difference in the properties of any molecule...'.

Maxwell's insight that identical atomicity had a very fundamental significance was to prove correct, though not in the way he interpreted it. What turned out to be indestructible was not the material molecules or atoms, but the atomicity of the changes in energy signifying changes of state. The indestructible atomicity was there, but it was in the laws of quantum theory.

Maxwell bequeathed to the Laboratory the problems of the origin of atoms, their nature and the apparent impossibility of their transformation. His equations for electromagnetic fields implied the equivalence of mass and energy. Planck's attempt to interpret black-body radiation in terms of them led him to invent the quantum theory.

The modern discoveries in the Cavendish Laboratory and elsewhere of the molecular mechanism of replication and the transmission of hereditary characters, have revealed the importance of integral quantities in a most direct way in the fundamental processes of life.

Besides having intuitions about the importance of integral quantities, he was also able to emancipate himself from subjection to physical models. He used these in aiding himself to discover the equations of the electromagnetic field, but when he had the equations he dispensed with the physical models.

This led Einstein to observe that: 'Since Maxwell's time, Physical Reality has been thought of as represented by continuous fields, and not capable of any mechanical interpretation. This change in the conception of reality is the most profound and the most fruitful that Physics has experienced since the time of Newton.'

Such, then, were the nature and quality of the thoughts with which Maxwell endowed the Cavendish Laboratory.

3

Building the Laboratory

While conceiving the intellectual tradition of the Laboratory, Maxwell also attended to its design and building. The University appointed a syndicate on 2 March 1871 to find a site and commission plans and estimates. The members were H. W. Cookson, the Vice-Chancellor and Master of Peterhouse, W. H. Bateson, Master of St John's College, Professors Adams, Humphry, Liveing, Maxwell and Miller, J. W. Clark and Coutts Trotter.

Maxwell visited Thomson's laboratory at Glasgow and Clifton's at Oxford to see what features of their arrangements might be adopted, and to examine their methods of teaching practical physics. Coutts Trotter also visited a number of laboratories for the same purpose.

The choice of site and the business negotiations in acquiring it were quite complicated. The syndicate had recommended the site in Free School Lane because it had easy access, but was sufficiently far from the main street to be comparatively free from vibration. But at the other side of the Lane, and facing where the Laboratory would be, was a side of Corpus Christi College. The College pointed out that their rights of 'ancient lights' would be infringed. £600 was paid to the College in lieu of these rights — nearly ten per cent of the cost of the whole Laboratory.

Maxwell indicated the general requirements to the architect, W. M. Fawcett of Jesus College. Fawcett produced a design, which proved more costly than the original estimate. The lowest tender for carrying out the design was by John Loveday of Kibworth, which was for £8450. After the Duke had seen the plans and heard of the extra cost, he wrote that he wished to meet this also, and 'present the building complete to the University'.

Loveday's tender was accepted on 12 March 1872, and building

47

soon commenced. While it was going up Maxwell lectured where he could. He wrote from Glenlair on 19 October 1872: 'Laboratory rising, I hear, but I have no place to erect my chair, but move about like the cuckoo, depositing my notions in the Chemical Lecture-room 1st term; in the Botanical in Lent, and in Comparative Anatomy in Easter'.

It was nearly complete by the autumn of 1873. The lecture-room and laboratory for students were ready, and students began to receive instruction in them during that term. The building was completed by Easter 1874, and formally presented to the University on 16 June 1874.

Hitherto, the Laboratory had usually been referred to as the Devonshire Laboratory. A Latin letter of thanks to the Duke as Chancellor of the University was read by R. C. Jebb, the Public Orator, in the Senate House. Besides thanking the Duke for his munificence, the contributions of his relative, Henry Cavendish, to the science of electricity were referred to, and the suggestion was made that the Laboratory should be given the name of the Cavendish family.

The Duke replied splendidly, in Latin, indicating his appreciation of the proposal, and then as Chancellor led the way to the new Laboratory. He unlocked the door, and handed the key to the Vice-Chancellor. The Laboratory was then inspected. Among those who took part were Leverrier from France, Stoletow from Moscow, Balfour Stewart and H. E. Roscoe.

The Duke's Latin reply, beautifully written out in his own hand, was subsequently handled by the Vice-Chancellor, Cookson, who cut it into pieces of a convenient size for filing. This is the form in which it is now to be seen in the University Archives.

To most Cambridge University men and Cambridge scientists the creation of the Cavendish Laboratory was seen and described entirely as a product of the internal development of the University and Cambridge science. Yet it was but one aspect of the general adaptation of British science to the new industrial society, which had been pursued during the previous half-century by many far-sighted men, from Babbage and Brewster to Playfair and Grove, and Tyndall and Huxley. In the 1860s the most sagacious of all was the physiographer, astronomer and instrument-inventor Alexander Strange, who as early as 1868 had clearly conceived the need for a ministry of science, advisory councils for civil and military science, and a national physical laboratory.

Mainly owing to his agitation, the Government set up on 18 May 1870 the Royal Commission to enquire into Scientific Instruction and the Advancement of Science. The Duke of Devonshire was appointed chairman, and in June 1870 started the sittings of this body, generally referred to as the Devonshire Commission. It continued for five years and published some four million words, containing a wealth of information and opinions, on the contemporary state and development of science in Britain, and what should be done to improve it.

The Secretary of the Commission was J. N. Lockyer, the civil servant and astrophysicist, who had been one of the most active promoters of the movement for the review and development of science in Britain. In 1869 under his editorship the journal *Nature* had been founded to assist in the promotion of this movement.

It was therefore not surprising that the editor of *Nature* was specially interested in the new Laboratory, which had been presented to the University by the Chairman of the Royal Commission of which he was Secretary.

On 25 June 1874, nine days after the opening, a detailed description of The New Physical Laboratory of the University of Cambridge was published in *Nature*. It opened with the observation that: 'The genius for research possessed by Professor Clerk Maxwell and the fact that it is open to all students of the University of Cambridge for researches, will, if we mistake not, make this before long a building very noteworthy in English science.' The forecast has indeed been brilliantly confirmed.

Because of its significance, *Nature* said that it proposed to 'put before our readers, as prominently as we can, a description of it'. The article is a striking example of scientific journalism taking deliberate action on the basis of insight.

It was illustrated with three figures of plans of the ground, first and second floors of the building, on the scale of 32 feet to the inch. The west front was built entirely of Ancaster stone, and the only ornate part of the building was the great gateway, thirteen feet high, and marked X on the figure of the ground-floor plan (see figure 3). The doors were very massive, and beautifully carved in oak. They bore the inscription *Magna opera Domini exquisita in omnes voluntates ejus*, from Psalm 101, Verse 2. In the Basic English invented by that eminent Cambridge man, C. K. Ogden, the translation is: 'The works of the Lord are Great, searched out by all those who have delight in them.'

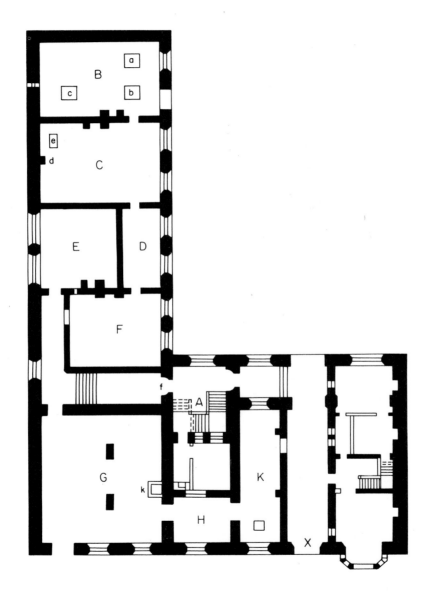

3   *Ground-floor plan of the Old Cavendish*

The arms of the Duke of Devonshire were depicted on the left-hand side over the gateway, and those of the University on the right. The motto of the Cavendish family, *Cavendo tutus*, 'Safe by being cautious', occupied the centre. The whole was surmounted by a beautifully carved statue of the Duke, in his robes as Chancellor of the University, and bearing in his hand the Cavendish Laboratory.

The rooms on the ground floor to the right of the entrance, consisting of a sitting room, scullery and kitchen, were occupied by the resident attendant. The external walls were two feet thick, and the foundations fifteen feet below the surface. With the exception of the west front, the tower and part of the lecture-room, they were built of brick, with Ancaster stone dressings. The tower, fifty-nine feet high, marked A in the plans, was about seventeen by fourteen and a half feet internally, and contained a very handsome stone staircase with carved oak balustrades. All the rooms on the ground and first floors, with the exception of the lecture-room were about fifteen feet high.

On the ground floor the room marked B was set apart for magnetic and other observations, requiring great steadiness and freedom from disturbance. It contained a brick pier eighteen inches high, at place *a*, with a stone top about four feet square. The pier was built separately from the tiled pavement of the room, commencing at a depth of eighteen inches below the pavement, and resting on a concrete foundation eighteen inches thick. The great electro-dynamometer of the British Association, the two large coils of which were each about half a metre in diameter, and containing about 225 turns of copper wire, was placed on this pedestal. The diameter of each circle of wire, and the distance between the two bobbins, which was about equal to their radius, had been accurately measured. The resistance of each coil had also been determined, so that all the electrical constants of the instrument were known with great accuracy. The electrical constants of all the other electromagnetic apparatus in the laboratory would be determined by comparison with this instrument. As the magnitude and position of each circle of wire in each coil was known, the coefficient of induction of one coil on the other could be determined.

The room contained two stone slabs each about four feet square, also on isolated piers. On one, marked *b*, was a unifilar magnetometer of the design adopted at Kew. In the upper part of the north wall of the room was a small window, to enable the direction of the meridian to be determined by astronomical observations. When the direction of the meridian had been determined, vertical mirrors would be placed opposite each other on the walls, each mirror being supported by screws so that it could be directly collimated. Three mirrors would be placed respectively on the north, east and south walls, but the fourth was to be placed on the west wall of room F, so as to be visible through the doorway from the mirror on the north wall of room B.

Room C was the clock and pendulum room. It had an isolated stone pier d, on which the principal clock was to stand. This would control electrically the other clocks in the building, and would be compared with the clock at the Astronomical Observatory. There was also a massive stone frame e, to carry an experimental pendulum.

The rooms B and C were each about thirty feet by twenty feet. The windows were furnished with wooden shutters, so that they could be darkened. Each window had a large stone shelf both inside and out, so that instruments could be set up partly inside and partly out. A small channel was provided to allow for the escape of rainwater.

The room marked E had two large windows on the north side, and was to be used exclusively for balances. The best balance they had at that time was an Oertling. It was robustly made, and sensitive to a difference of a milligram when each pan carried about two kilograms. This was considered sufficiently delicate for most physical purposes.

Room F on the ground floor was to be devoted to calorimetry and other heat experiments. It contained the apparatus devised by Maxwell for determining the viscosity of air. Three glass plates were caused to vibrate by means of a steel torsion wire between four parallel fixed plates in an airtight receiver. The amplitude of the vibration was measured by viewing the image of a graduated scale through a telescope, in a mirror attached to the vibrating plates.

Room G on the ground floor was used as a store room, where apparatus was brought directly in from the street, and unpacked. A lift k enabled pieces to be raised to the floor above. The room H was a workshop, furnished with a carpenter's bench, two vices, tools, etc. A five-inch self-acting screw-cutting lathe was to be added. The means would then be available for adjusting and repairing most of the apparatus required in physical research. Behind this minute workshop was a lavatory of about the same size.

Room K was the battery room, immediately below the lecture theatre on the first floor, so that wires could easily be carried to it through small hatches in the floor. The battery was to be of William Thomson's tray design, in which zinc plates are supported on porcelain cubes of one-inch edge. The interval resistance of each of these cells was about 160 mΩ.

A gas-holder containing oxygen was to be kept in this room. From it pipes would convey the gas to the lecture-room, so that oxy-coal-gas limelight would always be available. Looking back on this arrangement, it might have been thought that it was not particularly safe in view of the possibility of combustible mixtures of gases arising from the proximity of the battery and the gas-holder.

The south wall of room K was eighteen inches thick. It passed into the lecture-room above, independently of the floor, and carried the lecture table, which was accordingly free from floor vibrations. The floor was supported on two brick piers, isolated from the wall.

A long line was to be carefully measured on the stone pavement of the ground floor, with which other measures used in the laboratory were to be compared from time to time.

At f on the ground floor, an ancient stone gateway of the sixteenth century, which formerly served as the entrance to the Science Schools, was preserved.

On the first floor (see the plan shown in figure 4) at the east end, was the large laboratory, room L, intended for the general use of students. It contained ten large tables, and two more were to be added. Each of these tables, as in all the rooms on the first and second floors, was supported independently of the floor on beams resting on brackets in the walls. A standpipe passing through the centre of each table carried gas for four Bunsen or other burners. A closet with a good draught into the chimney was to be erected at the east end of the laboratory, for experiments producing objectionable fumes.

Most rooms were provided with fireplaces and a ventilator leading into the chimney near the ceiling. Water was laid on in all rooms, together with leaden sinks. A plentiful supply of rubber tubing lined with canvas was to be available in case of fire.

The Professor's private room was at M on the first floor. It was provided with two hatches communicating with the general students' laboratory, which could be readily opened or closed. Presumably, the professor would glance through them, from time to time, to see what was going on. The Thomson quadrant electrometer, made by White of Glasgow, was kept in the Professor's room.

The large apparatus room N was to be furnished with glass cases

53

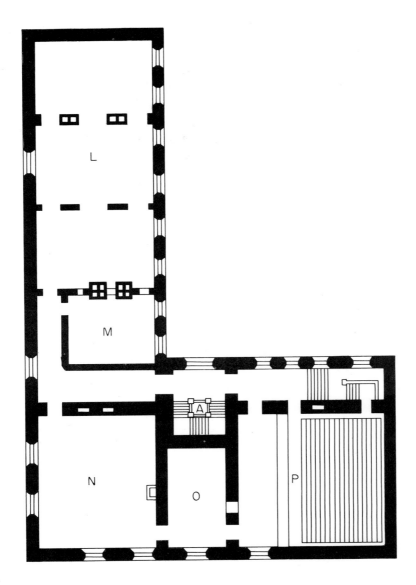

and cabinets, in which apparatus not in use was to be kept. A number of classical instruments were also to be kept there, such as the British Association's instruments for determining the original standard unit of electrical resistance.

Next to this was room O, the 'preparation room', communicating through a hatch with the lecture-room. In this, the preparatory arrangements for experiments illustrating lectures were to be carried out.

The lecture-room P was about thirty-eight feet long by thirty-five feet wide, and twenty-eight feet high, with seating rising at an angle of 30 degrees, for about 180 students. Three doors were provided, to give the audience adequate means of leaving. The room was panelled to a height of nine feet, above which the brick walls were relieved by handsome pillars. The oak lecture-table extended across the room.

The three windows were provided with wooden shutters, which folded together, and could completely darken the room. The shutters on the upper windows were operated by means of endless screws on a horizontal shaft. The ceiling consisted of wooden panels, those nearest the walls being perforated and communicating with horizontal shafts, to provide adequate ventilation. Three of the panels over the lecture-table were removable, so that a Foucault's pendulum, or other heavy apparatus, could be suspended over the table. Panels adjoining the north wall could be removed for the suspension of diagrams.

On the other three sides of the room the ceiling did not abut directly on the wall, but was covered in the form of a quadrant of a circle. This gave the whole room a very beautiful appearance. The lecture-room was in every respect a model of its kind.

On the second floor of the Laboratory (see the plan shown in figure 5), which in the *Nature* article is described as the third, the main rooms were intended for acoustics (Q), calculating and drawing (R), radiant heat (S), optics (T and U) and electricity (V).

The air in the electricity room V was to be kept dry by Latimer Clark's contrivance. This consisted of an endless flannel band, passing over a copper roller. This was heated by gas, so that the moisture in the flannel absorbed from the moist air in the room was driven off, and carried by the current of air feeding the burners inside the roller, together with the products of combustion, into a flue. Such was the device used for drying the air, and providing good insulation for the electrical instruments.

When the air in the lecture-room was too damp for good electrical

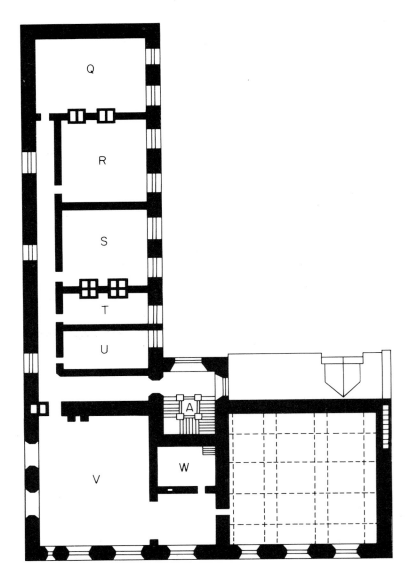

5 *Second-floor plan*

experiments, the electrical machine would be driven in the dry room on the second floor, and current conveyed from it by a wire through a small door in the wall of the lecture-room, seventeen feet from the floor.

A small dark room marked W was for photography and the development of negatives.

A small window in the west wall of the electricity room containing a heliostat enabled a ray of light one hundred and twenty feet long to be used for diffraction and other experiments.

All rooms were heated by hot-water pipes connected to a boiler in the basement. In room B on the ground floor, and the rooms near it on other floors, copper pipes were used for the heating, so as not to disturb the magnetic instruments.

A flight of steps in the tower led into the roof above the lecture room, and a few more steps into the highest room in the building, fifty feet above the ground. A Bunsen's water-pump was to be placed in this room, giving the water a fall of more than fifty feet. It would exhaust a large receiver, from which pipes would run to the various rooms. The rough vacuum provided could then be obtained in any of them by connecting to the vacuum pipe. If a better vacuum was required, further exhaustion could be secured by a Sprengel or other air-pump. Glass gauges on every floor would indicate the degree of vacuum available.

The top of the tower was to be equipped with a wooden mast carrying a pointed metal rod for collecting atmospheric electricity. This would be conveyed by a wire to the interior for experiments.

The floors of the laboratory were supplied with many hatches about eight inches square, so that wires could easily be suspended through the whole height of the building.

Besides the apparatus already mentioned, the Laboratory also possessed resistance coils of up to $100 \, \text{k}\Omega$, three mirror galvanometers of various construction, a three-feet six-inches glass-plate electric machine, a thirty-inch ebonite electric machine and a Holtz's electric machine. Finally, it possessed a hydraulic press of peculiar construction, made by Ladd & Co.

In the month following the opening of the Laboratory the Duke wrote to the Vice-Chancellor that it was his 'wish to provide all instruments for the Cavendish Laboratory which Professor Maxwell may consider to be immediately required, either in his lectures or otherwise'.

Maxwell had during the initial furnishing of the Laboratory kept the Duke carefully informed of what equipment he thought necessary. He provided lists of instruments, arranged according to the places and rooms in the Laboratory.

As the special duty of the professor was to teach the sciences of heat and electricity, and encourage research, the Laboratory would have to contain the equipment appropriate for these subjects, and also for whatever physical research seemed most important or most promising. He discussed with the Duke what was required, and explained that the complete furnishing of the Laboratory would take several years. He would himself contribute any instruments he had made in former years, and which might still be found useful; and also others which might be procured for special researches.

In his annual report in 1877 Maxwell announced that the Duke had now 'completed his gift to the University by furnishing the Cavendish Laboratory with apparatus suited to the present state of science'.

Nevertheless, the stock of apparatus was still small. Maxwell bought only the finest instruments by the best makers, and he also felt that the Duke should not be called upon too much.

But his interest in the Laboratory never declined. R. T. Glazebrook, who was appointed demonstrator in 1878, was also serving as Proctor in the University in 1887, the year of Queen Victoria's Jubilee, and had, in his latter capacity, to accompany the Duke as Chancellor of the University on an official visit to Windsor to present an address to the Queen. The Duke, who was then seventy-nine, took the train from Paddington to Windsor. Glazebrook met him at the station, and was at once closely questioned on the progress of the Laboratory, the number of students, and the work being done. The Duke was evidently fully acquainted with recent progress.

During Maxwell's professorship the number of students studying experimental physics was small. The Mathematical Tripos tradition was still very strong, and in spite of appropriate questions on mathematical physics being set, few candidates considered it worth while spending time on experimental physics. The situation was made worse in 1877–8, by the transfer of physics from the first part of the Tripos, to be combined with the more advanced parts of astronomy and pure mathematics.

Under this arrangement, very few students chose to read physics. The situation remained unremedied until 1892. New regulations then placed heat, elementary hydrodynamics and sound, and the simpler parts of electricity and magnetism into Part I. More candidates came to the Laboratory to prepare for this part, though still not in very large numbers. The more advanced mathematics and physics were placed in Part II.

In 1874 only seventeen men took the Natural Sciences Tripos, and few of these did physics. Twenty years later the number had much increased. In 1894 one hundred and three men took Part I of this Tripos, and sixteen Part II.

R. T. Glazebrook was one of those who came to work in experimental physics after they had taken the Mathematical Tripos. He entered the Laboratory in 1876, and the first experiments he attempted were on the measurement of electrical resistance. Not having had any previous experimental instruction, he did not know the principle of Wheatstone's bridge. Maxwell explained it to him. After these experiments Maxwell asked him to make some measurements on the Thomson tray form of Daniell's cells, which G. Chrystal and S. A. Saunder were then using in their research on the verification of Ohm's law.

Then Maxwell set him on the investigation of a stratified dielectric, consisting of layers of paraffined paper and mica. From these various experiments Glazebrook learned how to use different kinds of apparatus.

Maxwell did not conduct regular classes with a set drill of experiments arranged for examination purposes. He welcomed anyone who wished to work, but they were largely left to themselves to find out what apparatus to use, and how to use it.

Maxwell's approach to teaching was more like the modern individual method, in which the pupil is assisted to learn by attempting to find the answer to questions in which he is interested.

As Maxwell went round, inspecting experiments in progress, he was accompanied by his dog Tobi, which listened intently to the comments Maxwell murmured in undertones, as he looked down telescopes. Tobi was sometimes left in the Professor's room, while Maxwell went to Trinity, where dogs were not allowed. Tobi did not like this, and sometimes made so much noise that students in the adjacent room were distracted.

Glazebrook said that while Maxwell was always ready to listen, the answer did not always come at once. On one occasion he described a complicated difficulty to him. Maxwell replied: 'Well, Chrystal has been talking to me, and Garnett and Schuster have been asking questions, and all this has formed a good thick crust round my brain. What you have said will take some time to soak through, but we will see about it.' He came back a few days later. 'I have been thinking over what you said the other day, and if you do so-and-so it will be all right.'

Glazebrook was interested in optics, so Maxwell suggested he should study the optical properties of certain crystals. Glazebrook decided to investigate the wave surface of a biaxial crystal. Maxwell came in nearly every day to see how he was progressing. Later on, when Glazebrook was working regularly at the Laboratory, he saw him almost daily during term time for one and a half years.

4

The Early Atmosphere

Maxwell was Cavendish Professor for eight years. Most of his effort for the first half of this period was devoted to the planning and building of the Laboratory, and the organisation of his new department of experimental physics. During the second four years he had more time for personal researches, but already, in the last two of these, his health had begun to fail. Before this disaster he had endowed the Laboratory with the atmosphere of genius. He did this by the quality of his own contribution, and his selection of men and students to work in the Laboratory. In the brief time left he did not assemble many, but all were devoted. Maxwell always secured his effects by mind rather than quantity or force.

The aim of the physicists to get questions on the theory of heat, electricity and magnetism systematically introduced into the Mathematical Tripos was achieved in the new regulations for that examination established in 1873. Maxwell was brought in as an additional examiner, especially for the physical subjects. He was impressed by William Garnett, who passed as Fourth Wrangler, and whose papers exhibited an excellent knowledge of physics.

He offered the demonstratorship to Garnett, who was appointed early in 1874, before the Laboratory had been formally opened. Garnett served as Demonstrator until 1880.

The first student to work in the Cavendish Laboratory was W. M. Hicks, who became professor of physics at Sheffield, and the first Vice-Chancellor, after the Firth College was converted into a university.

Hicks said that the only regular experiments he made were measurements of electrical resistance and the use of the Kew magnetometer, to secure the best results for the measurement of the strength of the earth's horizontal magnetic field. He made an electrometer on the lines of that recently described by Lippmann. He found that trying to get it to work properly 'was worth any amount of routine measurement'.

Following Maxwell's development of the theory of electromagnetic waves, Hicks tried to measure their velocity experimentally. He invented a piece of apparatus, consisting of two coils, one large and the other very small. They were in the same circuit, and arranged to act on a light magnetic needle, where their combined force was zero. Hicks thought that as the large coil was much further from the needle than the small one, the effect of suddenly starting or stopping the current from the respective coils would take different times to reach the needle, hence causing it to move. The experiment could not succeed, but Hicks learned much in attempting it. He said that it was typical of one aspect of Maxwell's method.

The next student was J. E. H. Gordon. He worked under Maxwell on the accurate determination of electrical constants. He was the first to submit for publication the account of experimental work done in the Cavendish Laboratory. He read a paper to the Royal Society on 30 April 1875, on the rotation of the plane of polarisation of light in water, in a magnetic field of unit strength.

In June 1875 Gordon sent a letter to *Nature* on the recently discovered effect of light on the electrical conductivity of selenium. He found that not all specimens of selenium exhibited the effect, so that it must depend on the molecular state of the substance.

In 1879 he published a paper on the specific inductive capacities of various substances, especially transparent dielectrics. His aim was to test Maxwell's deduction from his electromagnetic theory of light, that the square of the refractive index is equal to the product of the dielectric capacity and the magnetic permeability. Gordon did not obtain an entirely satisfactory agreement with theory. He worked with great energy, making a laboratory at his home, where he continued the experiments Maxwell had inspired. He became a noted electrical engineer and pioneer of electric lighting.

Another early worker in the Laboratory was George Chrystal, the mathematician, and later professor of mathematics at Edinburgh. He had been Second Wrangler in 1875, and began research on the verification of Ohm's law. It was the first time that the law was submitted to a severe test.

In 1874 Arthur Schuster, just before his twenty-third birthday, read a paper to the British Association at Belfast on the accuracy of the evidence for Ohm's law. Schuster, who was based in Manchester, had been discussing the properties of electric currents with Balfour Stewart. This led him to examine the experimental evidence for Ohm's law, which he found to be very slender. While at Göttingen, Schuster had measured the deflections of a galvanometer magnet by a very weak current on which a much stronger alternating current could be superposed. He found that the latter always caused an increase in the deflection. He reported this result in his British Association paper. A committee was appointed to consider the result and investigate experimentally the accuracy of Ohm's law. It consisted of Maxwell, Everett and Schuster himself. As a consequence, Maxwell had set Chrystal to make an accurate experimental test of Ohm's law. His results showed that it was true to a high degree of accuracy. Maxwell commented that:

It is seldom, if ever, that so searching a test has been applied to a law which was originally established by experiment, and which must still be considered a purely empirical law, as it has not hitherto been deduced from the fundamental principles of dynamics. But the mode in which it has borne this test not only warrants our entire reliance on its accuracy within the limit of ordinary experimental work, but encourages us to believe that the simplicity of an experimental law may be an argument for its exactness, even when we are not able to show that the law is a consequence of elementary dynamical principles.

Crystal explained Schuster's anomalous result as due to a change in the longitudinal magnetisation of the galvanometer needle, caused by the alternating current in the galvanometer coil. This was not the only example of Schuster's peculiar gift for making faulty experiments and proposing fallacious theories, which nevertheless stimulated researches of great importance. Recently, his notion that large rotating solid bodies generate magnetic fields in space was revived by Blackett, and though in itself erroneous, it has led to major developments in geomagnetism, revealing new information about the geological history of the earth, the movement of strata and the processes of evolution in the earth's crust.

After the opening of the Laboratory the British Association's coils, prepared as standards of resistance to which the ohm could be referred, were transferred there from Kew Observatory. Chrystal and S. A. Saunder compared their resistances at frequent intervals, to determine whether they were subject to change. This work was taken up by J. A. Fleming and R. T. Glazebrook. It led to the general series

63

of investigations organised by Rayleigh, covering the whole field of electrical measurements. This work made the Laboratory the chief centre for the establishment and maintenance of accurate electrical units, until it was taken over by the National Physical Laboratory, formally opened under Glazebrook in 1902.

Schuster's contact with Maxwell inspired him with the desire to work under his influence. In 1876 he wrote to him for permission to work in the Cavendish Laboratory. Schuster was then twenty-four. Maxwell replied:

May 3, 1876

Dear Sir,
    It would do us all great good if you were to come and work in the Cavendish Laboratory. The very prospect of your coming has caused all our pulses to beat about one per minute quicker.
    The Schuster effect and the anti-Schuster effect have long been the objects of our regard, but we look forward to the time when these terms will have lost their significance as applied to the phenomena of electric conduction, and when every department of physics will have a recognised Schusterismus...

Maxwell even found a name for Schuster's pregnant errors! With his characteristic insight, Maxwell observed that 'Schusterismus' was worth noticing.

Later on Schuster came near to discovering the electron, and in 1898 wrote in an article in *Nature*. 'Astronomy, the oldest and yet most juvenile of sciences, may still have some surprises in store. May anti-matter be commended to its care!'

One of the points about Schuster was that he came of a wealthy family, and could afford to do and say what he liked, and not what was conventionally acceptable.

Maxwell continued: 'I know of no rule which would interfere with your working here (provided you do not let the gas escape), and members of the electrical committee of the B.A. have, by the desire of the Founder (the present Duke of Devonshire), liberty to make electrical experiments...'

He then described 'a very beautiful resistance coil of German silver wire 1/500 inch diameter', made by Garnett 'to test the Schuster effect'.

Maxwell concluded his letter with the remark that he was at present 'occupied with electric conduction through hot air, metallic vapours, flames, &c...'

64

Maxwell started the Cavendish Laboratory with an informal intellectual atmosphere, putting young men at their ease by amiable humour. Schuster went to Cambridge to discuss arrangements for coming to work in the Laboratory, and Maxwell invited him to stay in his house for a few days.

Schuster began research in October 1876, on a systematic investigation of diamagnetism. He had a vague idea that Weber's theory of molecular currents might be connected with the spectra of gases. This did not lead anywhere, and Schuster concentrated his interest more on spectra. He observed that the characteristic spectrum of the negative glow of oxygen persisted in the neighbourhood of the electrode for a short time after reversal of the current. He showed the effect to Maxwell, who was interested by it. However, others, and finally Schuster himself, failed to reproduce it; another of his 'Schusterisms'.

Sedley Taylor was one of the earliest to attend Maxwell's lectures. Taylor constructed a kaleidophone, containing a soap film which acted as an elastic membrane. When sound waves fell on the membrane they produced vibrations which caused the thickness of the soap film to vary. This, in turn, produced striking optical effects.

John Ambrose Fleming, later celebrated as the inventor of the electronic valve, gave up his appointment as science master at Cheltenham College, and in 1878 came to study under Maxwell. Fleming recorded that only three or four students attended his lectures at that time. He usually gave two courses a year, and Fleming was one of the very few to leave notes of the lectures on electricity and magnetism, which he described as 'splendid'. Fleming said he was surprised by this neglect of a teacher whom he regarded as in the very forefront of knowledge. During one term there was only one other attendant besides himself, Middleton, of St John's College.

In experimental research Fleming took up Chrystal's work on the comparison of the British Association unit of electrical resistance. He invented a special electrical bridge for determining the most probable value of the ohm from the British Association Committee's original measurements.

R. T. Glazebrook investigated the form of the wave surface in biaxial crystals, and confirmed that Fresnel's equations correctly represented the form of the wave surface. The issue of the *Philosophical Transactions* containing this paper also included two by Maxwell, the three Cavendish papers making up nearly one third of the whole volume. This showed that the Laboratory had already begun to establish itself as a centre of research.

A feature of the researches under Maxwell's direction was their wide range. Schuster expressed the opinion that the concentration of research after Maxwell's time along more particular lines was not altogether advantageous. Maxwell had a comprehensive interest in physical nature, any aspect of which might engage his attention, and to which he might direct the attention of others.

For example, in 1879 the Meteorological Council passed a resolution that a series of experiments should be made to test Dines's hygrometer, and compare it with the wet and dry-bulb thermometers, Regnault's hygrometer and de Saussure's hair hygrometer, for the determination of the absolute amount of water in air. The Chairman of the Council was the mathematician H. J. S. Smith, and Stokes was a member. They were instructed to arrange that the experiments should be carried out. Stokes approached Maxwell with the aim of their being done in the Cavendish. Maxwell, who was already suffering from his fatal illness, wrote to W. N. Shaw, who was working in Berlin, suggesting that he should undertake the experiments. Shaw became one of the most eminent meteorologists, and Director of the Meteorological Office.

J. H. Poynting, who had been working at Manchester with Balfour Stewart, included in his varied researches experiments on the measurement of the gravitational constant. He tried to detect a difference in the weight of a lead disc, when it was weighed in the vertical and in the horizontal positions. Poynting calculated that the effect might be magnified sufficiently to be experimentally detectable, if a large weight were placed under the balance pan. Maxwell became keenly interested in the research, and Poynting resigned his position at Manchester in order to continue his experiments in the Cavendish. Maxwell died soon after Poynting arrived. Schuster recalled that Joule had quoted a letter from Maxwell, in which he referred to the determination of the gravitational constant by weighing, remarking: 'You see that the age of heroic experiments is not yet past.'

Maxwell regarded the fostering of research in directions advanced by Henry Cavendish as one of the duties of a Cavendish professor. Cavendish had made the classical experiments on the determination of the gravitational constant, so, for this reason, work on that topic was worthy of special consideration, besides its intrinsic interest.

Maxwell used to go round the laboratory daily, saying a few words, or making suggestions. Schuster commented that he was most at ease in discussing the topic which was in his mind, and on this he would speak very freely. At the time when he was much concerned with Boltzmann's theorem on the equipartition of energy, he spoke on this to Schuster several times, remarking that if it were true for gases, it ought to be true for solids and liquids.

When others spoke to him, he often showed an apparent absent-mindedness. He might say nothing, so that it was impossible to tell whether he had taken the point, or he might make a remark that seemed to have no connection with it. On the following day, however, he might refer to the question, showing that he had been thinking about it. Schuster could never quite decide whether the question had remained unconsciously dormant in his mind until something reminded him of it, or whether he had consciously put it aside for future consideration. He often began a conversation with the remark: 'You asked me a question the other day, and I have been thinking about it.' This generally led to an interesting and original development of the subject.

Maxwell had an imaginative power of illustrating an argument by things which he knew were particularly familiar to his questioner. Once, when asked whether a certain so-called physical law was true or not, he replied: 'It is true in the three days but not in the five.' This was a reference to the Mathematical Tripos examinations, which at that time devoted three days to elementary papers and five days to more advanced subjects.

On one occasion Schuster remarked to Maxwell that he did not consider a certain paper had been worth publishing. Maxwell observed: 'The question whether a piece of work is worth publishing or not depends on the ratio of the ingenuity displayed in the work to the total ingenuity of the author.'

Maxwell devoted much of his experimental energies in his last years to repeating and elucidating Henry Cavendish's electrical researches. He was particularly fascinated by Cavendish's use of the human body for measuring electric currents before the galvanometer had been invented. The strength of the current was estimated from the intensity of the shock when it was suddenly passed through the body. Maxwell set up the appropriate apparatus, and everyone who came to the laboratory had to submit himself to the electric shocks, and be convinced that the method was sufficiently accurate to give consistent results. Not all visitors appreciated the test!

Schuster described how he had met Samuel Pierpont Langley, the inventor of the bolometer and pioneer of flying machines, during observations on Pike's Peak in the United States in 1878, in connection with a solar eclipse. In the following year Langley visited England, and expressed a wish to meet Maxwell. Schuster was working in the Cavendish at the time, and said he was sure Maxwell would welcome a visit, for he had spoken very highly of Langley's method of eliminating the personal equation in transit observations. When Langley arrived Schuster took him to the room where Maxwell was working in his shirt-sleeves, with each hand in a basin of water, comparing the intensities of two electric currents by sending them through his own body, and estimating the comparative strengths of the respective sensations. 'Every man his own galvanometer', said Maxwell, and would talk of nothing but Henry Cavendish. He tried to persuade Langley, unsuccessfully, to take his own coat off and try it, assuming that everyone would share his own intense interest in Cavendish's work and experiments.

Langley, who wanted to talk about his own scientific problems, could not get a word in. He was severely disappointed by this, and the affront, as he saw it, to his dignity. As he left the Laboratory he said angrily to Schuster: 'When an English man of science comes to the United States we do not treat him like that.'

Maxwell devised an improved form of Cavendish's experiment for proving that the whole charge of an electrified spherical conductor resides in its surface. The experiments were carried out by Donald MacAlister, who was Senior Wrangler in 1877, and later Principal of Glasgow University.

J. J. Thomson heard Maxwell lecture only once, and not in any of his courses. This was his Rede Lecture of 1878 on *The Telephone*, delivered in the Senate House. The apparatus was prepared in the Laboratory, where wires were laid between the basement and the attics. Maxwell said that the main difficulty was preventing the demonstrator from speaking so loudly that his voice, coming through the brick walls, drowned the reproduction in the receiver.

In the Lecture the telephone was demonstrated by a tune played in the Geological Museum and received in the Senate House; the distance between transmitter and receiver was less than twenty yards. J. J. said it was a very pleasant lecture to listen to. The audience were entertained by many flashes of the kind of humour characteristic of Maxwell's light verse.

In his vote of thanks to the lecturer the Vice-Chancellor explained how they had asked everyone they could think of to give it. All had refused, and he did not know what would have happened if at the last moment someone had not suggested Professor Maxwell.

William Garnett acted as demonstrator during the whole period of Maxwell's directorship. He superintended the ordinary laboratory practice. Maxwell frequently instructed students, especially those embarking on research after having received some training in physical measurements.

Garnett said that Maxwell's idea at first was to attach to the Laboratory a small band of graduates, each of whom would undertake a definite piece of work after having received his training in measurement. He regarded the Kew magnetometer as an excellent instrument with which to gain this experience, since it involved reading scales, making time observations and counting the beats of a watch against the vibrations of the magnet. All students on first entering the Laboratory were invited to measure the horizontal component of the earth's magnetic field.

At this time it was not contemplated that the Laboratory would be much used by undergraduates. The practical examination in physics for the Natural Science Tripos had not yet been introduced. About three years after the Laboratory was opened, elementary courses for medical and other students were started. These were given by Garnett and soon became popular, forming some of the largest science classes in the University. He was a good amateur carpenter and mechanic, and he helped students with their apparatus.

The first undergraduate student in the Laboratory was H. F. Newall, who later became professor of astrophysics at Cambridge. In his first term, against the advice of his tutor, he decided to work in the Laboratory. He afterwards told his close friend J. J. Thomson that he did not find the going easy. He entered the Laboratory in 1876. His tutor had tried to dissuade him on the ground that he could not hope to benefit from it. In his first week at Trinity in October 1876, Coutts Trotter told him that the Cavendish Laboratory was intended for the research work of graduates, and there was as yet no accommodation for young students. This was all the more striking because Coutts Trotter had been one of the Cambridge men who had done most to promote the foundation of the Cavendish Laboratory.

Newall, however, was an unusual young man. He had come from Rugby School, where the nowadays much criticised Head Master,

Dr Temple, had promoted the teaching of experimental science. Rugby became one of the pioneer schools in linking the learning of science with experimental teaching. The boys were not restricted to sitting in classrooms listening to unillustrated lectures and reading textbooks. They were given lectures by able science teachers, illustrated by striking experiments, and science laboratories, built in 1869, were provided, where they could themselves make experiments. Newall said that two terms' work in the new Science Schools early in 1876 under an able teacher was sufficient to inspire him with hope that the opportunities he would find in Cambridge would be fruitful.

He was told that there was no immediate opportunity for him. Writing to one of his friends, who was interested in the expanding opportunities for experimental physics at Cambridge, he said that he could not see how they could 'ever hope to get old students, unless they provide accommodation for young ones'. This incident illustrates how the development of science teaching in schools was a factor in creating the atmosphere that led to the university teaching of experimental physics and the foundation of the Cavendish Laboratory.

However, when Newall went to the Laboratory and saw Garnett, the demonstrator went to a good deal of trouble to find simple apparatus for him. Maxwell frequently came into the Laboratory, but the young Newall was in such awe of him that he used to hide motionless behind one of the brick pillars, lest Maxwell should find him puzzling over things that were made clear in the most elementary books. One day, however, Maxwell came on him unawares. He was struggling with experiments on the caustics and focal lines of a concave mirror in a mahogany frame, a piece of apparatus that became very well known to Cavendish students.

Newall felt embarrassed and suspected that Maxwell did, too. Maxwell was, however, apparently well-acquainted with the vagaries of this particular piece of apparatus, and commented: 'Yes, I have a shaving glass at home that performs much better. I'll bring it you.' Maxwell forgot to bring it, but, said Newall, he had in a moment by his simple understanding of the situation turned his 'awe of his power into enthusiasm for his quiet human helpfulness'.

In order to secure practical acquaintance with physical methods the young student was allowed to use a delicate balance in a room where J. H. Poynting was making his experiments preparatory to the re-determination of the density of the earth. As the apparatus was very sensitive it was important to see, before entering the room, whether the operator was using it; 'stalking his balance' was the phrase used. The balance was observed through a hole in the wall by means of a telescope, and had to be protected both from the gravitational pull of the mass of the operator's body, and the effects of his body's temperature.

Newall recalled how the original uses of some of the rooms were soon changed. Pieces of iron and electromagnets accumulated in the room on the ground floor specially built free from iron for magnetic observations. Disturbances of galvanometers in the room above, thought to be due to terrestrial magnetic storms, were traced to the magnetic apparatus in the supposedly non-magnetic room below.

Newall attended a course of Maxwell's lectures in 1878. The *Elementary Treatise on Electricity*, edited by Garnett in 1881, followed them very closely. Newall found the lectures very attractive, and he was deeply impressed by the contrast between the careful precision with which Maxwell chose his words in defining terms, and the apparently rambling remarks he made in explaining their use. 'He spoke quite informally as if conversing with a friend, and sometimes even as if speaking to himself. Every now and then a humorous remark would fall from him to the obvious bewilderment of some in the small audience, but much to the unrestrained amusement of Garnett, who nearly always attended the lectures, sitting in a chair on Maxwell's side of the lecture table.'

Newall recalled 'the sudden illuminating effect' of Maxwell's few words in explaining the definition of a magnetic field. He was very fond of using such phrases as 'the quality or peculiarity in virtue of which...' For example, he would define electromotive force as the quality of a battery or generator, in virtue of which it tends to do work by the transfer of electricity from one point to another. He cautioned against thinking that a magnetic field with a strength of so many c.g.s. units meant so many dynes. The field was 'there', whether or not an isolated unit magnetic pole was set in it experiencing the force of so many dynes. The dynes came in because of the arbitrary choice of a special aspect of the peculiarity of the field. The choice had to be consistent with other relations, such as the rotation of the polarisation of light. His words made 'the impression of an illuminating flash', and 'the raw student was made in an instant to wonder whether the entry of dynes into the measure of gravity did not leave the door open for glimpses of similar outstanding differences in peculiarities in the field of gravitation'. The pregnancy of Maxwell's thought is to be seen in many places in his writings; Newall described how it emerged in his speech.

Newall considered that Maxwell's precision of thought and clearness of statement were particularly exemplified in Appendix C to the Second Report of the British Association Committee on Electrical Standards (1863). It was signed by Maxwell and Fleeming Jenkin, but there was little doubt who wrote the vital parts. Newall said that 'it would be difficult to find so marked an example of the kind of spirit, which Maxwell's influence left as a heritage specially to be cherished in the Cavendish Laboratory'. The Appendix C opened with the Statement:

1. Object of the Treatise

The progress and extension of the electric telegraph has made a practical knowledge of electric and magnetic phenomena necessary to a large number of persons who are more or less occupied in the construction and working of the lines, and interesting to many others who are unwilling to be ignorant of the use of the network of wires which surrounds them. . .

Between the student's mere knowledge of the history of discovery and the workman's practical familiarity with particular operations which can only be communicated to others by direct imitation, we are in want of a set of rules or rather principles by which the laws remembered in this abstract form can be applied to estimate the forces required to effect any given practical result.

We may be called upon to construct electrical apparatus for a particular purpose. . . If we are unable to make any estimate of what is required before constructing the apparatus, we may have to encounter numerous failures, which might have been avoided if we had known how to make a proper use of existing data.

All exact knowledge is founded on the comparison of one quantity with another. In many experimental researches conducted by single individuals, the absolute values of those quantities are of no importance; but whenever many persons are to act together it is necessary that they should have a common understanding of the measures to be employed. The object of the present treatise is to assist in attaining this common understanding as to electrical measurements.

Newall considered the thirty-five page treatise 'as brilliant a piece of elementary teaching as was ever published'. One could not have had a more telling expression of faith from a master of science, 'showing in a really vital way his view of the importance of getting practical engineers to base their procedure on a real understanding of elementary scientific principles'.

Maxwell was about thirty-two when he expressed these views. His grasp of the technological need to pursue teaching and research in science was characteristically profound and mature. It was such qualities that enabled him to inspire the body of thought leading to the foundation of the Cavendish Laboratory, and the character of its programme and achievement.

While Newall pursued his study of experimental physics as a more or less lone undergraduate, he was surrounded by Chrystal, Schuster, Gordon, Poynting, Glazebrook, Fleming and other senior research workers, pursuing advanced experimental investigations. He relieved his scientific solitude by occasional envious visits to one of these.

He began to realise the wisdom of Coutts Trotter's unpalatable advice to wait a year or two until the regular classes for elementary students had been organised, in spite of the pleasure of trying to teach himself physics by practical experiments. He began to wish that all Tripos examinations were at the bottom of the sea, and the energy so released might then have been used for developing a course of experiments for the illustration of methods of physical measurement.

When he looked back in later years he could only hope that 'the mere presence of a forlorn undergraduate in the Laboratory in 1876 and 1877 may have had some influence in hastening the organization'.

The first organised course of practical experiments was advertised in the *University Reporter* in 1879. Newall did not attend it, though he was working in the Laboratory during the term. He could recall the profound impression caused by the news of Maxwell's illness, and the widespread mourning for his death.

In the first year of Rayleigh's professorship, Glazebrook and Shaw established the practical course in its classical style, and the number of students taking it rapidly increased.

Newall had meanwhile graduated, and returned to the Laboratory as a senior student. He recalled that his attitude then changed. His satisfaction at the establishment of the courses was tempered by the inconvenience caused to himself and others by the appropriation to the class-teaching of many instruments that they would have found very useful for their researches. Those who were bold enough borrowed pieces of the classroom apparatus, and often while it was in use in a private research, a messenger would arrive from the classroom, presenting compliments, and requiring its return. If the researcher did not then arrange his experiment at a time that did not clash with the class, a senior person would reprimand him!

Newall left Cambridge in 1881. When he returned in 1885 he found the development of teaching strikingly complete, both in the elementary and advanced classes.

Garnett introduced into the Laboratory workshop a few engineers' and joiners' tools, and engaged two instrument-makers to work there occasionally. When James Stuart was elected Professor of Mechanism in 1875, the workshops of the Engineering Department were started. The two instrument-makers were taken over, and the more important instrument work went with them. Garnett claimed that 'The little workshop in the Cavendish Laboratory was the starting-point of the University Mechanical Engineering Department.' He said that, as demonstrator

My own work was mainly to lecture to the more elementary students, and render mechanical assistance to the Laboratory students, and to those engaged in their own researches. I was occupied a good deal in devising experiments for lecture illustration and making suitable apparatus. Whenever I came upon a new piece of work I showed it to the students, but did not publish it; and while I assisted in a good many of the investigations which Maxwell suggested, I never embarked on a continuous research of my own.

Garnett later became Educational Adviser to the London County Council.

For several years Maxwell would not open the Laboratory to women, but when at last he consented, women were admitted while he was away on Long Vacation in Scotland. Garnett taught a class who were determined to go through a complete course in electrical measurement during the few weeks that the Laboratory was open to them.

Laboratory examination in practical physics was introduced tentatively in 1874. In that year candidates were presented with questions on chemistry, mineralogy and physics; Maxwell was the examiner in physics. The physics involved such questions as the determination of the focal length of a lens. Three similar simple questions were set in 1875. In 1876 the Natural Sciences Tripos was divided into two parts, one taken in June, and the other in December. In the second part a separate Laboratory paper was set, containing eight questions, of which not more than four were to be attempted.

Students were left largely to cope with their difficulties unaided. This stimulated independence and experience, but it was not always appreciated by those who were apprehensive about their examinations. Some thirty years later Schuster contrasted this with the 'over-instruction which is current in many laboratories at the present time'. Since the period of Rutherford's professorship there has been a change rather in Maxwell's direction, with more variety of research, and devotion to theory as well as experiment. In 1970 the change was marked by the alteration in the title of the professor, who, 'from the end of Professor Mott's tenure' was to become the 'Cavendish Professor of Physics', 'experimental' being left out.

A. P. Trotter, who became electrical adviser to the Board of Trade, reminded Schuster that when he was a student

Either I or someone else made a light-hearted attempt at measuring the current of the Holtz machine by means of a high-resistance Elliott galvanometer. Somebody scolded the culprit, but I could not see the enormity of the offence. Among the apparatus that I can remember using was a small spectroscope with which I measured the angles and refractive index of a prism. I had a book of trigonometrical tables in my hands for the first time. I knew all that the extras of the Little-go required as to trigonometry, but I could not understand why the sines and cosines stopped at 45°. I wanted an angle of about 60° and thought it must be in a second volume! I remember you allowing me to see the double sodium line in your big spectroscope, as a treat. I spent several days with the Kew magnetometer, and Garnett explained how to calculate the moment of inertia of the magnet—a quantity of which I had never heard.

I remember trying to measure a low resistance by the logarithmic decrement of a low-resistance Elliott galvanometer. The galvos were ordinary Thomson type. There was an optical bench for experimenting with lenses, and a good heliostat. The big cathetometer with a 24–25 vernier bothered me a good deal. There was a gramme dynamo with an ingenious driving arrangement, the construction of which I did not understand until two or three years after I went down.

The early years in the Cavendish Laboratory correspond, in a way, with those in the Boulton factory at the beginning of the Industrial Revolution. They did not mark the introduction of mass-production methods into research, just as Boulton did not introduce them into engineering production. The Cavendish Laboratory started as a kind of factory where individuals were assembled to do research, and assisted with facilities. Boulton assembled skilled craftsmen, providing them with one roof and better surroundings and facilities than they had possessed in their former individual workshops.

When Schuster arrived with the intention of doing research under the professor, there was no recognised place for him in the Cambridge university system. He was unwilling to sit for the Cambridge University graduate examination, since he had already graduated elsewhere. William Garnett presently succeeded in getting him admitted as a fellow-commoner at St John's, though not in any recognised category. Schuster thus became the first research student in the modern sense at the Cavendish.

75

Schuster said that it was difficult to convey 'how powerfully Maxwell's personality dominated the Laboratory and united those who worked in it'. He did not find it easy to define Maxwell's position as a teacher.

His influence on scientific thought was mainly exerted through his writings, and reached its full effect only after his death... The power which he exercised by personal contact was principally a moral one. Having originated new and fertile ideas in all branches of Physics, Maxwell might easily have found students eager to work out in detail some problem arising out of his theoretical investigations. This would have been the recognized method of a teacher anxious to found a 'school'; but it was not Maxwell's method. He considered it best both for the advance of science, and for the training of the student's mind, that everyone should follow his own path. His sympathy with all scientific inquiries, whether they touched points of fundamental importance or minor details, seemed inexhaustible; he was always encouraging, even when he thought a student was on a wrong track. 'I never try to dissuade a man from trying an experiment,' he once told me, 'if he does not find what he wants, he may find out something else'. It was the seriousness with which he discussed all ideas put before him by his students that, from the beginning, gave the Laboratory its atmosphere of pure and unselfish research'.

Schuster's view is particularly valuable because he later proved himself one of the greatest creators of a laboratory and a tradition. He planned and built the new Physical Laboratory at Manchester, and then retired at fifty-six in order to secure Rutherford from Canada as its new director. There Rutherford founded nuclear physics. Schuster's experience under Maxwell was a major factor in the development of his own genius as a laboratory planner.

His view of Maxwell was also particularly valuable because he had studied previously at Heidelberg, Göttingen and Berlin. At Manchester he had taught as well as studied, and had given a systematic course on Maxwell's theory of electricity and magnetism, before he went to Cambridge to work under Maxwell. One of his three students at Manchester was J. J. Thomson, still a youth in his teens.

Maxwell's teaching changed the Cambridge tradition of treating electricity as a subject of applied mathematics. It directed more attention to its physical nature, and thus to physical and experimental investigation.

Schuster described Maxwell as 'the most enlightened and progressive philosopher then living'. He thought it was impossible to say whether Maxwell was satisfied with the progress of the Laboratory. It grew steadily but slowly.

Schuster also discussed why the young physicists who surrounded

Maxwell in the Laboratory made little attempt to prove the existence of electromagnetic waves. He thought it largely due to Maxwell's letting his students go their own way and choose their own problems. Schuster knew from personal knowledge that the desirability of experimental proof of Maxwell's theory was realised by Cambridge men, but they were daunted by the apparent difficulty of the experiments, especially the weak intensity of the electromagnetic waves at a distance, and the delicacy of the measurements needed to detect them.

Another aspect of the question was the prestige of mathematics at Cambridge. A theory that explained so much must be true, whether or not there was experimental proof. The appeal to intellectual elegance was, to many, decisive.

Maxwell was one of those geniuses who were interested in everything human. He could attend equally to the curious mistakes of the beginner and the deepest problems of natural philosophy.

While he was teaching, building, experimenting and composing profound works he also wrote numerous shorter articles of outstanding quality. Among these were many for the ninth edition of the *Encyclopaedia Britannica*. Nowadays one does not look for the beginning of new knowledge in such works, but several of Maxwell's articles were to prove classical. In the article on *Ether*, under the heading 'Relative motion of the ether', he remarks: 'If it were possible to determine the velocity of light by observing the time it takes to travel between one station and another on the earth's surface, we might, by comparing the observed velocities in opposite directions, determine the velocity of the ether with respect to these terrestrial stations'.

This paragraph inspired the Michelson–Morley experiment. Its result, together with Maxwell's own electromagnetic theory of light, provided the material for the deduction of the theory of relativity.

In another *Encyclopaedia Brittanica* article Maxwell gave the first international recognition to the importance of Willard Gibbs, and his work on thermodynamics and heterogeneous equilibria. Maxwell made models with his own hands of Gibbs's thermodynamic surfaces. He sent one to Gibbs, and another is in the Cavendish Laboratory's museum (see figure 6).

Maxwell's researches on the mathematical description of electromagnetic forces in space led him to study the geometry of space, or topology. He made a number of contributions on this subject, together with P. G. Tait, to the London Mathematical Society, of which he was a keen supporter. Henri Poincaré, at the beginning of the

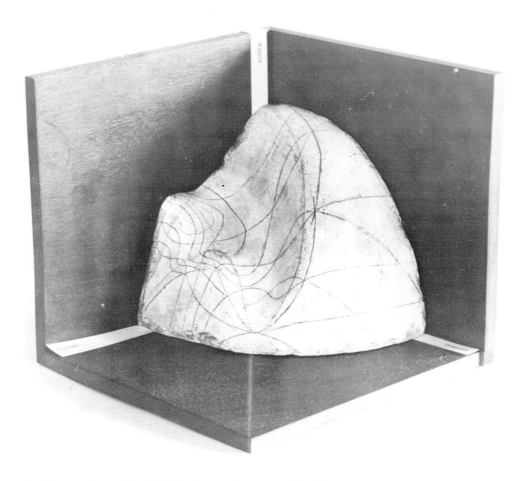

6   *Maxwell's model of Gibbs's thermodynamic surface for water*

twentieth century, sent papers on topology to be published in England in honour of the topological work of Maxwell and Tait.

Maxwell added an engineering sense to his other gifts. Besides his fundamental work on the measurement of strains in structures, he has been saluted by Norbert Wiener as the founder of modern cybernetics. Maxwell's theories were elucidated in his paper *On Governors* (both of steam-engines and scientific instruments), published in 1868.

Maxwell's concrete achievements were immense, and yet his premonitions were even greater. But his greatest gift and most important was an unsurpassable standard of intellectual inspiration. For Max

78

Planck, Maxwell's task in physics had been 'to build and complete the classical theory, and in doing so he achieved greatness unequalled'.

Yet to that must also be added his greatness as the designer, launcher and inspirer of the Cavendish Laboratory.

5

Rayleigh

Maxwell died in 1879, attended in his final illness by his favourite physician, Sir George Paget, the future father-in-law of J. J. Thomson. The Cavendish Laboratory grew in a highly complex and multiply connected society and social life, the product of much more than a chain of scientific ideas.

As the professorship had in the first instance been established only for the tenure of its first holder, University action was necessary to continue it. Those concerned with physics were therefore anxious to secure a candidate who would be a worthy successor.

The first thought was again of Sir William Thomson. The outstanding young man of ability was John William Strutt (see figure 7), who in 1877 had succeeded his father as the third Lord Rayleigh. Strutt, who was just under thirty-seven years of age when Maxwell died, had already been suggested by some in 1871 as a candidate for the original appointment. He had added his voice to those who had successfully persuaded Maxwell to stand. Now an influential group of Cambridge men presented a memorial to Rayleigh asking him to stand as Maxwell's successor.

We the undersigned members of the Senate of the University of Cambridge, whose names are on the Electoral Roll, are of opinion that in the probable event of the Establishment of the Professorship of Experimental Physics which has been terminated by the lamented death of Professor Clerk Maxwell, it would tend greatly to the advance of Physical Science and the advantage of the University that Lord Rayleigh should occupy the Chair.

The signatories included J. C. Adams, A. Cayley, J. Challis, G. H. Darwin, J. Dewar, R. T. Glazebrook, W. Hicks, W. D. Niven, E. J. Routh, W. N. Shaw, G. G. Stokes and I. Todhunter.

The Duke of Devonshire wrote to him on 15 November 1879:

7 *Rayleigh*

I understand there is a strong wish among the Cambridge residents originating with those who take a special interest in the experimental along with the mathematical treatment of Physics that your Lordship would consent to accept the chair of Experimental Physics, which has become vacant by the death of the much lamented Professor Clerk Maxwell.

Though it is perhaps somewhat unreasonable to ask you to undertake duties the discharge of which would involve heavy demands on your time, and might very probably be attended with no small personal inconvenience, I feel so strongly the advantage the University would derive from your acceptance of the office, that I hope you will allow me as Chancellor of the University, and also as taking a special interest in the Professorship, to support the appeal which I am told is about to be made to you, and to express a hope that you will consent to take the proposal into your favourable consideration.

Meanwhile, Rayleigh wrote to Sir William Thomson to ascertain whether he would again decline to stand. Thomson replied that he had been strongly attracted to Cambridge in 1871, but had decided then to stay in Glasgow; now he felt that his destiny would be in Glasgow for the rest of his life.

He thought that if Rayleigh could 'see his way' to take the chair, it would be for the benefit of the university and of science too, for the Cavendish Laboratory would provide him with 'means of experimenting and zealous and duly instructed assistants and volunteers', which would enable him to do more than if he were independent. If, however, he felt that the chair might be an uncongenial burden, and cause him to occupy time that might be more productive in independent research, he advised him to hold it only 'till some of the younger men might show suitable qualifications'.

Thomson viewed research colleagues and the professorship as possibly a hindrance to personal activity, and a burden to an otherwise independent gentleman.

On 17 November 1879 Rayleigh wrote to his mother, who did not like the idea of his becoming a professor, that neither he nor his wife were attracted to living in Cambridge, but in view of the agricultural depression, which had limited the family's means, perhaps he ought to take the post for three or four years, if they could not find anyone else fit for it.

Stokes thought there was no objection to his taking the post for a short time. Rayleigh's brother-in-law, Arthur James Balfour, was very much pleased with the idea that he should take it for a limited period. Lord Salisbury was of the opinion that if he failed with the professorship his scientific reputation would be lost.

On 1 December 1879 he wrote to his mother that he had made up his mind to stand. He felt there was a great deal to be said against it, but if he refused merely on grounds of convenience, he would feel ashamed afterwards. It was doubtful whether in any case he could have gone on living as formerly, owing to his increasing financial difficulties.

When it had become clear that there would be a suitable candidate for the professorship, the University continued it, taking the necessary action on 20 November 1879. Rayleigh was elected to the chair on 1 December 1879. He retained it until 1884, by which time his financial situation had improved, and a suitable successor had appeared.

When Rayleigh had inherited the family estates in 1873 they were comparatively prosperous. He had envisaged a future of personal research in a laboratory with several assistants in his grounds. Then a disastrous fall in the price of wheat caused him financial embarrassment, and became an important factor in making him a professional scientist for a short period.

The fall was due to the opening-up of the wheat lands of the American Middle West and export of large quantities of wheat to Britain. The economic pressure forced Rayleigh to attend to the unwelcome task of better management of the estates, which were turned over to prosperous dairy farming to supply milk to London. His younger brother was given a thorough agricultural education to qualify him to take over and develop this family economic interest. This left Rayleigh freer to devote himself to science as he pleased. He established a private laboratory on his estate, on a modest and economical scale, where he pursued his researches after 1884.

Rayleigh's forced attention to economics and management, though he escaped from them personally as quickly as he could, was good for the Cavendish Laboratory. When he became professor, he at once appreciated the need for improving its organisation, which had languished during Maxwell's illness.

Maxwell, the Scottish laird, and Rayleigh, the English landed aristocrat, were detached in their attitude to the chair and direction of the Laboratory. If they had been better off financially at the time they might not have accepted them. The independence of a gentleman was still socially more important than the most eminent scientific post.

In considering Rayleigh's part in the Cavendish Laboratory and in science, his background is of great importance. He was born on 12 November 1842, the son of the second Lord Rayleigh of Terling in

Essex. He came of a family of Strutts, who first became known in 1660 as corn-millers. They acquired the water-driven corn mills in the neighbourhood of Maldon, and invested their profits in land. When steam superseded water power in the Industrial Revolution they were no longer dependent on milling, and could live on the income from their estates; they were transformed from tradesmen into country gentlemen. The water mill on the river Ter, beside which the Strutts had their mansion, was pulled down, and the mill pond converted into an ornamental lake.

Rayleigh later used water power from this lake in his early researches on hydrodynamics and wave-motion. He used the echo from the echelon of steps to the kitchen garden to illustrate the principle of a diffraction grating. In his researches on the resolving power of optical instruments he made pin-hole photographs of the steps, obtaining results of basic importance for the design of the big modern telescopes, such as that at Mount Wilson.

Rayleigh, like the Duke of Devonshire, was a relative of the Boyles, his paternal grandmother belonging to that family. His father was forty-six when he married the seventeen-year-old daughter of an engineer officer early in 1842. Rayleigh was born near the end of that year.

His ancestors had included craftsmen, traders and engineers. He was born prematurely and was delicate throughout his youth. He was not sent to school until he was ten, then successively to Eton, Wimbledon, Harrow and a school at Torquay. At the latter there was good mathematics teaching, the best pupil in this subject being H. M. Hyndman, who became a well-known acquaintance of Karl Marx.

At about fifteen Rayleigh became interested in photography, which led to his researches in optics. He was then a tall, delicate youth who did not play outdoor games, but was personally popular. He competed for a scholarship at Trinity in 1860, but was unsuccessful. He entered the college as a fellow-commoner in 1861. As the eldest son of a peer he was entitled to dine with the dons at high table; he appreciated the privilege of social intercourse and conversation with them. His great abilities came out only after being taught by Routh. He was Senior Wrangler in 1865, second being Alfred Marshall, the economist. After this, he gained the First Smith's Prize. He attributed his success to his ability to identify the essential point of a problem, and attack it directly and quickly. This was precisely the quality that later distinguished his research method.

After these successes he set about trying to become a scientist. He did not find it easy, for there was no Cavendish Laboratory in which to gain a training. Stokes was the first eminent physicist whom he came to know personally. Stokes made brilliant experimental optical researches in his study, but did not invite Rayleigh to assist in them. He was a kind but reserved man, and Rayleigh was unable to acquire experimental technique from him, nor vital details, such as where scientific instruments could be obtained. He learned from sharp personal experience the need for systematic teaching in practical physics. He never had any instruction in the use of simple tools.

Rayleigh competed for a fellowship at Trinity, in spite of his father's doubts about the social propriety of the eldest son of a peer doing such a thing; the Duke of Devonshire and Lord Lyttleton had not condescended to it.

At Trinity he became a close friend of Arthur James Balfour, and presently married his sister Evelyn. Balfour's younger brother, the embryologist F. M. Balfour, became one of his ardent persuaders in getting him to stand for the Cavendish professorship. He said his management of the Laboratory was necessary in the interests of the University.

Rayleigh read Maxwell's papers on electromagnetism soon after they came out. He appreciated their importance immediately, and began to correspond with Maxwell in 1870. He did everything he could to cultivate his acquaintance, at British Association meetings and when Maxwell was examining in Cambridge. He extended Maxwell's experiment with the colour top, and developed his analogy between the self-induction of currents and the inertia of matter in motion. In one of his letters Maxwell described his 'finite being' which could separate the slow from the fast molecules, diverting them into separate containers, like a pointsman in a railway goods yard shunting some trucks along one line, and other trucks along another. He later published this as his famous 'sorting demon'.

Maxwell thanked Rayleigh for various improvements and small corrections in his forthcoming *Electricity and Magnetism*.

If Maxwell had lived, and there had been no agricultural recession, the Cavendish Laboratory would have developed along Maxwell's personal lines, while Rayleigh would have become a wealthy amateur scientist with a handsome private laboratory at Terling.

In the event Rayleigh was forced to become an unwilling but efficient professional, who had learned through his private affairs how to review organisational matters realistically. When he became the director of the Laboratory, he was immediately impressed by the shortage of instruments and equipment. It had been generally believed that the Duke of Devonshire had provided everything that was necessary. He had certainly provided a great deal, but Maxwell thought he was being called on for too much, and felt that the sending of bills to him as a matter of course should stop. Maxwell himself personally paid for a considerable amount of new apparatus.

When Rayleigh saw that much more was needed, he started an apparatus fund to which he contributed £500, the Duke another £500, and various others together subscribed a further £500. Rayleigh also contributed to the fund the fees paid by students to the professor. He aimed at developing the systematic teaching of elementary practical physics to a larger number of students. For this purpose large numbers of galvanometers and other apparatus were required. Inexpensive simple standardised instruments, manufactured in quantity, were not then available, for there was as yet no demand for them. James Stuart, the Professor of Mechanism, had on his staff two instrument-makers who had worked part-time in the Cavendish. He assisted by having some of the required instruments made in his workshops.

Only one laboratory assistant remained from the Maxwell period. He was unable to do much, for he was ill and died soon. Rayleigh advertised for a new assistant. Among the many applicants was George Gordon, who had been a shipwright at Liverpool, and was a competent craftsman in wood and metal; he had a keen amateur interest in science. When Rayleigh retired from the Cavendish, Gordon accompanied him as his personal assistant.

When Maxwell was ill, William Garnett had acted as his deputy, lecturing on his behalf. He had not organised the systematic teaching of practical physics to classes of elementary students, which Rayleigh felt was so necessary. This kind of teaching had already been carried a considerable way by Helmholtz in Berlin. Garnett had personal difficulties, and Schuster was among those with whom Rayleigh discussed the situation. On 8 January 1880 he wrote to Rayleigh:

I have always taken a great interest in the teaching of practical physics and the methods pursued in different laboratories.

There are no doubt great difficulties and very few even of the long established laboratories on the continent are in thorough good working order. Yet from what I have seen of Cambridge I should say the difficulties there, were less important than in any other place except perhaps Berlin.

In the complete teaching of physics two different classes of men have of course to be provided for, those undergraduates who cannot spend much time in the Laboratory and only want to know something about the theory of instruments and methods of observation and those more advanced students who are desirous of repeating some standard measurements or who are able to do original work.

With regard to the first class of students Kirchhoff of Heidelberg, and Quincke, arranged that twenty students at a time could be taught by his methods. Refractive indices, electric resistance, magnetic force, Kater's pendulum, were among the subjects. The stress and adjustment of instruments were explained. Each student sent in a report on his results and the methods he had used. This plan could easily be adopted at Cambridge.

Many undergraduates who are now kept altogether away because they cannot afford much time, would willingly hear one lecture a week, and spend one morning or afternoon in the laboratory. . .

I have very often suggested this plan and during the last vacation arrangements were made by Mr. Garnett and Glazebrook. Owing to Mr. Garnett's indisposition and subsequent marriage the work was almost exclusively done by Glazebrook. . .

With this exception no attempt was made for a systematic teaching of Physics. Students used to drop in at irregular hours and if the Demonstrator was there and not otherwise engaged he gave them some work. . .

Beginners of ordinary ability were often disappointed and discouraged by this regime, and of the number who started at the beginning of term, only a few enthusiasts remained at the end.

6

Systematising Teaching and Research

Rayleigh's first aim was to make the organisation of the Laboratory, and especially the teaching, more systematic. He saw the need for it, and did it well, in spite of having little taste for it. As is often the case with people who perform duties they see to be socially necessary but do not particularly enjoy, their correct behaviour is rewarded with important unforeseen results. In Rayleigh's case his sense of organisation, besides consolidating the Cavendish Laboratory, later enabled him to perform major services to the nation, as chairman of government committees on explosives and aeronautics. These were the first scientific advisory committees in the modern style appointed in Britain. In a sense, what Rayleigh did for the Cavendish Laboratory, he repeated for the nation.

Garnett resigned from the demonstratorship soon after Rayleigh became professor, and in 1880 Rayleigh appointed two new demonstrators, R. T. Glazebrook and W. N. Shaw. In the *University Reporter* of 13 April 1880 a notice was published under the heading *Lord Rayleigh's Lectures*, which ran:

The Professor of Experimental Physics will give a course of Elementary Experimental Lectures on Electricity and Magnetism in the Lecture Room of the Cavendish Laboratory on Monday, Wednesday and Friday during the present term, of twelve, commencing Wednesday 14 April. Gentlemen wishing to attend this course are requested to leave their name with Messrs. Deighton, Bell & Co., to whom the fee of one guinea should be paid. The Cavendish Laboratory will open daily from 12 April for the use of students, from 11 am until 5 pm, and the Professor or Demonstrators will attend daily to give instruction in *Practical Physics*. The fee for the use of the Laboratory will be two guineas per term.

Application should be made at the Laboratory between the hours of 11 am and 1 pm.

The private rooms at the Cavendish Laboratory will accommodate a limited number of students who have had the necessary training, and are desirous of prosecuting original research.

All applications for permission to prosecute research must be made to the Professor, and all investigations undertaken must be subject to his approval.

6 April, 1880.

Deighton, Bell was one of the well-known Cambridge bookshops in those days. The notion of collecting Cavendish Laboratory Lecture fees in a private bookshop contrasts with modern ideas of accounting, which are subject to more formal arrangements because of the large scale of laboratory finance; this comes mainly from the State.

According to the regulations the Professor had to reside in the University for eighteen weeks during the academic year. He had to deliver two courses of lectures, one in each of two terms; the total amounted to at least forty lectures during the year.

Rayleigh's first class in 1880 was attended by sixteen, including five of the first six Wranglers in the Mathematical Tripos of 1881; ambitious students already made a point of listening to the lectures, in case these provided pointers to the forthcoming examination. One or two of the audience were senior members of the University, who did not hesitate to ask questions on appropriate points. This contributed to the informality and illumination, but it became impractical with the larger classes of later years. Rayleigh's classes remained about the same size throughout his professorship.

In the autumn of 1880 it was announced that:

The Professor of Experimental Physics will lecture on Galvanic Electricity and Electro-Magnetism, Dr. Schuster will give a weekly course on Radiation, Mr. Glazebrook will give an elementary course of demonstrations on Electricity and Magnetism, Mr. Shaw will give a course of demonstrations on the Principles of Measurement and the Physical Properties of Bodies.

The Laboratory will be open daily from 10 to 4 for more advanced practical work under the supervision of the Professor and Demonstrators for those who have had the necessary training. Courses of demonstrations will be given during the Lent Term on Heat and Advanced Electricity and Magnetism, during the Easter Term on Light, Elasticity, and Sound.

Demonstrations lasted two hours, and were given on three days a week. In the Lent Term of 1880, eighteen students attended the elementary demonstrations on heat, and fourteen the advanced demonstrations on electricity and magnetism. The number gradually increased, so that by 1884 it became necessary to duplicate the elementary courses. Two assistant demonstrators were appointed in 1884. In the Easter Term of 1884 Rayleigh delivered his last course of lectures as Professor to fourteen students. Forty-seven attended Glazebrook's course, and twenty-three Shaw's.

During Rayleigh's first term in 1880 five courses of physics lectures were delivered by himself and others. In his last year, in the same term in 1884, the number of courses of lectures had risen to ten.

The University decided in 1881 to divide the Natural Sciences Tripos into two parts. In Part I candidates took the elementary parts of three or more subjects, while in Part II they were expected to specialise in one subject. Part II required a more advanced knowledge of physics, and the teaching for this had not yet been provided. With Rayleigh, Glazebrook and Shaw worked out the course of instruction in practical physics, which continued unchanged in principle for more than fifty years, and formed one of the strongest and most lasting features of the Cavendish system.

In each term there were demonstration classes in a branch of elementary physics, each lasting two hours, three times a week. The subjects for Part I of the Natural Sciences Tripos were covered in three terms. The student was required to complete each experiment, and write out an account of it in his notebook, which had to be passed by the demonstrator before he could proceed to the next experiment. Advanced subjects were treated in the same way, but the student was allowed more than two hours to deal with each experiment, if he found it necessary. J. J. Thomson said that one of the merits of the system was that the student thought he was writing for posterity, so he took particular care to make his account as good as he could. It was an excellent training in clear thinking and writing.

In the Natural Sciences Tripos of 1879, just after Maxwell's death, twenty-five candidates passed. By 1884, at the end of Rayleigh's professorship, there were sixty successful candidates in Part I of that Tripos and twenty in Part II.

Rayleigh regarded himself as to the right of centre in politics, but on some questions he differed sharply from conventional opinion. He was influenced by the ideas of John Stuart Mill, and believed that women should be given equal intellectual opportunities with men. His acquaintance with the intellectual ladies of the Balfour family strengthened this opinion.

Rayleigh's social acquaintance, which was extending at this time, enabled him to make some illuminating observations on the scientific capacities of eminent statesmen. Lord Salisbury asked his advice on magnetic experiments. After seeing him in his laboratory he wrote to his mother that he was 'too awkward to succeed well as an experimenter'.

Then he went to stay with the Gladstones at Harwarden. He at once felt the power of Gladstone's personality, but found him incapable of thinking scientifically. He persisted in contending that snow had more power than water of penetrating leather. Rayleigh suggested that the snow had probably fallen inside the top of the boots, and in any case, it penetrated them in the form of water. Gladstone would not accept this; he repeated with emphasis that snow had a *peculiar power of penetration.*

Rayleigh became engaged and married to Evelyn Balfour in 1871. In the following year he had a serious attack of rheumatic fever. He had previously been a slender, youthful-looking man; after it his figure and appearance were middle-aged. For his convalescence he and his wife spent the following winter in Egypt, being accompanied by Eleanor Balfour, with whom Rayleigh talked mathematics and science. She married Henry Sidgwick, the Cambridge moral philosopher. After she had settled in Cambridge she assisted Rayleigh throughout the whole period of his professorship. She was patient, accurate and neat. She kept notes on measurements, and checked the lengthy computation of results.

Eleanor Sidgwick became a leading feminist and pioneer of higher education for women. She joined in the founding of Girton, and became Principal of Newnham. Her opinions were rather too advanced for Rayleigh, but he respected her great ability. She said that Rayleigh inspired and led others by his sympathetic interest. He had an even temper, and was never aggressive or ambitious. She considered him an excellent teacher, who contributed much to the development of the Laboratory as a teaching institution. As a lecturer and director of research, his first reflections on a new idea or problem were particularly impressive. His audience could see how his mind worked, piercing immediately to the point.

In 1882 Rayleigh opened the classes in the Cavendish Laboratory to women, on equal terms with men. Mrs Sidgwick collaborated with Rayleigh especially in the re-determination of the absolute value of the ohm, the electrochemical equivalent of silver, and the e.m.f. of the Clark cell.

Seeing Rayleigh as a collaborator and a relative, Mrs Sidgwick said that he always tackled a problem from 'the big end'. He made patient and profound efforts to form a physical picture of a phenomenon before he made any calculations. He said that 'a good instinct and a little mathematics is often better than a lot of calculations'.

The method of class teaching of practical physics developed under

Rayleigh was adopted from that which had been established by Pickering in the Massachusetts Institute of Technology. In 1873 Pickering had published the textbook *Elements of Physical Manipulation*, to be used in conjunction with his course. He had stated in the preface to his book that the 'rapid spread of the laboratory system of teaching Physics, both in this country and abroad, seems to render imperative the demand for a special text-book to be used by the students'. He explained that according to his method

> Each experiment is assigned to a table on which the necessary apparatus is kept and where it is always used. A board called an indicator is hung on the wall of the room, and carries two sets of cards opposite each other, the one bearing the names of the experiments, the other those of the students. When the class enters the laboratory each member goes to the indicator, sees what experiment is assigned to him, then to the proper table where he finds the instruments required, and with the aid of the book performs the experiment.

Instructions were given on how the results should be calculated and entered up. Pickering said that, 'By following this plan an instructor can readily superintend classes of about twenty at a time and is free to pass from one to another answering questions and seeing that no mistakes are made.'

The first thoroughly worked out system of class teaching in experimental physics in the Cavendish Laboratory was inspired by this American development, which shows marks of American thinking on factory organisation.

Glazebrook, who had perhaps the chief part in adapting this system in the Cavendish, said that most experimental physicists in the last quarter of the nineteenth century in Britain as well as in America were brought up on this system. He thought that the plan had drawbacks as well as merits; there was a tendency for the student to be assisted too much. The apparatus on the table was so well set up and adjusted that all the student had to do was, as it were, press a button. But if a large class of students was to be taught, all of whom have to learn something about the facts and laws of nature in a limited time, a system of this kind was absolutely essential.

The alternative method, by which he had himself learned three years before by trying to work out his own experiments, was more interesting, but it was organisationally impossible for large classes.

Glazebrook as a beginner had been given a very brief instruction by Maxwell on how to measure electrical resistance and electromotive force. He had then been told to measure these quantities for a set of large-gravity Daniell cells which Chrystal and Saunders were using in their determination of Ohm's law. The instrument that was assigned to Glazebrook for measuring the e.m.f. was a large-pattern Thomson quadrant electrometer. This type of instrument, on which Rutherford was later to make crushing comments, is notoriously difficult to operate, and Glazebrook recorded that his own difficulties, 'at any rate for a beginner, were not small'.

Glazebrook and Shaw found their work as demonstrators most interesting. One of their first jobs was to find a laboratory attendant. It was Glazebrook who heard of George Gordon and recommended his appointment to Rayleigh. One of Glazebrook's friends in Liverpool had been teaching a Science and Art Department class, in which Gordon had enrolled. Gordon made much of the apparatus used in the Laboratory. He was not a skilled instrument-maker, but under Rayleigh's guidance he acquired a capacity for attending to essentials, and his apparatus served its purpose admirably.

Shaw had worked under Warburg at Freiburg, and Helmholtz in Berlin. Glazebrook's own experience had been in the Cavendish under Maxwell. Both drew on Pickering's textbook, and on Kohlrausch's book on physical measurements, but neither book was entirely suited to the Cavendish needs. They started a series of manuscript notebooks, each dealing with one experiment, describing the apparatus which would be used, the method of using it, and the way in which the results should be written up.

The notebooks were continually revised, and became the basis for the famous textbook *Practical Physics* by Glazebrook and Shaw. They found that in preparing it, there was no comparable textbook to which the reader could be referred for expositions of the relevant theoretical physics, so they included a good deal of theory in it.

The advanced practical teaching was based on broadly similar principles, but students were from the beginning introduced to original work by the repetition of important researches that could be repeated with the means available in the laboratory. The students were expected to read the original memoirs in which the work was described, and to reach a high order of accuracy in their own repetition of the work. The notebooks in which they wrote out their results became, in time, a valuable collection. The teaching of these able men was a great pleasure for the demonstrators.

Glazebrook's advanced class in electricity and magnetism in the Lent term of 1881 contained fourteen students, including Edward

Hopkinson, A. R. Forsyth and Richard Threlfall. He and Shaw had the assistance of Rayleigh in what they considered to be the best of all ways. The syllabus and plan of the course was arranged with him after the most thorough consideration and discussion. When they found themselves in a difficulty, they could go to him for advice. This was always given with sympathy, and ready assistance was provided when necessary. Otherwise, they were left free to teach as they thought best, and develop the work in the directions that seemed to be most appropriate.

The two demonstrators were each on duty three days a week. On those days the teaching kept them fully occupied, but on the other days they were free to pursue their own researches, college duties or private teaching. They found the condition of work delightful. They had the pleasure of developing a comparatively new line of teaching experimental physics; they had the sympathy and support of the Professor, and were aware of the importance of the work and the future discoveries it promised. All this combined to make the years 1880–4 a period of particular happiness for the Laboratory staff.

Besides the system of teaching, the system of examining in experimental physics required development, under the stimulus of the increased number of students. Rayleigh examined in 1882 and 1883, and Shaw in 1884 and 1885. Besides the examining conducted in the Laboratory for the University, examining was also done for the annual College and Scholarship examinations. These grew in size and importance as the number of science candidates increased. It provided an opportunity for identifying talent in boys arriving from school, and thus guiding them to a scientific career from an early stage. It was appreciated that early recognition and encouragement, putting young students onto the right lines, is of major importance in conserving talent, and making the most efficient use of it. Young scientists who find the right line of work from the beginning of their career can achieve a great deal very quickly, and even men of modest talent can achieve much if they find an appropriate line early, and stick to it.

The development of examining involved careful organisation both with regard to the nature of the questions, and the method and apparatus required for their solution. The co-ordination of the conduct of the examination and the normal work of the Laboratory was also necessary, so that it was not upset.

As examinees had in general at that time little knowledge of how to attack questions, they were required to write an account of their method before they began their practical work. With the growth in preliminary science teaching this became less necessary.

They had also to work out how a large number of candidates could be examined simultaneously, when only a limited amount of apparatus was available, and they had to try to find a satisfactory way, which they did not find easy, of distributing marks between the written answers and the practical work itself.

During the summer vacation the Laboratory was more or less closed to the elementary students. The demonstrators and the advanced students concentrated on their own research work. But there was one very important teaching course carried out during the vacation: a special class was organised for teaching physics to mathematicians.

Those who had graduated in the Mathematical Tripos and some others were put through an intensive course to enable them to enter the advanced students' class in experimental physics in the following autumn term. With a small group of selected men it was possible to get through a great deal of work in an interesting way in a short time.

When Rayleigh began his professorship he did not immediately introduce a new plan of research for himself and the Laboratory. He continued with the researches he had been pursuing privately in his home at Terling. These concerned the effect of electricity on gases. Mrs Sidgwick assisted him in these.

He then considered what general topic of research might be introduced, in which he and other research workers might collaborate in an organised way. He decided to take up Maxwell's suggestion in his Introductory Lecture that the Laboratory should pursue exact physical measurements systematically, develop the work of Gauss and Weber on the theory and measurement of physical units, and extend the work on the determination of physical units undertaken by the British Association.

The first clear ideas on physical units arose from the development of astronomy, in which it came to be seen that all quantities in science could be expressed in terms of length, time and mass. Each physical quantity, such as length, is measured in a unit of its own kind. Thus a moving object has unit velocity when it covers unit distance in unit time. The unit of force is that which, when applied to a unit of mass, increases its velocity in a unit of time by the unit of velocity. This system of dynamical units is called an absolute system of units.

Gauss and Weber showed that the absolute system of dynamical units could be extended to electrical and magnetic phenomena. Ohm showed that the current in a conductor is proportional to the electromotive force causing the current, such as that provided by an electric battery.

It became necessary to relate electromotive force, current and resistance to the fundamental units of dynamics. The units so obtained were inconvenient in practice, since the unit of current was far too great, while that of electromotive force was far too small. The development of electrical engineering made it essential to adopt more practical units. These were named the ampere for current, the volt for electromotive force and the ohm for resistance. They were one-tenth, one hundred million times and one thousand million times the respective values in the centimetre–gram–second system of absolute units.

The introduction of this system was due to the British Association Committee on Standards of Electrical Resistance, which, from 1862 onwards, had a leading part in establishing electricity both as an exact and a practical science. The Committee had been formed under the inspiration of William Thomson, whose work on the design and construction of the Atlantic Cable had impressed on him the need for a precise absolute system of electrical measurement. Until this period, each worker defined his own units. For example, Lenz in 1838 defined his unit of electrical resistance as that of one foot of Number 11 copper wire.

When the electric telegraph was introduced, the English electrical engineers used the resistance of one mile of Number 16 copper wire as standard, while in Germany the resistance of one mile of Number 8 iron wire, and in France of one kilometre of four-millimetre iron wire were adopted as units. It was presently realised that the electrical conductivity of the metal in the different wires varied, owing to impurities. Werner von Siemens then suggested that the resistance of a column of mercury one metre long and one square millimetre in cross-section, at the temperature of melting ice, should be the standard.

The British Association Committee pointed out that all of these variations should be abolished by establishing an absolute system, not dependent on the physical properties of particular substances. Weber had already produced resistance coils with their values expressed in absolute units, so the Committee decided to check his results. This was done by spinning-coil experiments, carried out for the Committee by Maxwell and Fleeming Jenkin at King's College London.

Rayleigh's programme of research at the Cavendish cleared up much of the confusion and uncertainty, and made it possible to measure to a high degree of accuracy the strength of a current in amperes, an electromotive force in volts and a resistance in ohms.

One of Rayleigh's earliest courses of lectures was on the history of electrical measurements. As the original British Association apparatus was in the Laboratory, he set it up again with the aid of Schuster and Mrs Sidgwick and improved it to obtain greater accuracy. He used the stroboscopic method for measuring the rate of rotation more accurately.

When Rayleigh started on these researches the value of the ohm was known to an accuracy of only 4 per cent, and the electrochemical equivalent of silver, or amount deposited by a standard current, was known only to within about 2 per cent. After his experience with the original spinning-coil apparatus, he constructed a second coil, strengthening it, and determining its dimensions and constants with great care. Then he repeated the determination by a method invented by Lorenz. This led to the publication of a paper on the comparison of methods of determining resistances in absolute measure. The various sources of error in the different methods were carefully analysed, and their effect on the reliability of results estimated.

In 1881 the resistance of a column of mercury 106 centimetres long, and one square millimetre in cross-section, at the temperature of melting ice, had been adopted as the legal and commercial ohm. Rayleigh carefully measured the resistance of such a column of mercury in absolute c.g.s. units by his rotating-coil method, obtaining for the value of the ohm the resistance of 106.31 centimetres of mercury.

After the determination of the unit of resistance, the unit of current was determined by electrolysis. He followed the method previously used by Mascart. Rayleigh and Mrs Sidgwick obtained a result that differed slightly from his. Glazebrook said that shortly before Mascart died he told him that the difference was due to his use of a standard metre which turned out to have been inaccurate.

Rayleigh also determined the electromotive force of the Clark cell, much used as a standard in laboratory experiments.

The development of electrical engineering made it essential that the chaos in electrical measurement and standards should be eliminated, and a legal code for them established by the Government. In 1890 the Board of Trade appointed a committee to consider the question. Making use of the Cavendish investigations it reported in 1891, and its figures were adopted by the International Electrical Congress at Chicago in 1892. They became the foundation of legislation on electrical standards throughout the world. They were an indispensable basis for the development of the modern technological age, in which exact electrical measurement is necessary not only in production, but in control.

In social manner and intellectual attitude Rayleigh and Mrs Sidgwick could scarcely have looked less like industrial pioneers. Yet they and the recently founded Cavendish Laboratory provided not only abstract ideas, which might ultimately be of fundamental industrial importance, but a whole class of immediately necessary fundamental data, without which industrial development could not have proceeded.

Glazebrook remarked that all Rayleigh's writings on the fundamentals of measurement were

marked by the same characteristics: perfect clearness and lucidity, a firm grasp on the essentials of the problem, and a neglect of the unimportant. The apparatus throughout was rough and ready, except where nicety of workmanship or skill in construction was needed to obtain the result; but the methods of the experiments, the possible sources of error, and the conditions necessary to success were thought out in advance and every precaution taken to secure a high accuracy and a definite result.

Glazebrook held that the stream of discoveries that came from the Laboratory in later years sprang from seeds sown by Rayleigh, and in his own experience his years under him were perhaps the happiest he ever spent.

In the period 1881–7 Rayleigh published sixty-two papers, nearly all of which dealt with researches he had carried out during the five years of his professorship. A score of these were on the measurement of electrical units and problems involved with it. The rest included his personal researches in optics, sound, hydrodynamics, aerodynamics and general physics, besides Presidential Addresses to the British Association and the Physics Section of the British Association.

Characteristic of his topics in his short, pointed papers was his treatment of 'The Use of Telescopes on Dark Nights'; 'On the Invisibility of Small Objects in a Bad Light'; 'On the Theory of Illumination in a Fog'; and 'On the Colours of Thin Plates'. His papers on sound included 'Acoustical Observations on the Pitch of Organ Pipes'; 'Estimation of the Direction of Sounds with one Ear'; 'Rapid Fatigue of the Ear'; 'Sensitive Flowers'; and related topics. His hydrodynamical papers included 'The Form of Standing Waves on the Surface of Running Water'; 'On the Vibrations of a Cylindrical Vessel Containing Liquid'; 'The Crispations of Fluid resting upon a Vibrating Support'; 'On Porous Bodies in Relation to Sound'; and 'The Equilibrium of Liquid Masses charged with Electricity'.

His aerodynamics papers included one on 'The Soaring of Birds', and another on 'An Instrument Capable of Measuring the Intensity of Aerial Vibrations'. His general papers included one 'On a New Form of Gas Battery'; 'The Distribution of Energy in the Spectrum'; and a number of papers on electromagnetic topics.

J. J. Thomson said that the paper that pleased Rayleigh most was that on the soaring of birds. In it he explained why birds, such as albatrosses and gulls, can soar round and round, without moving their wings. Even when the wind was horizontal they might be able to rise. In general, the bird went downwards when flying with the wind, and upwards when flying against it. Rayleigh discussed the paper with J. J. before publication, hesitating to send it to the press because the argument was so simple and beautiful that he feared there must be a snag in it somewhere. It is one of the classics of aerodynamics, associating Rayleigh with Leonardo da Vinci, who also looked directly and elegantly at this particular phenomenon.

His paper on a new form of gas battery was only fifteen lines long, but it contained the suggestion that platinum gauze should be substituted for platinum plate in the Grove gas battery. This improvement has contributed to the development of the fuel cell as a supplier of electrical power in modern space vehicles.

Some years later he made another prescient technical suggestion, when he showed that a turbine expansion engine could provide a particularly efficient method of producing very low temperatures.

His discovery of the distribution of energy in the spectrum contributed to the evidence that led Planck to propose the quantum theory. Rayleigh accepted the mathematical results of the quantum theory, but hesitated about the theory itself, since he was unable to form physical pictures in his imagination of what happened in quantum action.

His scientific fertility and influence continued unabated. In 1919, at the age of seventy-seven, he gave a theory of the reflections from multilayer structures, which explains the recently discovered phenomenon of the sea-scallop, which has eyes that do not contain lenses but mirrors. Its eyes work on the same principle as the Schmidt telescope, which was utilised by nature two hundred million years before Schmidt.

As Rayleigh's five years of professorship progressed he felt that his constitution, which was not robust, could not sustain the great intensity of his intellectual, teaching and directive work. His financial situation had improved, owing to the general recovery of trade and the skilful management of his estate. He longed to return to individual research in his private laboratory at Terling. This had become practicable because several experienced and talented possible successors were in sight. He resigned in 1884, and, free from academic administration and the preoccupations of laboratory direction, concentrated on personal researches. His pursuit of Maxwell's principle of continually increasing the accuracy of experiment as a means to revealing quite new and unexpected facts led to the discovery of argon and the noble gases. In 1895 Rayleigh even suspected that argon might react with fluorine, as was proven in 1961.

Rayleigh brought the self-confidence and social power of the British imperial and ruling class to bear on, and advance science. A powerful ruling class takes it for granted that its desires will be studied and met; it expects that nature, when called upon to answer its scientific queries, will duly answer.

Rayleigh resigned from the Cavendish professorship as early as he could. His personal contribution was by far the greatest part of the Cavendish achievement during his professorship. The number of research workers had increased during his direction, but they were still small, and none of them had yet become very prominent as investigators. The electors of his successor were confronted with one of the most difficult problems in making their choice: were they to elect one of the most experienced men, or one of the most promising?

In 1881 a new name appeared: that of the research student J. J. Thomson, who was then twenty-five years old. He published two papers in that year, one experimental and the other theoretical. The first was on electromagnetic properties of open circuits, and the second was 'On the Electric and Magnetic Effects Produced by the Motion of Electrified Bodies'. This was a very remarkable paper. In it, Thomson showed how, according to Maxwell's theory of electromagnetism, a moving electrically charged body must increase in mass with increase in speed. This follows at once from the Theory of Relativity, which was not announced until 1905, twenty-four years later, when it was formulated by Einstein.

Besides producing this classic theoretical paper, J. J. Thomson had published useful experimental work; as a physicist he became armed on both sides. In 1882 he published a paper on 'The Vibrations of a Vortex Ring, and the Action upon each other of two Vortices in a Perfect Fluid'. Helmholtz, William Thomson and Maxwell had already considered the conception of the atom as a vortex. This had introduced the notion of rotation as an element in the structure and operation of atoms, underlying some of their essential properties. This was at the backs of the minds of those who later explained atomic properties in terms of rotating electrons.

In the same year J. J. Thomson also published two papers on the dimensional theory of magnetic poles. These were in line with Rayleigh's stimulation of the development of dimensional theory of units. In 1883 Thomson published another paper on electrical units, and one 'On a Theory of Electric Discharge in Gases', a significant new direction in his research. In Rayleigh's last year he published two more papers, one on an electrical subject, and the other 'On the Chemical Combination of Gases'.

It was evident that the young man, in 1884 still only twenty-seven, might have the intellectual ability to be a worthy successor to Maxwell and Rayleigh. His work was brilliant and varied; he was contributing to several fundamental aspects of physics, and he was interested in experiment as well as theory. But could he keep it up, and could he direct a laboratory with a staff older than himself, and an increasing number of students? The young Thomson had been far too busy with personal research to have much opportunity for acquiring directive experience.

Again, Sir William Thomson was approached, but once more refused to stand. Schuster, and Osborne Reynolds, the professor of engineering at Manchester, applied. They had both been J. J.'s teachers at Manchester. Other applicants were Glazebrook and Garnett, who had been his teachers in Cambridge, and had carried much of the burden of organising and running the Cavendish. There was also the distinguished senior physicist G. F. Fitzgerald of Trinity College, Dublin.

The Electors were the Vice-Chancellor, Dr Ferrers, Professor Clifton, Professor G. H. Darwin, Sir William Grove, Professor Liveing, Professor W. D. Niven, Professor Stokes, Professor James Stuart and Sir William Thomson.

They chose the young man. It was a bold scientific judgment, and an act of administrative genius, possibly the most crucial event in the history of the Cavendish, but several of the older men who had been passed over felt their disappointment.

Under J. J. Thomson the Cavendish was to become acknowledged as the leading laboratory of experimental physics in the world. The appointment was a manifestation of the quality and health of the Cambridge scientific environment in the 1880s. The achievement of the Cavendish was much more than the addition of the efforts of three outstanding professors. This succession of appointments was not a result of the chance appearance of three outstanding men within a generation; it was an outcome of the contemporary state of science, and the social attitude to science at Cambridge. Compared with the more rationally organised development of academic science in Germany and France, this was odd. Yet, out of the peculiarities of the Mathematical Tripos, the college system and other historical circumstances, Cambridge managed to produce not merely a system that served, but one that was to be outstandingly effective. In this respect, the Cavendish Laboratory has had some of the features of the British Parliament: a mixture of empiricism and rationality. Behind it has been the power and inspiration of history as a whole.

It was to be the outstanding product in the realm of physical science of three centuries of industrial and territorial expansion. It was this power which ultimately led to the inspiration of talent and correct decisions.

7

J. J. Thomson: Early Days

Clerk Maxwell brought the genius of the Scottish Celtic lairds and Rayleigh that of the English landed aristocracy to bear on the creation of the Cavendish Laboratory. J. J. Thomson added to this the potent spirit of the nineteenth-century British middle class. He had great scientific gifts, but his shrewdness, industry, common sense and membership of that class of men who dominated the Victorian period was also very important. He was better equipped than Maxwell and Rayleigh for developing the Cavendish Laboratory after it had been started and consolidated. He was personally closer than his predecessors to the increasing number of research students who necessarily came more and more from less privileged social strata. His manifold qualities enabled him to bring the Cavendish Laboratory to the summit of scientific fame in less than twenty years.

Joseph John Thomson (see figure 8) was born in 1856 in Manchester, the son of a bookseller, at the time when 'What Manchester thinks today the world thinks tomorrow'. The city was then setting a world pattern for capitalist development, and had a cultural life that paralleled its social vigour. Its Literary and Philosophical Society was the first of its kind, and the headquarters of Dalton and Joule. J. J.'s father was a member, and introduced him when a boy to Joule, saying that someday he would be proud to have met that gentleman.

Owens College, the forerunner of Manchester University had been founded in 1851. J. J.'s father wanted him to be an engineer, and as there was a long waiting list for apprentices he was advised to send him to Owens, where he would receive some preliminary education, which would make a good foundation for an engineering training. J. J. accordingly entered the college when he was fourteen. In three years he secured a certificate in engineering, and then concentrated for two more years on physics and mathematics, which had interested him most.

J. J.'s early orientation to engineering was important, for it strengthened his practical approach and handling of physical matters. It helped him to add a shrewd earthy sense to the intuitions of Maxwell and the distinction of Rayleigh in the growth of the Cavendish tradition.

Owens, reflecting the cultural inspiration of the contemporary Manchester, had started with a brilliant staff, including Balfour Stewart in physics and Osborne Reynolds in engineering. The former was a highly original thinker, and the latter a creator of scientific engineering. Thomson had the opportunity of learning in his youth the nature of genius.

However, the man from whom he learned most was Thomas Barker, the professor of mathematics, who was an excellent teacher. A typical

8 *J. J. Thomson investigating the conduction of electricity in gases.*

Mancunian of the period, he was a successful investor and man of independent means. He indulged his fancy, which ran to botanical research in his spare time. J. J. Thomson used to say in his old age that he might perhaps have preferred to have been a botanist rather than a physicist, if the circumstances of his life had been different.

Thomson, who started his academic career as a poor man, also became a shrewd investor, and left a fortune of £82 000.

Barker was a Trinity man, and advised J. J. to compete for a mathematical scholarship at Trinity. Owing to Barker, he arrived at Cambridge with a knowledge of quaternions before he had learned analytical geometry.

At Manchester J. J. had been one of Balfour Stewart's half-dozen research students, and benefited from the influence of his physical imagination. The conditions in which he studied under Balfour Stewart were not unlike those introduced by Maxwell in the Cavendish Laboratory. The number of students was small, and they were allowed a good deal of choice in their experiments. They had to set up their own apparatus, and they were free to spend as much time as they pleased on any point of interest that turned up. J. J. thought that this was more interesting and educational than the highly organised systems that became necessary when there were many students.

J. J. was awarded a mathematical scholarship at Manchester in 1874, and published his first paper, 'On Contact Electricity of Insulators', in the *Proceedings of the Royal Society*. In spite of this, he was unsuccessful when he competed for a scholarship at Trinity. In the following year he was awarded a minor scholarship, and started training for the Mathematical Tripos, graduating as Second Wrangler in 1880.

He was one of the audience of three who attended Cayley's lectures. The other two were senior members of the University, and therefore sat beside Cayley. Being only an undergraduate, he had to sit facing the lecturer, and was able to observe his method of work upside-down, which was very instructive. Cayley apparently never stopped to think, filling large sheets with jumbles of symbols, which he gradually reduced to most elegant conclusions.

J. C. Adams read his lectures from beautifully written scripts, 'which he brought into the lecture-room in small calico bags'. The lectures he enjoyed most were those of Stokes. They were much more concerned with physical ideas, and illustrated by experiments which succeeded with a precision J. J. grew to envy, when he had to give similar lectures.

He was 'coached' by E. J. Routh. This consisted essentially of a series of very well-arranged and exceedingly clear lectures, which were attended by much larger audiences than most of those of the official University lectures. They were most ingeniously devised to cope with the many peculiar features of the Tripos examination.

Such was the kind of instruction available to an able Mathematical Tripos candidate in the period when the Cavendish Laboratory was being built up. J. J. said that the preparation for the Mathematical Tripos left him with no leisure for experimental work. He preferred this old system because it had compelled him to read much more widely in pure mathematics than he would have done under the more specialised system that superseded it. He had found this very valuable later, for 'even a superficial acquaintance with the higher parts of pure mathematics is that it prevents one from being deterred by the formidable aspect which pure mathematicians, by their nomenclature, delight to impart to the simplest and most evident theorems'.

J. J. soon became popular, and universally known by his initials. Even his son, G. P. Thomson, when he grew up, always referred to him as 'J. J.'.

He did not attend Maxwell's physical lectures, and did not enter the Cavendish Laboratory until one month after Maxwell had died, and four years after he had first come into residence. He chose as his research an experimental comparison of the ratio of the electromagnetic and electrostatic units of electricity, corresponding roughly to the units for electricity in motion, and at rest. According to Maxwell's theory the ratio should be equal to the velocity of light.

His results were not very good, and the experiment was repeated more successfully in later years with the collaboration of G. F. C. Searle. His next research, made at the suggestion of Rayleigh, was on the measurement of electrical units.

In 1881 he applied unsuccessfully for the new chair of Applied Mechanics at Manchester. His old teacher Schuster, who was five years his senior, was preferred.

J. J. was, however, elected a Fellow of Trinity, which gave him the opportunity of devoting all his time to research for the next seven years. He looked for an electrical subject to which he could apply both his experimental and mathematical technique.

Maxwell had written in *Electricity and Magnetism*, published in 1873, of the processes involved in the conduction of electricity in gases. He had observed that 'These and many other phenomena of electrical discharge are exceedingly important, and when they are better understood they will probably throw great light on the nature of electricity as well as on the nature of gases and of the medium pervading space. At present, however, they must be considered outside the domain of the mathematical theory of electricity.'

Since then, William Crookes had published in 1879 researches on the rays emitted from the cathode in tubes in which electric discharges were passed through gas at low pressure. E. Goldstein of Berlin, who had investigated these rays at the suggestion of Helmholtz, had discovered many of their properties. He named them 'cathode rays', and published impressive papers on them in the English *Philosophical Magazine* in 1880. Goldstein was convinced that the cathode rays were waves, because they threw sharp shadows of objects placed in front of them, whereas Crookes was as firmly convinced that they were streams of electrified particles. The difference was influenced by national modes of thought; nearly all the German investigators were sure they were waves, while the British were equally sure that they were particles.

Thus, in 1880, the nature of the cathode rays was one of the central problems of physics. As beams of cathode rays repel each other, J. J. decided to calculate the magnitude of their mutual interaction, on the assumption that they were electrified particles, and obeyed the laws of Maxwell's electromagnetic theory. This led to his famous paper of 1881 on the electric and magnetic effects produced by the motion of electrified bodies, which contributed much to his election as Rayleigh's successor.

His deduction that the mass of the electrified particles increased with their velocity was confirmed by Kaufmann twenty years later. Among other results he explained the green phosphorescence seen in highly exhausted vacuum tubes, as due to the sudden stoppage of cathode particles.

This paper, written when he was a Cavendish research student, was perhaps the most remarkable he ever wrote, and contained the germs of his major achievements.

His research on vortex rings, published in 1883, had implications for the problem of atomic structure. His book *Applications of Dynamics to Physics and Chemistry*, published in 1888, made him one of the founders of the new science of theoretical physical chemistry. If he had continued in it, he might have become as distinguished in physical chemistry as he became in physics.

He independently discovered Le Chatelier's principle, and re-discovered some of the results of Willard Gibbs. His intense concentration on the mathematical theory of the dissociation of chemical molecules probably prepared his mind for the idea of atomic dissociation, or the emission of electrons from atoms by the absorption of energy.

J. J. had a wide range of physical interests, combined with mathematical skill and experimental understanding. In this respect he was like Helmholtz. He had not the intuitive theoretical genius of Maxwell, and did not discover the quantum theory of Planck, the relativity theory of Einstein or de Broglie's wave theory of matter. But he did have the requisite combination of theoretical and practical gifts to achieve the decisive discovery of the electron. He was not mathematically minded, and did not extend his mathematical equipment after his student days. He used it more and more as a tool for exploring the implications of experiments, and the physical ideas conceived to explain new experimental data.

J. J. brought a professional middle-class attitude to both research and teaching. It was a contribution of critical importance, which made the Cavendish Laboratory as useful as it was brilliant, and completed the establishment of the characteristics for which it became famous; under his guidance the Laboratory adapted to the scientific and social climate of the age.

8

Building on the Foundations

After Rayleigh resigned in 1884 J. J. applied for the chair, without dreaming that he might be successful, and did not think seriously about the responsibilities and problems of direction. He was greatly surprised by his election, and so were many others. Glazebrook, who had done so much of the early work in establishing the Laboratory was deeply hurt, and was unable for some time to control his feelings, and offer his congratulations.

J. J. saw the situation in terms of the contemporary social idiom. He reported that one college tutor had remarked that things had come to a pretty pass when mere boys were appointed professors. He described himself as feeling like a fisherman with a light tackle, who had casually thrown in his line at an unlikely spot, and found he had hooked a very large fish too heavy to land.

He said that his want of directive experience was less harmful than it might have been, owing to the loyal support of Glazebrook and Shaw. They continued the practical classes they had organised. Early in 1884 two assistant demonstrators had been appointed to strengthen the teaching.

One of J. J.'s first tasks was the appointment of an assistant to superintend the workshops. Gordon, who had been appointed by Rayleigh, left with him to be his private assistant. The efficiency and success of laboratories depends very much on the quality of the workshop assistance. This was particularly true of the Cavendish Laboratory, owing to the high degree of identification of its research staff with the educated middle and upper classes. (See figure 9, which shows the assistants in 1900.)

It is not insignificant that J. J. started his survey of the first twenty-five years of his professorship with an account of the workshop

109

assistants he had had. He said that, naturally, they had to be skilled workmen. They had also to be men of strong character, to maintain discipline among the younger assistants, and, he might have added, among the research students competing for apparatus and attention. They had to exercise diplomatic tact in sharing the workshop resources between those engaged in teaching and in research. The technician in charge had to be business-like, since he was entrusted with the stores and the purchase of materials. J. J. held that the smooth running of the Laboratory depended almost more on him than on anyone else. He first engaged a young Scotsman, D. S. Sinclair, who was a skilled mechanic. In 1887 Sinclair left for an appointment in India and was succeeded by A. T. Bartlett, who served until 1892, when he left to engage in electrical engineering. He became head of the research department of the English General Electric Company.

After him, the superintendent was W. G. Pye, who served until

9 *The laboratory assistants in* 1900

1899, when he left to found a scientific instrument firm. It specialised in making good apparatus at a reasonable price for practical teaching in schools and universities. It was an early example of Cambridge physics stimulating industrial production. A branch of the firm became Pye Radio. Pye's father had been superintendent of the Physiological Laboratory's workshop, from which Horace Darwin had developed the Cambridge Scientific Instrument Company.

Pye was succeeded in the Cavendish by F. Lincoln, who had joined the Laboratory staff in 1892 as a boy so small that he had had to stand on a box to reach the bench.

J. J. appointed a chief lecture assistant in his first term as professor. This was W. H. Hayles, who was a very skilful photographer and maker of lantern slides, and devoted himself to the perfecting of the lecture experiments and illustrations. J. J. also had James Rolph who had been appointed by Rayleigh.

J. J. appointed E. Everett as his private assistant. Everett gave him invaluable help in his most famous experimental investigations, providing, among other qualities, the physical manipulative skill that J. J. lacked.

Hayles served the Laboratory for more than fifty years, and Lincoln, Everett, and their successors also for comparable periods. The continuity of the workshop characteristics was one of the strands in the Cavendish tradition.

The strain on J. J. at the beginning of his professorship was great. Besides adjusting to the psychological problem of directing scientists who had been his seniors and teachers, and coping with problems of management, of which he had no previous experience, he was profoundly conscious of the intellectual responsibility of succeeding Maxwell and Rayleigh. When he gave the first of his first course of lectures as Cavendish professor in 1885, he nearly fainted, and had to dismiss the class. He never suffered from such an attack again. His success in mastering these early difficulties illustrated the toughness of his mental and physical fibre, in addition to his intellectual gifts.

Among J. J.'s early students in the Cavendish was W. H. Bragg. After graduating in mathematics as Third Wrangler in 1884 Bragg spent part of 1885 in the Cavendish, but before the end of that year he was appointed Professor of Mathematics and Physics in Adelaide University, South Australia. J. J. had been asked to recommend a candidate, and happening to meet Bragg on Kings Parade in Cambridge, he asked him whether he would like it. In those days not all young men of talent were willing to go to Adelaide, but W. H. Bragg joyfully accepted the offer.

J. J.'s first experimental research after his election was on the discharge of electricity in gases. His close friend Richard Threlfall collaborated in the beginning of this research, which ultimately led to J. J.'s classic discoveries. He had formed the opinion that whenever a gas conducts electricity, its molecules must have been split up, and that it was the molecules in this modified condition that enabled the gas to conduct electricity. When the molecules were in their normal, whole condition, the gas was a perfect insulator. He thought that some of the molecules were split into two atoms, one being positively, the other negatively electrified. His first experiments were aimed at testing this idea.

Threlfall, who particularly appealed to J. J., was appointed assistant demonstrator. He was appointed to the chair of physics at Sydney, New South Wales, in 1886, but soon returned to England, and became a leader in scientific industry. He was prominent in armament developments in the First World War, and became an important link between J. J. and military and industrial science. J. J. was appointed a member of Lord Fisher's Board of Invention and Research in 1915, and Threlfall later joined the panel of scientific advisers. He developed phosphorus smoke screens for concealing ships at sea, and the manufacture of phosphorus bombs. He was the first to suggest that helium should be used in airships, instead of hydrogen.

Another early research student of whom J. J. was particularly proud was H. L. Callendar. J. J.'s interest in and understanding of scientific men who became prominent in business and engineering was a reflection of his practical managerial sense. He looked and behaved externally like an absent-minded professor, but this was far from the case. Besides his advisory work in military science, he became chairman of an influential government committee on the position of science in the British educational system.

J. J.'s first foreign research students from Europe were K. Olearski of Lemberg and W. Natanson of Cracow. The former arrived in 1885, and worked on the dielectric strength of mixed gases. Natanson investigated the dissociation of nitrogen oxides.

J. J. himself studied the electrical discharge through gases in a uniform field, in order to discover the state of the field before the discharge occurred. The distinction between the steady conduction of electricity in the discharge and the disruptive effects of the sparks was not yet clearly recognised.

Accordingly J. J. began investigations with the collaboration of Newall, another member of his first group of research students, on the leakage of electricity through bad conductors such as benzene, paraffin, oil and carbon disulphide. They found that at low voltages it obeyed Ohm's law. It had previously been shown by Quincke that at high voltages of the order of 30000 the leak was proportionately larger. This was evidently because the electric field had had to do work in splitting up the molecules.

W. Cassie was the first to hold the Clerk Maxwell Studentship, founded in 1890 by funds bequeathed to the University by Mrs Clerk Maxwell on her death in 1887. One of the conditions of the award was that it should not depend on the result of an examination, but on promise in research. This kind of senior research studentship was an important addition to the Cavendish Laboratory's facilities.

There were then very few opportunities for a senior research student to devote sustained attention to an important piece of work, unless he had private means. Many able men had to find remunerative work, such as teaching, soon after graduating, because of lack of other means of support. Few English schools, unlike some of those in Germany, left teachers with the energy and time for research.

Cassie worked on experimental implications of the electromagnetic theory of light, in particular on the relation between the refractive index of a transparent medium and its specific inductive capacity. It was part of that branch of Cavendish research which aimed at confirming Maxwell's electromagnetic theory of light.

J. J. recorded that Cassie, owing to circumstances beyond his control, had not been able to complete his research as well as he would have wished. He insisted on returning his stipend to the Laboratory for the purchase of apparatus, for that period during which, he felt, he had not been pulling his full weight. He was a comparatively poor man, and the sum represented a large part of his income. The incident illustrates both the high principles of Cassie, and the strong feeling of a Cavendish man not to let the Laboratory down by the imperfection of his work.

L. R. Wilberforce experimented with a delicate interference fringe method for the detection of a translation motion in the electromagnetic medium caused by a displacement current. The experimental proof of the Maxwellian electromagnetic theory was naturally a major concern of the Cavendish experimenters. J. J. thought about and attacked the problem from many different approaches.

113

However, the experimental proof came not from the Cavendish, but from Hertz, in 1887, when he demonstrated the existence of electromagnetic waves longer than those of visible light. The experiments made in the Cavendish and elsewhere had failed because the methods of detection used had not been sufficiently sensitive.

Hertz's results were received with enthusiasm in the Cavendish. The younger mathematical physicists had been brought up on Maxwell, and were inclined to accept his theory without experimental proof, because it was theoretically so elegant and simple. Hertz's beautiful experiments confirmed their ideas; for them it did not come as a revolutionary discovery. But on the continent of Europe, where Maxwell's theory had carried much less conviction, it came with revolutionary force, both in the theory of electromagnetism and in the wider field of physics.

The failure of the Cavendish experimenters to prove the existence of electromagnetic waves experimentally has significance for the condition of research in the Cavendish up to about 1887. Maxwell's own lack of urgent concern about the experimental proof of his theory was a factor in the failure. He had a passionate conviction about ideas, manifested in all his thinking. He may be seen as a scientific poet, who could create theories more real than common knowledge, in the sense of Shelley's poet:

> '. . . create he can
> forms more real than living man.'

J. J. was a great scientist, but experimentation was not the chief factor in his genius. If Rutherford had been ten years older, and had arrived at the Cavendish ten years earlier, he would probably have been the first to demonstrate the existence of electromagnetic waves.

While Hertz made the crucial discovery confirming Maxwell's theory, J. J. kept busily on with his multifarious researches in many directions, and with increasing numbers of research students. In some ways he owed more to Balfour Stewart and Manchester. His own greatest discoveries were to come more from the mode of thought of his *Applications of Dynamics to Physics and Chemistry*, inspired by the ideas of Balfour Stewart, who had concentrated more on atoms and molecules, and their dissociation. J. J. was in Cambridge for four years without getting to know Maxwell.

The investigation of the discharge of electricity through gases appeared to be more difficult than the conduction of electricity through liquids and solids, even though the properties of gases are simpler than those of liquids and solids. Because of this, it is in principle easier to form a mental picture of what is going on in a gas. The study of gases was of prime importance to the advance of physics in the seventeenth century, because they present matter in its simplest form.

In spite of the obscurity of the phenomena of the electrical discharge in gases, history was to repeat itself in the nineteenth century. The next great advance in knowledge of the constitution of matter was again to come from the study of gases, but this was not commonly apparent in the 1870s and 80s. A large part of the researches at the Cavendish were concerned with the electrical properties of liquids and solids. Owing to their inherent difficulty they produced many incomplete and inconclusive results, creating an impression of a degree of energetic chaos when looked at as a whole.

J. C. McConnel, who had been appointed as assistant demonstrator by Rayleigh in 1884, had extended optical researches in the line of Stokes and Glazebrook. His health broke down, and he retired to Switzerland. He studied meteorological phenomena, observing iridescent clouds, the colours of which he explained as due to diffraction by ice crystals. He related cloud appearances to the size of the smallest ice crystals, as they had been measured on Ben Nevis. He died in 1890, in his thirtieth year. McConnel's research interests and approach were not unlike those of C. T. R. Wilson, who was inspired in 1894 by cloud appearances on Ben Nevis.

In the attempts to confirm Maxwell's electromagnetic theory, J. J. and H. F. Newall started an experimental research in 1886 on the continuity of displacement currents. Hertz's discovery in the following year removed its point, but the preliminary researches revealed information of interest. They intended to use a very strong magnetic field, and found that they could attain much higher inductions in iron than any that had been published, and higher than the maximum attributed by Rowland for iron. They then learned that Ewing had obtained inductions still higher than their highest. Quite a number of Cavendish researches were overtaken by others.

Investigations on general properties of matter were pursued. A. M. Worthington found that the coefficient of extensibility of liquids remained constant at pressures up to 17 atmospheres.

S. Skinner experimented on the compressibility of liquids, and in particular of that of solutions compared with solvents. Electrolytic solutions were found to be considerably less compressible, whereas non-electrolytic solutions showed little or no diminution. The results were in agreement with J. J.'s theory of osmotic pressure.

The most notable work on heat was done by H. L. Callendar. J. J. held that his career in the Laboratory was in some respects the most interesting in all his experience. Callendar had read classics at school, and had not done any physics. He graduated in 1884 with a First Class in classics, and in the following year with a First Class in mathematics. When he came to work in the Laboratory, he had not had any previous experience of practical work, and no training in theory except for the ideas he had acquired in casual reading.

J. J. pondered on what topic he should propose to him for research, which would provide scope for his gifts, and would be least hindered by his lack of experience. He concluded that the most suitable would be the accurate measurement of electrical resistance, and the application of this technique to the measurement of temperature.

Werner von Siemens had devised a pyrometer on this principle. It consisted of a platinum wire wound round clay, and enclosed in an iron tube. A British Association report in 1874 had indicated, however, that it was inaccurate.

The simplicity and convenience of using a wire for measuring temperature appealed very strongly to J. J., so he suggested to Callendar that he should investigate whether the weakness of Siemens's instrument was inherent in the principle, or due to a defect in design.

Beginning research in 1887, Callendar attacked the problem enthusiastically, and within eight months showed that if the wire was kept free from strain and contamination, it could serve as a very reliable as well as convenient thermometer.

Callendar foresaw the importance of a suitable notation and mechanical system for recording and communicating thermometric observations, an approach which has become very important with the development of automation and the computer. His research caused a revolution both in high- and low-temperature thermometry.

J. J. looked in on him from day to day to see how the research was going, and encourage him when he had difficulties. After only a few months of research, he had made a new device of absolutely first-class importance in physical research, and in technological processes utilising regular and exact control of temperature.

Callendar submitted his research as a dissertation for a fellowship at Trinity College, and was elected at his first candidature, which was unusual. J. J. thought that Callendar's case raised doubts about the need for the large amount of time devoted to advanced courses in practical physics. He observed that Joule, Stokes, Kelvin, Rayleigh and Maxwell never attended any courses in practical physics.

Callendar reported in 1899 that his experiments suggested that the resistance of certain metals vanishes at a temperature appreciably above absolute zero. He did not pursue this suggestion, and so did not open up the field of very low-temperature research, or discover superconductivity.

He was appointed Macdonald Professor at Montreal at an early age, and was succeeded there by Rutherford. When Rutherford arrived, he was rather irked by the extremely high esteem in which Callendar had been held, and had the impression that he was expected to try and live up to him. In 1898 he wrote to his fiancée: 'This extra-ordinary belief in Callendar is a little unfortunate for me, as however well I do I will never rise to the acme of greatness attained by him.' Less than a year later, however, he wrote again: 'As a matter of fact, I don't quite class myself in the same order as Callendar, who was more an engineering type than a physicist, and who took more pride in making a piece of apparatus than in discovering a new scientific truth — but this between ourselves.'

The difference in attitude of J. J. and Rutherford to Callendar reflected the difference in their scientific minds.

During the 1880s and 90s there was a good deal of research in the Cavendish on electrolysis, as part of the general concern with the nature of electricity. W. N. Shaw wrote a report on the subject for the British Association in 1890. T. C. Fitzpatrick investigated the conductivity of electrolytes to alternating currents, and W. Coldridge experimented on the properties of stannic chloride, a liquid that Faraday had found to be a non-conductor.

Interest in electrolysis was stimulated by Arrhenius's development of the Clausius–Williamson hypothesis of the dynamical equilibrium of chemical compounds, based on the continual dissociation and re-formation of molecules. The Cavendish experiments confirmed their ideas, and corroborated Kohlrausch's notion of specific ionic velocities.

W. C. D. Whetham continued research on the velocity of ions in electrolytes during the period 1890–4. He devised a new method depending on the use of coloured salts, and obtained results which were in excellent agreement with Kohlrausch.

W. N. Shaw pursued researches on hygrometry, which made him an authority on ventilation. During the period 1889–90 he carried out, with the collaboration of R. S. Cole, research on an analogue of electrical and air resistance, on which his important work on ventilation was based.

Besides this varied experimental physics, the Cavendish men were also publishing mathematical papers.

J. J. very energetically pursued a variety of researches during the first decade of his professorship. He was, as it were, ploughing the soil. The most important line of research, however, was on the passage of electricity through gases. Following the indication of Maxwell, J. J. had explored the theoretical implications of the researches of Crookes, Goldstein and their predecessors, and the ideas of Clausius and Williamson. He viewed them in the light of the generalised mode of thinking he had acquired from Balfour Stewart, as well as from Maxwell's theory, and in 1883 he had announced that the dissociation connected with the electrical discharge through gases was not an accident attendant on it, but an essential feature of the discharge, without which it could not occur.

In 1889 J. J. showed that if a permanent gas is heated in a closed vessel up to 300° C the potential required to produce a discharge does not change. If the vessel was left open, the potential fell, due to the change of density in the gas caused by the rise in temperature. If the temperature was raised so high as to cause dissociation, then the potential required fell very low or to zero.

In the same year he studied the resistance of electrolytes to alternating currents of very high frequency. They are as good conductors for currents alternating a hundred million times a second as for direct currents. As they are transparent to light they should according to Maxwell's theory become insulators for currents alternating with the frequency of light vibrations. J. J. inferred from this that the molecular processes on which conduction in electrolytes depend must occur at rates of 10^9 to 10^{15} per second.

In 1890 J. J. started his investigations on the discharge of electricity through hot gases. He found that dissociable gases became good conductors, and obeyed Ohm's law, while gases that did not dissociate remained poor conductors. He discovered the importance of glowing terminals in promoting the discharge through a gas.

In the following year he investigated discharges produced without the involvement of electrodes. He showed that it was due to an electromagnetic effect, involving a discharge through gas and accompanied by a brilliant luminosity.

J. J. found that it was often very difficult to initiate the electrodeless discharge. Though the phenomena involved were extremely interesting, he laid the topic aside, because of the experimental problems and the difficulty of making quantitative measurements. The capacity to relinquish attractive but not immediately profitable researches in favour of more immediately promising ones was one of J. J.'s gifts.

After the difficulties with these researches, he was pleased with the experiments of J. B. Peace, who discovered in 1892 that if the pressure in a tube was kept constant, the potential required to produce a discharge diminished as the distance between the plates diminished. However, when the distance was reduced to less than a certain figure, the required potential increased.

In 1891 J. J. determined the rate of propagation of the luminous discharge by means of a rotating mirror. He showed that the luminosity started from the positive electrode, and travelled with a velocity of rather more than half the speed of light.

In 1893 he studied the effect of electrification on a jet of steam, and showed that negative is more effective than positive electrification. This was of significance for the line of research on cloud formation and electrification, which was to come soon afterwards from C. T. R. Wilson.

The same year saw the publication of J. J.'s third revised edition of Maxwell's *Treatise on Electricity and Magnetism*, and his own *Notes on Recent Researches*.

In 1894 he determined the velocity of the cathode rays, obtaining a value that was much greater than the average speed of molecules, and much less than that of the main discharge from the anode to the cathode. He found that it agreed very nearly with the velocity that a negatively charged atom of hydrogen would acquire under a potential fall near the cathode, if it had an electrolytic type of charge. In that case, the ratio of charge to mass would be about 1 to 10000. He pointed out that the deflection of the cathode rays by a magnetic field showed that the particles must have a velocity of this order.

Thus, in 1893 he had almost discovered the electron. Four years were to pass, however, before that major discovery was completely proved. The first ten years of J. J.'s professorship were marked by variety and industry in research, and almost, but not quite, culminated in major discovery.

While all this research activity was progressing, he was carrying out teaching duties, in addition to administering the Laboratory and taking part in University affairs. His lectures were much appreciated by the students, whether as teaching for the Tripos examinations, or as suggestions for post-graduate research. He had a gift for suddenly introducing old knowledge, familiar to his hearers, in a new light. This stimulated their interest and focused their attention. He never assumed that his audience knew much mathematics, and always liked to work things out from first principles. He made mathematical calculations on the blackboard with extraordinary speed and accuracy.

A young post-graduate wrote of J. J. and his lectures in the *Cambridge Review*:

'It is very apparent that you are always physicist first and mathematician second. For when in the course of some investigation a new function turns up which would keep some of your colleagues in Trinity contented and happy for months, you merely 'with a grave scornfulness' select such of its properties as you require and march straight on to the goal you have in view. And this accounts for what sometimes befalls you in lecture-room. For, though knowing well what is the result you wish to obtain, you have occasionally mislaid the envelope back containing the details of the investigation and are compelled to plunge at short notice into a sea of symbols. Yet when, since even Professors make slips sometimes, it becomes evident that the desired result is not coming and you find it necessary to apply an empirical correction to the work on the blackboard, the cool confidence with which you say 'Let's put in a plus' and the smile of cheery conviction with which you turn to your audience put to shame the incredulity of the most sceptical among them.'

Newall quoted a paragraph from J. J.'s *Applications of Dynamics*: 'The object of the following pages is to endeavour to see what results can be deduced by the aid of these purely dynamical principles without using the Second Law of Thermodynamics.' He thought that J. J.'s use of the word 'see' instead of 'show' illustrated his power as a teacher; it expressed his sympathy with the hearers, making himself one with them, and proposed that they should join him in seeing.

By the end of J. J.'s first decade as professor, research in the Cavendish had been established on a much wider foundation. During the period 1874–83, about 126 papers were published from the Laboratory; of these, 40 were by Maxwell, and 27 by Rayleigh, the number of contributors being 23. In the period 1884–93 there were about 124 papers of similar standard, of which 34 were from J. J., while the total number of contributors was 34.

Newall estimated that 220 papers were published in the period 1885–94, from about 36 contributors. He excluded papers of a theoretical nature which did not lead immediately to experimental research. General physics, heat and optics, accounted for 31 per cent; electricity and magnetism and conductivity in gases for 40 per cent; 10 per cent were non-experimental, and 19 per cent covered reports and miscellaneous topics.

Very broadly speaking, the volume of work and number of research workers doubled in J. J.'s first decade. J. J. published at about the same intensity and volume as Maxwell and Rayleigh, but his work formed a smaller proportion of the total output of the Laboratory. It was no longer crucially dependent on individual geniuses like Maxwell or Rayleigh, and consequently had been placed on a firmer foundation. A large number of research men had been trained, and the organisation had become more thoroughly established.

Besides the development of the formal aspects of organisation, the more subtle qualities of intellectual atmosphere were promoted. The Cavendish Physical Society, formed by J. J. in 1893, met fortnightly in term time, primarily to discuss recent advances in physics. It was quite informal, with no written rules or membership; anyone interested could come in. J. J. took the chair and led the discussions. As he was a fluent speaker, and the best and most widely informed, he did most of the talking.

Members and visitors gave reports of recent researches, both of their own and those in other places. This kept the Laboratory abreast of research, and gave members practice in lecturing, and expressing themselves. New work was reported on before it was published, and in the following discussions improvements were often suggested.

In 1895, on Mrs Thomson's suggestion, tea was provided before the meetings of the Society. She presided, and the arrangements were kept as simple as possible to preserve a workman-like atmosphere. During these teas research workers came to know each other better, and also made the acquaintance of visiting scientists. The Society helped to form a Cavendish type of personality.

In 1894, when J. J. was thirty-eight years old, the man and the Laboratory were well prepared to attack deep problems in physics. Then in 1895, however, they received a revolutionary increase in strength as a consequence of a change in University regulations, by which graduates of other universities were admitted as 'research students'. If, after two years' residence, they submitted an acceptable thesis on the research work they had carried out, they became entitled to a Cambridge University Master of Arts degree. Presently, the University replaced this by a specially instituted Doctorate of Philosophy, or Ph.D.

The first arrival under the new regulation was Ernest Rutherford from New Zealand. Several other new research students of outstanding ability soon followed. But the full effect of the regulations was far deeper; it became clear that they signified a change in the status of the Laboratory, and in that of the University itself.

Originally, Cambridge had been an English University. In Isaac Newton's time it had become the leading centre of English science; during the nineteenth century it became the leading centre of British science, signified by Clerk Maxwell's move from Scotland. Then at the end of the century it started to become the main centre of science in the British Empire and Commonwealth, signified by the arrival of Rutherford. Up to 1895 the Cavendish Laboratory had drawn its inspiration and resources from inside Britain; its achievements were essentially a reflection of the scientific aims of the British society of the nineteenth century, with its strengths and limitations. But in the twentieth century it became more of a world institution.

Before long these research students formed one of the most characteristic features of the Laboratory. They became numerous, and, at various times, came from nearly all of the important universities of the world, creating a very international atmosphere. Together with them were the research students from other British universities, who now found it easier to go to Cambridge to do research.

Professors from other universities in the world came to work there during their sabbatical years. This extension of the range of research workers, with their different points of view, training and temperament, was highly stimulating, broadening and educative, especially for the students who had first graduated at Cambridge before starting on research, knowing only the Cambridge outlook. Besides gaining a better appreciation of the work done in other countries, they learned also about their political and social problems. When the Laboratory workers met for tea in the afternoon, the leading events of the day were often discussed, and if they had occurred in a foreign country, there was generally someone present who had personal knowledge of the conditions where they had happened.

J. J. said that he had listened during Laboratory teas to American Republicans and Democrats fighting their battles over again, and had learned far more from them about American politics than from the reports of the special correspondents.

The research students instituted an annual dinner, to which J. J.

and one or two other eminent persons were invited. It was a rollicking affair. Langevin sang the Marseillaise with such fervour that a French waiter embraced him. Later on, light verse was composed for reciting or singing to popular tunes. Much of it was written by the mathematical physicist A. A. Robb, but others helped with odd lines and verses. A collection was printed under the title *The Post-Prandial Proceedings of the Cavendish Society* in 1904; it was revised up to the sixth edition, which appeared in 1926.

One of these compositions gave a happy depiction of J. J.'s behaviour just after he had discovered something:

> When the professor has solved
>   a new riddle,
> Or found a fresh fact, he's
>   fit as a fiddle,
> He goes to the tea-room and
>   sits in the middle
> And jokes about everything under the sun.

Verses on *The Revolution of the Corpuscle* (the electron) were sung to the tune of *The Interfering Parrot* from the light opera *The Geisha*. J. J. was devoted to Gilbert and Sullivan operas, and prided himself on being present at their first nights. 'Some professional agitators only holler till they're hoarse,' but the corpuscle was not like them. 'He struggled for freedom against a powerful foe...' He:

> 'radiated until he had conceived
> A plan by which his freedom
>   might be easily achieved.
> The atom and the corpuscle
>   each made a single charge,
> But the atom could not hold
>   him in subjection,
> Though something like
>   a thousand times as large.
>
> The corpuscle won the day,
> And in freedom went away.
> And became a cathode ray.'

Their most popular was *Ions Mine*, sung to the tune of *Clementine*:

> 'In the dusty lab'ratory,
> 'Mid the coils and wax and twine,
> There the atoms in their glory,
> Ionize and recombine.'
> . . . .

123

J. J. contributed a verse to this, his only known metrical composition:

> 'And with quite a small expansion,
> 1.8 or 1.9,
> You can get a cloud delightful,
> Which explains both snow and rain.'

Another of the songs was dedicated to J. J. and his assistants:

. . . .

> 'My name is J. J. Thomson, and my lab's
>   in Free School Lane,
> There's no professor like J. J. my students
>   all maintain,
> I've been here six-and-twenty years, and
>   here I shall remain,
> For all the boys just worship me at my
>   lab. in Free School Lane.
>   :
> The people are delighted with the wondrous
>   things we do,
> But few have any notion that we're such a
>   jolly crew,
> If some of them were here tonight I think
>   we'd make it plain
> We're not all just as dry as dust at the lab.
>   in Free School Lane!'

The *Post-Prandial Proceedings* reflect the informal, lively, strong community spirit of the Cavendish, but they also show that Cavendish men shared the period's middle-class taste in popular amusements and attitudes of mind. As a social institution, the Cavendish had distinct class features.

In the period 1895–8 less than half of the more prominent research workers in the Cavendish were English. The visitors from overseas included Rutherford from New Zealand, Townsend, McClelland and Henry from Ireland, Langevin from France, Zeleny from the United States, Novak from Austria and McLennan from Canada.

When the new regulation was introduced, the general attention was focused on solving the immediate problems of physics. The able band of research workers was welcomed as an acquisition of scientific strength, without much reflection on the significance of where they came from.

In 1877 the number of students working in the Cavendish had been about twenty. By 1882 the figure had risen to sixty-two, and in 1885, ninety. Of these, ten were engaged in research.

J. J. instituted practical classes for medical students in 1888. This raised the number of students using the Laboratory to 153 and made the problem of accommodation acute. Relief was obtained by transferring the medical classes to a room formerly used for anatomical dissection, but additional accommodation was obviously necessary.

In 1893 an adjoining site in Free School Lane, occupied by old houses, was assigned by the University for an extension to the Laboratory. A Syndicate consisting of the Vice-Chancellor, the Provost of King's, A. Austen Leigh, the Master of Christ's, J. Peile, Professor Liveing, Professor J. J. Thomson and Dr R. T. Glazebrook, was appointed to report on the building of an extension. It reported in November 1894 that a suitable three-floored extension could be erected for £10 000. However, in view of the University's financial stringency, it was recommended that only the ground floor should be erected immediately, and the second and third floors be added later.

J. J. offered £2000 towards the cost of this construction, which he had saved from the fees of students, and which had originally been intended for the provision of apparatus. A large ground-floor room was then built by Sindall of Cambridge, according to the designs of W. M. Fawcett, the architect of the original Laboratory. A small lecture-room, and a private room for the Professor were also added. The cost amounted to about £4000.

The extension was first used in 1896. The official announcement of the completion of these constructions appeared in the same report as that in which the new system of graduate research students was announced. It was a happy circumstance that the Laboratory should have been extended just when this highly significant influx began.

The extra accommodation was extremely welcome, but with the continually increasing numbers of undergraduate and research students, it soon became insufficient. Shortage of accommodation and equipment continued to be a feature in the Professor's annual reports.

9

Fruition

By 1895 a great deal of significant information on the discharge of electricity through gases had been discovered by the Cavendish and other investigators, but the phenomena were as yet too complicated to be explained by any concise idea. The gases were made conducting by passing sparks through them. The voltages needed to start the discharge were so high that a variety of effects was produced, which could not be satisfactorily disentangled. While the amount of information increased, its explanation advanced less quickly.

Faraday and Maxwell had pointed out that the facts of electrolysis could be explained if electricity itself was atomic, but both of them had given warning of the difficulties of extending this idea to the explanation of other electrical phenomena.

In 1884 Schuster had seen that if the cathode rays consisted of charged particles in motion, the ratio of their charge to their mass might be measured and in that way something might be discovered about the mass of the charged particles, in comparison with that of the hydrogen atom. He submitted a stream of cathode-ray particles, propelled by an electric field of known strength, to deflection by a magnetic field of known strength. From the data he calculated that the ratio was about one thousand times greater than for the hydrogen atom. He published this result in 1890, but the idea that the atom of electricity could exist independently of the atom of matter was still far too intellectually revolutionary to be entertained as a practical reality.

As Schuster was experimenting with gas at a not very low pressure, he assumed that the cathode particles must collide with many molecules, and that their velocity would therefore be much lower than if they were moving under an electric field in a high vacuum. Schuster's method was later used by Kaufmann in his accurate determination of the ratio of e/m for cathode-ray particles.

While the need for using high voltages was one obscuring factor in the research on the conduction of electricity through gases, another was the imprecise conception of the cathode-ray particle. Until this became clearer, it was difficult or impossible to make faster progress. A fact of fundamental importance was discovered by Hertz in 1892, when he showed that cathode rays were able to pass through very thin films of matter. This was pursued by Lenard, who showed that the cathode rays could pass through a small opening in a discharge tube, which enabled their properties to be examined outside. He demonstrated that their power of penetrating matter depended only on the mass of matter present, and was substantially independent of its chemical composition.

These discoveries were of great importance because they altered the mode of thinking about the penetration of matter by radiation. Hitherto this had been thought about in terms of the penetration of transparent material by light waves. The discovery that matter could be penetrated by material particles led to the application of the idea to the phenomena of the conduction of electricity in gases. Even more important, it prepared the scientific imagination to realise that there might be other kinds of radiation, besides light and certain material particles, that were capable of penetrating matter.

It then became more likely that any manifestation of penetration by radiation, whether wave or particle in nature, from electric-discharge tubes would be recognised and scrutinised. It was in this climate of scientific knowledge and thought that Röntgen, in 1895, recognised the existence of a highly penetrating radiation emitted by electric-discharge tubes. Within a few months, working in such secrecy that he did not even tell his wife, he discovered most of its fundamental properties, and then in November 1895, published his discovery, which he called X-rays.

Their extraordinary and spectacular properties multiplied many times the shock to the scientific imagination caused by the discoveries of Hertz and Lenard. They raised the temperature of the scientific mind and created an expectancy of radical discovery, which was realised by a swift succession of major experimental and theoretical developments.

Interest in the X-rays reinforced research on how they were produced in electric-discharge tubes. It soon became clear that when the cathode rays impinged on a metal object they produced X-rays. This increased interest even more in the nature of the cathode rays.

In 1896 J. J. discovered that X-rays made gases at ordinary pressures electrically conducting, and apparently produced in them large numbers of positively and negatively charged particles. If X-rays could do this at ordinary gas pressures, it seemed probable that a strong electric field would be able to do it at low gas pressures. This made it more likely that the cathode rays consisted of negatively charged particles, as Crookes had suggested. But the conclusion could not be certain until it had been shown that the cathode rays were deflected by electric as well as magnetic fields.

When X-rays were discovered in 1895 their extraordinary nature at once inspired numerous experiments and researches. These were along several lines. One was an application of their most spectacular property: their power of revealing the position and shape of bodies in opaque substances, such as bones in the human body. Another was on the nature of the rays: were they waves or particles? The first line was obviously of medical and industrial importance, while the second directly confronted the investigator with the problem of the fundamental nature of radiation, to which a quick answer was hardly to be expected.

J. J. pursued a third line: he at once sought to investigate whether he could use X-rays as a tool in advancing the research that he had already long been carrying out on the conduction of electricity in gases. He was very well prepared to make use of a good new tool, which was a more promising line of advance than asking the most difficult and fundamental questions about the new phenomenon.

One of the most important qualities in scientific genius is to ask questions that are soluble and not too far in front of their time, both in ideas and resources. J. J. showed this sense of strategy in researching the utilisation of X-rays.

At the beginning of the Easter term in 1896, Rutherford, who had just completed his researches on his magnetic detector for radio waves, joined J. J. in a systematic research on the nature of the process by which X-rays cause a gas to become an electrical conductor. It had been noticed that a gas, which had been rendered conducting by X-rays, retained its conductivity for a second or so after the rays had been cut off. A rapid current of air was passed through a long metal tube, and X-rays were directed into the air near the beginning of the tube. The conductivity caused by the rays was retained by the

128

air up to a distance of several feet from the radiating point. If the stream of conducting air was made to pass through a strong magnetic field the conductivity, which enabled it to discharge electrified objects, disappeared. A similar loss of discharging power occurred if the conducting air was passed through a plug of cotton wool, or bubbled through water.

It was evident that the radiation produced something that was removed by a magnetic field, or by filtering. The experimental facts could be explained if positively and negatively charged particles had been produced in the air.

The conductivity persisted for some time after the initial radiation, because the positively and negatively charged particles took some time to recombine after the radiation had stopped. The disappearance of the conductivity in strong magnetic or electric fields was due to the removal of the charged particles by the fields. The conducting gases did not obey Ohm's law, because strong fields removed the charged particles so quickly that most of them did not have time to recombine.

The current through the gas was at a maximum under these conditions, so long as the fields were not strong enough to produce new ionisation. This maximum or 'saturation' current was a characteristic property of all ionised gases not exposed to voltages high enough to produce spark discharges. It indicated that under defined conditions the X-rays produced conducting particles at a definite rate.

J. J. and Rutherford published a paper, 'On the Passage of Electricity through Gases exposed to Röntgen Rays', in November 1896, giving a simple mathematical theory of all the experimental facts, old and new, on the conduction of electricity through gases, without making any assumptions about the nature of the ions. It became the basis for all subsequent work in the subject.

Rutherford experimentally verified J. J.'s assumption that the rate of recombination of ions was proportional to the square of the number present. He discovered that the rate was much affected by the presence of dust or other nuclei in the gas. As the ions diffused through the gas, they came in contact with relatively large particles of dust, and stuck to them or transferred their electric charge to them; thus dust rapidly removed the ions.

Besides knowing the general features of the movement of ions, it was necessary to know the exact velocity with which they travelled in a field of known strength. In 1897 Rutherford measured the sum of the mobilities of positively and negatively charged ions in various gases at atmospheric temperature and pressure. In air the sum was found to be 3.2 centimetres per second, and in hydrogen, 10.4.

It followed from the kinetic theory of gases that a molecule of hydrogen, if it carried a charge equal to that of a hydrogen atom, should travel in hydrogen at the rate of 340 centimetres per second for a voltage gradient of 1 volt per centimetre, while the observed sum of the mobilities was less than one-thirtieth of this figure. This indicated that the mass of the ion was much greater than that of the molecule of the gas in which it was produced. One suggested explanation was that the ion attracted a cluster of neutral molecules, which slowed it down, but later it became clear that it was due more to the charge on the ion multiplying collisions, so that it was slowed down relative to the uncharged particles.

The discovery of X-rays and Rutherford's collaboration with J. J. gave a fresh impetus to the Cavendish researches on the electrical conduction of gases. Up to 1895 J. J. had been engaged in extensive explorations of both the theoretical and experimental aspects, but, like other leading investigators, he had not been able to reach definitive conclusions. This led to misunderstandings. Feeling that Schuster also had not yet reached definite conclusions, he did not read his papers thoroughly, and did not immediately appreciate what he had done.

After 1895 experimentation was facilitated by X-rays, and Rutherford in particular improved the precision of measurements. As J. J. later put it, 'Rutherford devised very ingenious methods for measuring various fundamental quantities connected with this subject, and obtained very valuable results which helped to make the subject 'metrical' whereas before it had only been descriptive.' Following these advances J. J. succeeded in 1897 in giving quantitatively precise conclusive proof of the existence of the electron and its fundamental properties. His success arose not so much directly out of these advances, but from the increased research confidence arising from the X-ray discoveries, and the growing possibility of making exact quantitative experiments.

Jean Perrin had shown in 1895 that if cathode rays were deflected through a suitable opening, they carried with them a large negative electric charge. This provided strong evidence that the rays consisted of negatively charged particles.

J. J. repeated the experiment in a more closely defined form. He projected a pencil of cathode rays between two parallel plates, which were kept at a constant potential difference by connection to a battery. When the gas pressure was very low the pencil was deflected towards the positively charged plate, which was strong evidence that the cathode rays consisted of negatively charged particles.

He found that when the pressure of the gas was higher, the charged carriers produced in the gas by the passage of the cathode rays masked their deflection by the electric field between the plates. Under these conditions the cathode rays appeared not to be deflected by an electric field. This explained how Hertz had been led to believe that they were wave-radiations; because, like light, they appeared not to be deflected by an electric field.

J. J. had now satisfactorily proved that the cathode rays consisted of streams of negatively charged particles. The next requirement was to obtain more information on their mass. He found the ratio between their charge and mass by a method similar to Schuster's, and was struck by the fact that the mass came out at about one-thousandth that of a hydrogen atom. He therefore devised an experiment that would conclusively determine the mass. A narrow pencil of cathode rays in the discharge tube was subjected to a magnetic field of known strength. The size of the deflection provided a measure of the product of the mass of the individual particles and their velocity, divided by the unit charge. The pencil of rays then fell on a thermocouple, which gave a measure of their kinetic energy. The total charge carried by the pencil was measured by collecting it in an insulated metal cylinder. From these three measurements the values of the velocity of the particles, and of the ratio e/m could be calculated.

Experiments were made with various gases. In all cases the ratio of charge to mass was the same within the limits of the experimental error; the velocity of the particles was about 10000 to 100000 kilometres per second.

J. J. then devised a simpler experiment, in which the pencil of cathode rays was subjected both to a magnetic and an electric deflection. He obtained a slightly lower result for the ratio of charge to mass. Since the ratio of charge to mass for the hydrogen atom is very much smaller, J. J. at first thought that the cathode-ray particle might carry a charge larger than that on the hydrogen atom, but concluded that the evidence was strongly in favour of the charge being the same, but the mass very much smaller.

He had now conclusively proved the existence of the cathode-ray particle, that it moved at an enormous speed, was very much smaller

131

than the hydrogen atom, and that it was of the same nature from whatever source it was obtained. He referred to the particle as a 'corpuscle', and continued to use this term for many years. It became plain, however, that the 'corpuscle' was the real object corresponding to the hypothetical unit charge of electricity implied by the phenomena of electrolysis. In 1891 G. J. Johnstone Stoney of Dublin had introduced the term 'electron' to describe this unit charge. It had been adopted by Larmor and Lorentz in their researches in electromagnetic theory and later became generally accepted.

J. J.'s proof of the existence of the electron was to have vast significance, not only for physical science but for industry and society. The development of radio and the computer are only two of the examples of the application of electronic discoveries that have revolutionised the conditions of human life.

In 1897 J. J. was forty years old; he had been Cavendish professor for thirteen years, and he was surrounded by talented research students from abroad as well as from Britain. He had organised the Laboratory in a characteristic running style, best suited to achieving the aims he desired. He had had time to build up those sides in which he was personally not at his strongest. He did not have the intuition of a Maxwell, but he made up for this by industriously working through every vague possibility, until ideas gradually became clarified.

He was clumsy with his hands and had had to train assistants who could carry out manipulations for him. He enjoyed the collaboration of research students such as Rutherford. The moment came when circumstances were in his favour, and he took advantage of them. Improving on Schuster and Perrin, he set out to devise experiments which were completely unambiguous in their implications, and finally succeeded.

Schuster, Wiechert and others had anticipated his results in varying degrees, but they had not provided the complete proof that the fundamental nature of the enquiry demanded. Wiechert's direct measurement of the speed of the particles was particularly important, because it confirmed the correctness of the electromagnetic theory that J. J. had invoked.

In his paper of 1897 announcing his results, J. J. indicated their philosophical consequences.

The explanation which seems to me to account in the most simple and straightforward manner for the facts is founded on a view of the constitution of the chemical elements which has been favourably entertained by many chemists. This view is

132

that the atoms of the different chemical elements are different aggregations of atoms of the same kind. In the form in which this hypothesis was enunciated by Prout, the atoms of the different elements were hydrogen atoms; in this precise form the hypothesis is not tenable, but if we substitute for hydrogen some unknown primordial substance x, there is nothing known which is inconsistent with this hypothesis, which is one that has been recently supported by Sir Norman Lockyer for reasons derived from the study of the stellar spectra.

If, in the very intense field in the neighbourhood of the cathode, the molecules of the gas are dissociated and are split up, not into the ordinary chemical atoms, but into these primordial atoms, which we shall call for brevity corpuscles; and if these corpuscles are charged with electricity and projected from the cathode by the electric field, they would behave exactly like the cathode rays... Thus in this view we have in the cathode rays matter in a new state, a state in which the subdivision of matter is carried very much further than in the ordinary gaseous state: a state in which all matter—that is, matter derived from different sources, such as hydrogen, oxygen, &c.—is of one and the same kind; this matter being the substance from which all the chemical elements are built up.

He pointed out that the mass of the electron might be electrical in origin, following his calculation of 1881, when he showed that the magnetic field set up when an electrified particle is in motion increases the effective mass of the particle.

J. J. gave the first public account of his results in a lecture at the Royal Institution on 30 April 1897, but it did not at first cause much comment. The material and ideas were probably too new for their significance to be grasped immediately.

The research gave J. J. intense joy. R. J. Strutt (the fourth Lord Rayleigh) was an undergraduate attending the Cavendish Laboratory at the time. During the summer of 1897 he happened to meet J. J. in King's Parade. He walked with him, while J. J. eagerly described his experiments and their significance. When they came to the entrance to Whewell's Court, where Strutt's rooms were, Strutt walked past the entrance so as not to interrupt J. J. They went on, past St John's College and the Round Church, to the other entrance in Whewell's Court, where J. J. left him, after standing talking to him for a few minutes.

Shortly afterwards J. J. gave an account to the Cavendish Physical Society. It was received with an enthusiastic awareness of its importance, and pride in the Laboratory's achievement. Those present included research students from various parts of the world, who looked on it as an international as well as a British and a Cambridge event. The strength of J. J.'s research and personality was manifested by the choice of able research men to work on problems bearing on

133

his electrical researches, and even to give up important work of their own in order to join in his.

Schuster and J. J. had observed that the negatively charged ions in electric discharges appeared to be more mobile than the positive. In 1898 the American Zeleny showed conclusively that the negative particles always moved faster than the positive ions in an electric field of given strength. He also showed that the flow of ions in an electric field set up convection currents in the gas.

When Townsend arrived in the Laboratory, he worked on the magnetisation of liquids. Then he began to investigate whether chemical actions are accompanied by electrical effects. Laplace and Lavoisier had noted in 1782 that the hydrogen evolved when a metal is dissolved in sulphuric acid is electrically charged. Townsend studied the gases liberated in electrolysis, and found that they were very highly charged. They could pass through cotton wool, or be bubbled through liquids without losing their charge. When the charged gas was bubbled through water, it produced a dense white cloud; but if the gas was passed through a drying agent, the cloud disappeared. However, if passed on into a moist atmosphere, the cloud re-appeared. Townsend drew the conclusion that the charge was not carried by the gas itself, but was due to the presence of charged particles. These were much heavier than ordinary ions, because they diffused slowly and moved slowly in an electric field.

These experiments led Townsend to develop the theory of the diffusion processes that were involved. He afterwards applied the theory in experiments on the measurement of the rate of diffusion of ions, which enabled him to prove that the ion carried the same charge as the hydrogen atom. Thus Townsend came in to add to the foundations of the general electrical research led by J. J.

Yet another even more remarkable example was provided by C. T. R. Wilson. He, like J. J., had studied at Owens College, Manchester, before coming to Cambridge. He had graduated with a First Class in zoology when he was eighteen, and was urged to compete for a scholarship at Cambridge. This he did, taking an examination in practical physics and chemistry, in which he had not previously received any instruction.

The examiner was the chemist F. H. Neville. Wilson went up to him and explained the situation. Neville was sympathetic, and told him to do as well as he could in the circumstances. Wilson did so well that he found himself awarded a scholarship in Neville's college, Sidney Sussex! He went up to Cambridge in 1888.

Wilson did not belong to the generation of research students who entered under the new regulations of 1895; he took a Cambridge degree in the ordinary way. He attended Glazebrook and Shaw's courses in the Cavendish Laboratory. Glazebrook always lectured in detail about particular experiments, while Shaw, who arrived at the lecture-room in morning coat and silk hat, liked to look at all sides of a question, and avoid committing himself.

Wilson graduated in 1892. He was the only student in Cambridge in that year taking physics as his main subject. He succeeded in making a living by some work as a demonstrator, and by private coaching. It left him with little time and energy for his research, which was on the comparison of the behaviour of substances in solutions by an optical method. Feeling his position was precarious, he became a science master at Bradford Grammar School, but this proved not to be his vocation. He embarked on the 'desperate venture' of returning to Cambridge, without clearly seeing how he might achieve his aspirations. His courage was rewarded by good fortune. The University's decision to extend the teaching of physics to medical students provided Wilson with a post as supervisor of the students' practical work in the physics laboratory.

Wilson had already received his scientific inspiration in Scotland. As a youth in the hills of Arran he had first recognised 'the beauty of the world', and acquired a permanent passion to understand nature. This led him in 1894 to apply for a post as temporary observer for a fortnight at the meteorological observatory on Ben Nevis, the highest mountain in Britain. Now he began to see nature in dazzling forms. Sometimes the visibility exceeded a hundred miles. Occasionally, the mountain snow was covered with insects blown from great distances. There were fogs, mists, rain, snow and violent thunderstorms. The scenes at dawn were breathtaking, the most striking being the observer's shadow thrown on the top of the cloud below, in the phenomenon called the Brocken spectre. The extraordinary cloud effects incited him to try to reproduce them in the laboratory.

On returning to Cambridge at the beginning of 1895 he made artificial clouds by the method of expansion of moist air invented by Coulier and John Aitken. By illuminating them he succeeded in reproducing in the laboratory the Brocken spectres, and the exquisitely coloured glories and coronas which he had seen on the mountains.

But besides satisfying his aesthetic sense, he immediately made a physical discovery. Aitken had been unable to produce cloud in his expansion chamber if the air was free from dust. Wilson discovered that cloud could indeed be produced in dust-free air if the expansion was big enough, and he also measured very exactly the degree of expansion that was necessary. Moreover, he found that after each expansion the number of drops produced did not decrease.

He published his discovery early in 1895. It indicated that there were nuclei in air around which water vapour could condense, which were not dust, and which were continually regenerated. As his experiments were further refined his results indicated that the nuclei were of a magnitude not greatly exceeding molecular dimensions, and 'at once suggested that we had a means of making visible and counting certain individual molecules or atoms which were at the moment in some exceptional condition. Could they be electrically charged atoms or ions?'

Wilson's meticulous research led him to discover that two different clouds could be produced in dust-free air by different degrees of expansion. These corresponded to two different kinds of nuclei. The first kind were few in number, while the second were very numerous, but required a much bigger expansion to become operative. It seemed obvious that the two kinds of nuclei were respectively positive and negative ions, of the kind that had been discovered in ionised gases.

When Wilson began these researches in 1894, he found the Cavendish Laboratory a comparatively quiet place. Most of the research was done by senior men in the time left after other duties. They were seldom in the laboratory for long periods, so no regulations were made on hours of work; anyone could work there at any time, if he chose. Wilson took advantage of this. During the summer vacation of 1895 he worked there every day, and was often the only person in the laboratory. There were no students whose specific duty was to do research.

This situation changed dramatically with the arrival of the new research students in October 1895. The first of them, Rutherford, Townsend and McClelland, created the cheerful, friendly atmosphere that became a characteristic of the Laboratory, and Wilson immediately felt the pleasure and stimulus of their companionship.

The contrast of atmosphere in 1896 with that experienced by Wilson in the summer of 1895 is brought out by Rutherford's correspondence with his fiancée. On October 30 he wrote. 'The

Cavendish is crammed with Research people and it is hard to find room for them. I am very glad I came last year, as I have got a good start.' In the following month he wrote that he was enjoying the unheard-of privilege of working in the Professor's room, where he found his assistant Ebenezer Everett 'very handy'.

The Laboratory was now full of 'researchers after truth', some of whom he had got to know very well. He reported that he and some other Trinity men had formed a new Physical Science Club. The first meeting was held in his rooms in the College, and, as was usual in these cases, he had supplied 'coffee, biscuits, baccy and cigarettes'. He had also delivered the inaugural paper.

The contrast between Wilson's and Rutherford's descriptions no doubt owes something to their very different temperaments: Wilson was the supreme individual artist in experiments, whereas Rutherford was the gregarious natural leader of the tribe. The arrival of the graduate research students in 1895 had certainly created a new atmosphere. It was essentially that of young men of various traditions and nationalities in their twenties and thirties. The veteran J. J. was still only thirty-eight.

Wilson felt the encouragement of the new atmosphere in his exacting research, which involved much difficult glass-blowing, with repeated failures. J. J.'s powers were at their height. Wilson, with his peculiarly acute observation, noted that the activity of J. J.'s mind was continually manifested in his expression.

It was in Great St Mary's church during the University Sermon that I remember on more than one occasion being able to study his countenance while we were both unusually free from distraction. His alert and eager expression seemed to indicate a mind that had already reached the solution of some great and fundamental problem and was now tremendously active in following out the many consequences of this solution.

Wilson thought that though J. J.'s great concentration did not hinder him from discussing others' ideas, it did prevent him sometimes from realising how much he owed to others. He had a remarkable instinct for knowing which experiments were worth while, and the best way of making them. Unlike Rayleigh, he was not interested in designing elegant experiments for their own sake. He did not try to influence Wilson in the design of his experiments, but indicated by his degree of amiability what he thought of them, and how he was doing them.

After the general stimulation caused by the arrival of the research

10 *Rutherford's magnetic detector for radio waves*

students in 1895 the physics meetings were held fortnightly. Stokes, who was then Secretary of the Royal Society, used to come to them. Wilson noted that his rather stern expression lighted up with a wonderful smile when something appealed to him, and he compared it with a similar transformation in Bohr's expression in 'his sudden change from solemnity to humour'.

Wilson also felt that the early Cavendish research students had unusual physical energy, especially Langevin, who 'had the most extraordinary vitality which even seemed to be proclaimed by the intense blue-black colour of his hair'. This remark of Wilson's is specially interesting, because Langevin was particularly aware of Rutherford's vitality. He and Rutherford shared a room for a time as research students. Many years later, when asked whether he had been very friendly with Rutherford at the Cavendish, he replied that their

138

relationship had been excellent, but one could not exactly be 'friendly' with a 'force of nature'.

Rutherford's physical energy was also exceptional, but Wilson did not feel the full impact of his genius until after he had left Cambridge, and accomplished his researches on radioactivity at Montreal. While Wilson's researches on clouds and their beauties flowed in the most unexpected and brilliant way into J. J.'s electrical investigations, Rutherford dived into J. J.'s stream in an entirely different manner. The point here is that J. J. was such a figure that he could attract these men of great and very different genius.

When Rutherford arrived at the Cavendish in 1895 he was twenty-four years old. He had already invented a magnetic detector for radio waves in New Zealand (see figure 10). It consisted of a magnetised needle partially demagnetised by the passage of electrical oscillations through a solenoid surrounding it. This proved a very sensitive device for detecting electrical waves. He developed it in Cambridge, and, using large Hertzian oscillators, the electrical waves emitted were detected at a distance of about half a mile, which, for a short time, was the long-distance record for radio communication. Rutherford wrote in 1910: 'These experiments were made before Marconi began his well-known investigations on signalling by electric waves. This effect of electric oscillation of altering the magnetism of iron is the basis of the magnetic 'detectors' developed by Marconi and others, which have proved one of the most sensitive and reliable of receivers in radio-telegraphy.'

The practical significance of his work was obvious, and he was taken up by leading figures in Cambridge society and London clubs, who perceived that he might revolutionise means of communication. J. J. and his wife treated the bluff, hearty, farmer-like young man from New Zealand with the utmost kindness. J. J. gave him great help in settling into an academic society new to him, and finding him jobs to provide extra money. Mrs Thomson personally discovered lodgings for him. Rutherford sent detailed accounts of what happened to him to his fiancée in New Zealand, expressing his profound appreciation and regard for the Thomsons.

As a research student, he pursued his radio investigations with an unparalleled combination of scientific and social success, yet after a short time he decided to discontinue them, and join J. J. in his research on the electrical discharge in gases. The decision was of great interest from several points of view. It showed his respect for J. J.;

his scientific judgment that J. J.'s line of research into the nature of electricity was more fundamental than his own brilliant researches; and his strength of character in giving up these researches to join in those of another which he regarded as more scientifically important.

The incident was of profound credit to them both, and like many moral intellectual judgments, was followed by its reward. The collaboration of J. J. and Rutherford probably provided just that edge to the Cavendish electrical researches which led to the definitive discovery of the electron.

After Rutherford had completed his research with J. J. on the ionisation produced by X-rays, he investigated the discharge of electrified bodies by ultraviolet light, which had been discovered by Hertz. He found that the negative ions produced by ultraviolet light were identical in mass with those produced by X-rays.

When in 1896 Becquerel announced his discovery of radioactive rays from uranium, Rutherford at once began a systematic examination of the conductivity produced by them in gases. He showed that the radiation ionised the gas throughout its volume, and that the ions were identical in character with the ions produced under similar conditions by X-rays. In the course of this investigation he made the significant discovery that the radiation from uranium and its compounds was of two distinct types; one, which was very easily absorbed, he called α-rays, and the other, more penetrating, β-rays. The α-rays were to prove to be the tool with which Rutherford founded nuclear physics, and disintegrated the atom. The research was published after he left Cambridge in 1898 to be professor in Montreal.

Thus, within two years, Rutherford twice changed his line of research, each time moving from a first-rate line of investigation to one yet more important. This capacity to judge and change was one of the most significant qualities in his genius.

The electrical theory of matter, and the age of electronics were brought to birth in the Cavendish Laboratory. J. J. took his place among the foremost founders of modern science, and the Cavendish Laboratory established its position and reputation in world science.

10

From Outside to Inside the Atom

When, in 1899, J. J. had obtained the same result for the direct, as for the indirect methods of measuring the ratio of charge to mass of the electron, he observed: '... we have clear proof that the ions have a very much smaller mass than ordinary atoms; so that in the convection of negative electricity at low pressures we have something smaller even than the atom, something which involves the splitting of the atom, in as much as we have taken from it a part, though only a small one, of its mass'.

Thus the famous words 'splitting the atom' appeared in the literature of science.

J. J. attended the Dover meeting of the British Association in 1899. Many members crossed the Channel to hold a joint session with the French Association for the Advancement of Science, which was meeting at the same time in Boulogne. J. J. gave a talk 'On the existence of masses smaller than atoms', in which he outlined the implications of the Cavendish researches. The notion that the atom was not a permanent particle was not well received by the chemists, but J. J. felt that for the first time the implications of the Cavendish researches on the electron, and its relation to the structure of atoms, began to be appreciated in circles wider than the Cavendish Laboratory and Cambridge.

The transference of attention from the world of impersonal, uniform atoms to that of individual atoms was expressed some years later by Rutherford in his remark to Eve that 'ions are jolly little beggars, you can almost see them'. It came to be said of Rutherford that every α-particle was one of his personal friends.

After 1897 a great deal of research effort in the Cavendish was concerned with getting to know the individual atoms and sub-atomic

141

particles better, to become acquainted with their individual visages and expressions, and not merely with their uniform properties as members of crowds.

Great progress had been made in explaining the phenomena of electricity in terms of ions; given some ions, much could be derived from them. But where did the ions come from in the first place?

This was explained by C. T. R. Wilson in 1900. He showed with his cloud-expansion chamber that there were always a few ions present in the air or gas, because he could always obtain a few drops with the appropriate expansion. The air should therefore always be slightly conducting. He tested this with a gold-leaf electroscope, finding that it always lost its charge, even if very slowly. Investigating the phenomenon precisely, he discovered that positive and negative ions were constantly being produced in the air in equal numbers, at the rate of about 14 per cubic centimetre of air per second.

This discovery was of major importance. Not only did it provide the explanation of how an electric discharge could pass through air, in the laboratory, or in the atmosphere in the form of lightning: it provided the starting point for the discovery of cosmic rays.

J. J. pointed out in 1898 that C. T. R. Wilson's condensation experiments indicated that it might be possible for ions of only one sign to be produced, and that this might have meteorological implications. It might, for instance, explain why the atmosphere is normally positively electrified, and a negative electric charge is carried down to the earth by rain and precipitation.

As a result of J. J.'s suggestion of its possible meteorological significance, C. T. R. was able to pursue this research after he had been appointed by the Meteorological Council to investigate the phenomena of atmospheric electricity.

In March 1903 J. J. published a new and more refined determination of the ionic charge. His result was about half of his earlier one, the difference being ascribed to the counting of only the negative ions in the earlier experiment; consequently, the charge ascribed to each ion had been twice what it should have been.

In the same year, a month later, H. A. Wilson, another of the Cavendish researchers, published an independent determination of the charge. His method was different from J. J.'s, though it also depended on making the position of the ions visible by condensing water on them, and measuring the rate of fall of the drops. H. A. Wilson's results agreed fairly well with J. J.'s, though later both were proved to be too low.

H. A. Wilson observed that many drops in the clouds carried more than a single ionic charge. They might have charges of two, three, or more, multiples of the ionic charge. R. A. Millikan subsequently used this multiple-drop phenomenon in his experiments on the measurement of the ionic charge.

The Cavendish research workers carried out a large number of fundamental experiments in a comparatively short time. The electron did not at first present itself in one simple, clear-cut guise. It was associated with various ions. It could be obtained from many different kinds of source, such as gases, and metals. Elster and Geitel in Germany had shown that red-hot carbon emitted a negative charge, which could be deflected in a magnetic field. J. J. suggested that this emission consisted of electrons, on the basis of the electronic theory of conduction, elaborated independently by himself, and Riecke and Drude in Germany.

According to this theory, a conductor contains a large number of loose electrons. When an electric force is applied to it, the electrons flow in a direction and strength that can be calculated, by treating them as particles obeying the kinetic theory of gases.

J. J. pointed out in 1900 that when the temperature of the carbon or metal was sufficiently high, some electrons would have enough energy to be able to jump through the surface, and constitute an emissive current.

O. W. Richardson, who entered the Laboratory in 1900, pursued the phenomenon of the emission of electrons from hot wires. He independently conceived the same idea as J. J., and investigated it experimentally as well as theoretically.

Within a period of five years C. T. R. Wilson had made the fundamental observation leading to the discovery of cosmic rays, and O. W. Richardson had provided the scientific basis for the development of thermionics.

The variety of electronic manifestations made it essential to track down their precise qualitative and quantitative features. Zeleny concentrated on accurate absolute determinations of the velocities of ions in an electric field. The method he used required large quantities of gas. Langevin devised an improved method, and made extensive measurements of ionic velocities, which he published in 1902.

McClelland pursued his studies of the electrical conductivity of air and other gases caused by drawing them past an electric arc or over an incandescent metal. He showed that it could be explained by the presence of slow moving ions, the negative moving faster than the positive. He found that when platinum wire was heated, positive ions appeared as soon as it became red, negative ions appearing only at considerably higher temperatures. He also investigated the effects of lowering the temperature of the gas around the hot wire. He found that at moderately low pressures they could be explained as due to ionisation by collision, but at very low pressures there was evidence that electrons were being emitted directly when the metal was at incandescent temperatures. He published these results at the same meeting of the Cambridge Philosophical Society, in November 1901, at which Richardson published his first paper on electronic emissions from hot metals.

The Cavendish investigators were pursuing many parallel and overlapping investigations, and sometimes had difficulty in deciding questions of priority among themselves, as well as in relation to researches made elsewhere. Priority was important to the young men, who had still to make their reputations. J. J., however, viewed what was going on in the Laboratory with the higher detachment of a commander-in-chief. He discussed individually with his research lieutenants their ideas and experiments, suggesting developments and ways out of their difficulties.

He would listen to a half-formed new idea being suggested for the first time by one of the talented young men, and immediately point out improvements on it; or he would listen, go away and return a fortnight later with the whole idea worked out, convinced that it was his own. Most of those who had this kind of experience had a sufficiently strong sense of their scientific and intellectual obligation to J. J. not to be much upset by it, but one or two felt permanently hurt. J. J. seems to have been aware that there were complaints, for he once jocularly remarked that in the Cavendish, there were two kinds of experimenters: those who made the discoveries, and those who got the credit for them.

He did not make adequate acknowledgement of his indebtedness to Townsend's work on ionisation by collision. Townsend was considerably upset by this. R. J. Strutt remembered being present when Townsend pointed out to J. J. the implications of ionisation by collision. But from the historical point of view, questions of priority were less important than the spectacle of a group of gifted young men enthusiastically and successfully exploring a new field of nature.

Besides discovering new knowledge some acquired the technique of discovery, which later became fertile in other directions. For example, R. S. Willows in 1900 studied the remarkable coloured striations that appear in the positive column of the discharge tube. He later became a leader of scientific research in the textile industry, whose group worked out the manufacture of non-shrinking cotton fabrics.

R. J. Strutt, who later became the biographer of J. J., began research in the Cavendish in 1899. In the following year he started to investigate the ionisation produced by the various radiations from radioactive substances. Rutherford had recognised and defined the α- and β-rays, and it was known that the latter were negatively charged particles. Strutt discovered that the ionising effects produced by both types of radiation resembled each other much more than those produced by X-rays. He threw out the suggestion that the α-rays might be positively charged particles, and that a sufficiently powerful magnetic field might deflect them in the opposite direction to that taken by the β-rays. Geitel in Germany, and C. T. R. Wilson showed almost simultaneously in November 1900 that a charged body suspended in a closed vessel gradually lost its charge, even when leakage through its support was prevented. The leakage through the air of the vessel showed the same properties as that of the conduction of electricity through an ionised gas. In order to test whether the ionisation was due to a very penetrating radiation passing through the atmosphere, Wilson in the following vacation compared the ionisation in a closed vessel when carried into a railway tunnel, with that produced when the vessel was in an ordinary room. He found no indication that such rays had been absorbed. In the following year he investigated this 'spontaneous' ionisation in various gases, and obtained results remarkably similar to those obtained by Strutt in his experiments on radioactive radiations. This suggested that the 'spontaneous' ionisation might be wholly or partly due to radiations from radioactive materials, such as those constituting the walls of the vessel, which might be slightly radioactive.

J. Patterson experimented with air in a large ionisation chamber, and concluded that the ionisation was due to a radiation from the walls, which was easily absorbed. Strutt, McLennan and E. F. Burton, and H. L. Cooke showed that the ionisation was partly due to radiation from the material of the walls, and partly to a much more penetrating radiation passing through the walls.

It was thought that the penetrating radiation was probably due to the presence of radioactive materials in the atmosphere and the earth. In 1903 J. J. himself noticed that when air was bubbled through water radiation effects were produced. He found that these were due to traces of radioactivity in Cambridge tap-water.

Elster and Geitel had shown that when a negatively charged body is exposed to the atmosphere, it becomes temporarily radioactive. This suggested that there was a radioactive material in the atmosphere. The observation led C. T. R. Wilson to suggest in 1902–3 that freshly fallen rain and snow might be found to be radioactive. This was investigated, and found to be so.

Another subject of investigation was the possibility that chemical combination might cause ionisation in gases.

C. G. Barkla, who had been investigating the velocity of electric waves along wires, and its dependence on the diameter and material, made use of Rutherford's magnetic detector in starting, at J. J.'s suggestion, on his important researches on secondary X-rays. He continued them after his appointment to a chair at Liverpool.

While the researches arising from the conduction of electricity in gases dominated the Laboratory's work, some research was done in other parts of physics. For example, in 1901 G. F. C. Searle and T. G. Bedford introduced a new method for measuring magnetic hysteresis. The old method was long and tedious. In theirs, an almost instantaneous determination was made by measuring the throw of a ballistic galvanometer. It greatly simplified the measurement of effects such as that of temperature and strain on hysteresis.

The importance and number of researches carried out in the Cavendish Laboratory at about the turn of the century was striking, yet in this period the average number of research students was only about twelve. The success and fame brought more research students; in the six years after 1902 the number rose to only about twenty-five, however. In the years before the First World War it had risen to more than thirty, then regarded as a large number.

J. J. inspired and led the activity without being domineering. He liked research students to propose their own problems, but he realised that only a few could do this successfully; the majority required a research problem to be suggested to them. J. J. had an inexhaustible supply of ideas, but he found it hard to suggest problems that were both easy and worth while. Like Maxwell he did not discourage a man from trying an experiment, however unpromising scientifically or technically. With talented men, the seemingly absurd sometimes proved to have more in it than at first appeared.

He had an immense enthusiasm, and thought about his problems almost continually. He suddenly jumped or quickened his step in the laboratory and in the street, when a new idea occurred to him. He was very widely read in physics, and had an exceptional memory. He would recommend a research student in difficulties with his experiment to read some paper which, when looked up, proved to be comparatively old, and at first sight neither important nor relevant; yet J. J. had stored it in his memory, and might cite the page and the place in it, where the point might be found. He was keenly interested in cricket and football, and had a similarly powerful memory for their statistics.

But his mind did not work like a machine; he did not memorise in exact detail. On the other hand he was very quick and effective at approximate mental calculations. His intense concentration made him absent minded. Sometimes he seemed obstinately to refuse to see the point of other people's ideas. Part of this occasional unperceptiveness may have been due to inattention caused by preoccupation.

His absent mindedness was noticeable in his own experimental work. He was heavily dependent on the manual skill of his assistant Everett (see figure 9), and sometimes forgot what he was doing with his fingers in manipulating apparatus, or even failed to see what was patently visible, because his mind became fixed on some other idea. He would then give wrong instructions which occasionally led to the destruction of the apparatus. He conveyed his admission of his mistake by surveying the debris dejectedly.

The brilliance of the researches led to a rapid change of the researchers. Rutherford had left in 1898, Townsend went to Oxford in 1900, R. J. Strutt to Imperial College in 1902, and others to a variety of chairs and senior appointments.

The Laboratory attracted an increasing number of research men from America and other countries. One of these was the American physicist H. A. Bumstead, who worked in the Laboratory in 1904–5. When asked to write on his impressions of the Cavendish Laboratory, as one accustomed to universities with a different tradition and organisation, he replied that the task was difficult, because the Cavendish Laboratory was 'a pretty complex social organisation', and also because he was so delighted with it that he feared that his praise might be indiscriminate. He thought that the University as a whole had skilfully utilised the survivals from the past. It had not swept them away in the fear that they might decay into abuses, but had adapted

147

them, creating a better instrument for teaching and research than they had been able to do in America up to that time, even with the ground clear for all modern improvements.

He thought that the college system was particularly advantageous for the students' and dons' social life, and their broad intellectual development.

The tendency in America was for the creation of schools of engineering, law, medicine and other specialities. The students and staff tended to keep within themselves, and lost the broadening influence of social integration with those studying and teaching over the whole spectrum of knowledge. He also admired the college fellowship system; he thought it the best way of promoting research and scholarship, though it was subject to some abuses. American benefactions of the day seemed to demand quick results from their endowments, which hindered the cultivation of deep research.

He thought the social life was much richer and fuller, partly because Americans were often too busy to enjoy each others' society. In America, a laboratory where few started work before 10 a.m. or worked later than 6 p.m. would have seemed disgracefully indolent, and yet the Cavendish men discovered more. He confessed he could not quite understand it.

The Laboratory was obviously dominated by J. J., yet he had never seen a laboratory where there was so much independence, and lack of restraint on ideas. He had been struck by the regard for J. J., but he had also observed that there was no subservience. He had noted a number of men following their own lines, and the frank discussion and criticism of the Professor's ideas at meetings of the Cavendish Society. He had noted, too, the mutual discussion among the research students. Frank criticism and demonstration of error was regarded as a friendly service. There were not many places where friendship could stand that kind of criticism, but Cavendish men were usually able to bear it.

The first twenty years of J. J.'s professorship had culminated in the discovery of the electron and the determination of its main properties. J. J. summarised the work of the period in his *Conduction of Electricity in Gases*, published in 1903. The next step was to investigate the place of the electron in the structure and mechanism of matter.

Physicists in many countries developed theories based on the new experimental information. J. J. naturally led the Cavendish researches in this direction. In the years 1903–6 he published papers on the structure of the atom, and its magnetic, optical and conductive properties, in terms of the electron. For the first time, even if only tentatively, electrical and thermal conduction, magnetism, radioactivity, the refraction and emission of light, and the chemical reactions governed by the notion of valency, were brought within the bounds of a single theory.

Experimental research was directed towards the testing of this theory, and the discovery of new facts on which it could be further developed.

On the first page of *The Corpuscular Theory of Matter*, published in 1907, J. J. stated that 'From the point of view of the physicists, a theory of matter is a policy rather than a creed; its object is to connect or to co-ordinate apparently diverse phenomena, and above all to suggest, stimulate, and direct experiment'.

As P. Zeeman observed, J. J. counted the number of electrons in atoms; arranged with his fertile imagination the electrons in coplanar rings inside a sphere of positive electricity; made suggestions towards the explanation of the periodic law of the chemical elements; offered views on the structure of light; and boldly attempted to explain great difficulties in the old theory by his conception of the speckled wavefront. In the midst of these difficult investigations, the discovery of the isotopes of neon by his parabola method appeared to Zeeman as a charming episode. Zeeman also greatly esteemed J. J.'s paper of 1881 on the increase of electromagnetic mass with velocity. Being published when he was only twenty-five, it was regarded as a most remarkable example of early genius, which unconsciously prepared him for his forthcoming electrical researches and the isolation of the electron. Zeeman thought, too, that his *Applications of Dynamics to Physics*, published in 1887, was a work of profound mathematical insight, and a forerunner of Hertz's *Principles of Dynamics*.

C. G. Barkla, continuing at Liverpool the researches that he had started in the Cavendish, discovered that when an element was bombarded with electrons, it emitted a series of X-radiations of specific wavelength, dependent on the energy of the bombarding particles. This discovery of the characteristic X-radiation of an atom was of fundamental importance. It provided basic data for Niels Bohr's theory of the electronic structure and quantised energy levels of the atom.

Insight into atomic structure was sought by detailed investigations of the absorption of radiations, and the secondary radiation they produced. J. A. Crowther studied the absorption of β-rays, and the ionisation and secondary radiation produced by X-rays in gases. He interpreted his results as giving support to J. J.'s model of the atom; the disproof of his deductions by Rutherford was an important stage in the advance to the correct theory of the structure of the atom. But fundamental progress on the investigation of the structure of the atom was to depend less on electrical and electronic researches, and more on radioactivity.

The lead in this field had been taken by J. J.'s pupil, Rutherford, who at Montreal had published in 1902–3, together with Soddy, an explanation of radioactivity based on the spontaneous disintegration of atoms.

The Cavendish investigators in radioactivity were inclined to the theory that all matter was more or less radioactive. Their researches helped to show that radioactivity was very widely distributed in the materials of the earth. A. Wood and N. R. Campbell discovered in 1906 that potassium and rubidium were quite considerably radioactive. In 1907 they detected a distinct diurnal periodicity in the intensity of the penetrating radiation from the earth and air.

Rutherford had learned to make use of radioactivity, which proved to be the most crucial phenomenon for elucidating atomic structure. J. J. had elaborated the provisional electronic theory of atomic structure that provided the model for Rutherford, who was now the best equipped to prove or disprove it.

Various researches were made on the nature of X-rays and the photoelectric effect, probing phenomena that were presently to be explained by Einstein and von Laue. The Cavendish investigators were not among the leaders of the advances that arose from quantum and relativity theories, nor were they as adept in interpreting wave phenomena as they were at particle phenomena.

In 1906 J. J. gave the theory of two methods for estimating the number of electrons in an atom. One consisted of the measurement of the diffusion of a parallel beam of electrons as they passed through thin sheets of matter. If the deflections of the electrons are due to collisions with electrons within the atom, then the way they are scattered should give some indication of the number of centres, or electrons, causing the scattering. In the other method, the intensity of the scattered X-radiation from a radiator subjected to a beam of primary X-rays was measured. This was an indication of the number of electron centres causing the scattering.

The experimental investigation utilising these methods was carried out by J. A. Crowther from 1906 until 1911. His results showed that the number of electrons in an atom was comparatively small, and they were interpreted as showing that the main mass of the atom must be associated with the positive charge. J. J. used Crowther's results in elaborating his theory of the atom as a region of positive electrification, in which electrons were distributed like plums in a pudding.

Rutherford closely analysed Crowther's experiments, and concluded that they did not provide valid evidence for this model.

J. J.'s own experimental work was concentrated, from 1908 until the outbreak of the First World War in 1914, on positive rays.

The phenomena of secondary radiation excited by X-rays and photoemission were investigated by Innes and Hughes, respectively. They became increasingly inexplicable in terms of the classical wave theory of light. J. J. suggested in 1907 that the wavefront of a beam of light might be discontinuous, and N. R. Campbell and G. I. Taylor investigated this experimentally in 1909. Then, in 1912, von Laue's discovery of the diffraction of X-rays by crystals created the possibility of revolutionary advances, both in the theory of radiation, and the application of radiation to the elucidation of atomic structures.

R. Whiddington's experiments in 1912 on the excitation of X-rays characteristic for an element, through bombardment by cathode rays, provided evidence for the quantum theory of X-rays.

The number of students and research workers, and the variety of researches increased the organisational and administrative problems of accommodation and teaching.

11

Accommodation and Teaching

The advance of science depends not merely on intellectual endeavour. It needs accommodation, teaching and equipment to enable scientific ability and genius to unfold. The growth of the Cavendish activity continually pressed on the provision of accommodation and apparatus. J. J. particularly mentioned them in his reports of 1903 and 1905.

Working conditions became extremely crowded. There was competition even for a single table, and some research men had to find laboratory space in other Cambridge institutions. Others left Cambridge prematurely, because they could find no proper place to work.

In 1904 Rayleigh was awarded the Nobel Prize for physics. He wrote at once to J. J., saying that he had probably noticed that he proposed to present the prize money to Cambridge, and asked him whether he had any suggestions to make on what form the gift should take. Physics would of course have the first claim, and he would like him to turn the matter over in his mind.

J. J. is reported to have urged the claims of the Cavendish, and especially for a new building free from radioactive contamination, caused by gaseous radon, and the solid deposit that it left on the walls and elsewhere. The radioactivity from these deposits discharged electrical instruments and hindered accurate experimental measurements.

Rayleigh assigned £5000 of the Prize to the Cavendish, and the rest to the University Library. J. J.'s suggestion that there should be an extension with frontage on Free School Lane was approved.

A syndicate was appointed to obtain plans and estimates. It reported in February 1907. W. M. Fawcett was again asked to design plans. He provided for a large basement and a number of small rooms for research, a library, a chemical room and a demonstrator's room.

This enabled a room being used as a library to be restored to the Porter's Lodge, whence it had been appropriated.

The cost was estimated between £7000 and £8000, or about ten-pence to one shilling per cubic foot. Besides Rayleigh's £5000, J. J. offered a second time to provide £2000 accumulated from students' fees, and he stated that he did not expect that the University would be called upon for further maintenance.

The report was approved in March 1907. Sindall of Cambridge made the lowest tender, which was for £7135, including £500 for provision of heating. The financial position was so good that it gave no ground for opposition from the University, and construction began in the autumn of 1907.

The building was completed early in 1908. Lord Rayleigh had just succeeded the Eighth Duke of Devonshire, the son of the founder of the Cavendish Laboratory, as Chancellor of the University. His first act in that position was to open the new wing of the Laboratory.

The extension was opened on 16 June 1908, the thirty-fourth anniversary of the opening of the original Laboratory. The Chancellor proceeded in state to the Laboratory, and declared the new extension open.

According to *The Times* Rayleigh said that his interest in the Laboratory went back to a date before the Laboratory existed, when there was little opportunity of acquiring the experimental art. The five years he had spent there as Maxwell's successor had been labor-ious, but had been lightened by the assistance of his colleagues. Much of his work had been devoted to the determination of electrical units. In those days he had began his lifelong friendship with Kelvin. Then, too, he met Helmholtz, the only physicist to be put in the same rank as Kelvin.

Since his retirement nearly a quarter of a century before, the fame of the Laboratory had spread far. The large number of young physicists who had been trained in it would all be outspoken in their recognition of what they owed in direct teaching, and in stimulus, to J. J. Thomson.

In his thanks to the Chancellor J. J. drew attention to the fact that the new wing was being opened on their founder's day, and he hoped that those using the Laboratory and its new wing would show by their work that they deserved the gift.

Rayleigh's son, R. J. Strutt, was present, and recorded that the remark of J. J.'s that struck him most was that, in his experience of many generations of students, he had known far more to fail from lack of grit and perseverance than from the want of what was commonly called cleverness.

Strutt, in his life of J. J., indicated that all was not plain sailing in the happy affair of the new Rayleigh wing. J. J. was exceedingly busy in the early years of the new century, and various details were overlooked. For instance, difficulties about ancient lights arose, and J. J.'s lecture-room assistant, W. H. Hayles, was dismayed to find that no blinds had been provided for darkening the lecture-room. When J. J. was told of this, he asked why they should want to darken it, momentarily forgetting, in his absent-minded way, that a lantern was in almost constant use.

Even at that time, the professor was burdened with management. It was becoming obvious that he should receive more assistance in that direction, as well as in finance. His provision of £4000 towards capital costs of the Laboratory from the accumulation of fees was a noble and economical act, but it also consolidated the practice of running the laboratory on an excessively small budget, which was not the best preparation of its research men for the coming age of large-scale expensive physics.

The new building came into full use in 1908, and the whole of the physics lectures in the University were, for the first time in many years, delivered within the Cavendish Laboratory. The teaching of practical physics was an extremely important feature of the work of the Cavendish. It was worked out and stabilised during the first quarter of a century of J. J.'s professorship. Owing to the nature of the subject, and conditions of Cambridge life, it was complex. Among the latter were the college system and the financial stringency of the University.

Teaching was provided partly by the college and partly by the University. This presented organisational difficulties, since the college tutor and the university lecturer did not necessarily agree on the order of importance or method of presentation of subjects. As the tutor was paid by the college, which might have the same level of finance at its disposal as the University, he could be very independent.

The merit of the tutorial system was that a student's difficulties could receive personal attention. If the tutor was able, experienced and sympathetic, as he often was, psychological problems of understanding, which a student would hesitate to reveal before a class, could be unravelled in front of a comfortable open fire in a private study. This was worth the complication of the need for co-ordinating the two sources of instruction.

154

The shortage of University finance had caused the teaching to be carried out in overcrowded laboratories and makeshift rooms. A considerable part of the complication of the Cavendish system arose from the need to fit classes in here and there, wherever accommodation could be found. It had less to do with physics than with the peculiarities of Cambridge educational life and University finance, but in developing it, some conclusions of general interest were noted.

The students in the advanced classes were supplied with very little explanatory material. The demonstrators selected a representative set of experiments, and gave lists of textbooks and references to original papers. The student had to collect the material he considered requisite for dealing with each experiment. This trained his habits of reading and judgment, and his capacity for organising experimental work.

In the elementary classes it was soon found from experience that the best way of training beginners was to require each to keep a notebook, in which he wrote an account of the experiments he had witnessed, and the measurements and results obtained. The demonstrator could see from reading his notebook how far he had understood what he had heard and seen, and give him help where needed.

In the advanced classes a significantly different system was adopted. The Laboratory kept a series of notebooks, which remained its property. Each student was required to enter in them an account of his methods and results. Students were concerned about what they left on permanent record, and worked hard, both at the experiment and at writing it up. The Laboratory acquired a historical record of how a number of able men tackled certain problems when they met them for the first time, which was instructive for their successors.

In these advanced demonstrations the student was not expected to attempt all the experiments. He was encouraged to think deeply about those he chose, learning the necessary technique, determining the sources and magnitude of error, and the degree of accuracy of the result. He might take several weeks on one experiment, his capacity and application tensely stretched, giving him excellent mental training.

J. J., following the practice of Rayleigh, started by giving two courses in the year, one on properties of matter, and the other on electricity and magnetism. They were not designed as preparatory for the Tripos examinations, but were intended to be generally stimulating. They attracted large audiences, and were attended by virtually all the workers in the Laboratory, and also students from other branches of learning, who wanted to know what was going on in science.

The elementary courses of lectures were presently supplemented by elementary practical classes. The majority of the men attending these classes were medical students. In the examination for the medical bachelorship candidates had formerly been set written questions only, and no oral or practical questions were asked. About 1887, however, a simple oral examination on the use of practical apparatus was introduced, with spectacular results. 'The memories of the first examiners almost collapsed under the strain of storing for future recital the immense collection of "howlers" perpetrated by their victims.'

To meet these deficiencies J. J. appointed T. C. Fitzpatrick in 1887 to organise practical classes for medical students. In the beginning the classes were not too large to be handled by one demonstrator. At that time the only place available for them was in the lecture-room. Foolproof apparatus was devised.

The numbers in the classes grew rapidly, and had to be taught in a laboratory that was already overcrowded. The students were divided into groups, and a sequence of times arranged, so that each group could make use of the small amount of space and apparatus available. Some classes were held so late that the porter made a formal complaint that he was being deprived of his proper hours of sleep.

The situation was relieved in 1890, when a corrugated iron shed became available; it had formerly been used as a dissecting room. The physics demonstrators were delighted with it, but had difficulty in getting rid of the odour traditional in such rooms. Calcium chloride and carbolic acid were not effective, and they did not succeed until they placed jampots containing salt and manganese dioxide on each table, poured sulphuric acid into each pot, ran out before they were overcome by the escaping chlorine, and then locked the room for two days. After this and a good airing, it became fit for use. But this was not the end of their trouble. They were given a small boy to act as a laboratory assistant. He could help during class time, but was terrified of entering the room alone between classes for fear of ghosts. He had to be sent to J. J., who tried to dispel his fears by reasoning.

The use of old medical rooms for other purposes is far from a joke. Hertz died at the age of thirty-seven from tuberculosis, probably contracted from the walls of his laboratory, which was in a converted hospital building.

156

The demonstrators who were expending energy and time in this way were able men. They soon found that much of their effort was being wasted, for many of these elementary students could not understand the instructions. Rather against their theoretical principles of teaching, they felt compelled to prepare descriptions of the experiments and forms on how their results should be drawn up. They issued these as a 'Laboratory Note-Book of Elementary Practical Physics'. While such a book could be used as a crib by lazy students, it was a real help to earnest beginners, and a necessity when dealing with numbers of students in a large class. The teaching of the elementary students was greatly helped by the opening of the extension of 1894.

As the number of candidates for the Natural Sciences Tripos Part I increased, a course less elementary than that for the medical students became desirable. These students had some background in physics, and the experiments were designed to deepen rather than widen their knowledge. Standardised apparatus of sufficiently advanced design was not yet manufactured in quantity, and much organisation was required to make the best use of the apparatus available.

G. F. C. Searle made a particularly big contribution to the development of Cavendish teaching. He continually devised new experiments, often of a highly original and ingenious character. He worked and wrote out the theory and experimental details with the utmost thoroughness. Many of these new experiments first appeared in the physics papers of Part I of the Natural Sciences Tripos examination when Searle was an examiner.

The feature of Searle's experiments was that they should illustrate points of theory, and especially mathematical theory, rather more deeply than was usual. They should be susceptible of very accurate performance, and be carried out with apparatus appropriately designed for them. Every detail was designed to illustrate some point of interest, and the geometry of movements and constraints was dealt with carefully.

Searle had personal qualities which secured the attention of students. He could be amusing; when one of his students, named Green, cut his lectures for some time, he eyed him on his return, went to the blackboard and announced: 'In honour of the return of Mr. Green, we will use green chalk only today.'

Financial independence and financial stringency greatly influenced the development of the Cavendish Laboratory. It was not merely a question of teaching experimental physics, and of exploring nature experimentally; all of this had to be done within the financial traditions of Cambridge University in the nineteenth and twentieth centuries. J. J.'s saving of fees, the ingenuity devoted to working in inadequate rooms with too little apparatus, were part of the price paid for the preservation of the traditional Cambridge and British social values. But the preservation of these values also gave a return in originality.

The arrival in 1895 of the research students from other universities had a considerable influence on Cavendish teaching. These students were directly supervised by J. J. from the start. It soon became evident that some of them brought a new outlook, knowledge and skill, and that others were no better prepared than the Cambridge candidates for Part II of the Tripos. It became possible to compare research as a method of teaching, with the instruction in the advanced courses, which had previously been worked out in the Cavendish. It appeared that there were advantages in both methods. Those who followed a carefully worked out course of advanced instruction acquired a sound and balanced knowledge of physics, and would be well prepared for any future research they might undertake. Those who engaged in research at too early a stage were liable to develop a mental short-sightedness, through too much concentration on a narrow field.

The advantage of research as a medium of teaching is that the student is conscious that he may discover new knowledge not only of interest to himself, but to the rest of the world. This is an immense spur to his keenness and enthusiasm.

When the organised advanced courses had been started, the demonstrators had not always been able to identify the causes of difficulties, but with experience they became able to tell if a student was failing to follow the instruction. This was often revealed by a decline in interest. They found that the best students worked best when they saw that the demonstrator himself was in difficulty.

A sustained piece of original research provided in various ways a better medium for mental training than organised exercises. It involved imagination and resourcefulness more, while the demand on observation and reasoning was at least as great.

Those who attended too much to the systematic acquirement of existing knowledge and technique tended to regard the perfection of their equipment as an end in itself, leaving them with insufficient energy for the strenuous work of advancing the frontiers of knowledge.

While the question of the best psychological moment for a student to change from instruction to research was debatable, there was no doubt that the vitality of the Laboratory was increased by the new research students. Their arrival again produced an accommodation crisis. It was said that more physics was done in the Cavendish per square centimetre than in any other laboratory in the world. During this period men appropriated scarce apparatus when their users were not looking. The situation was relieved for a time by the opening of the Rayleigh wing in 1908.

The physics training had a complicated and long evolution, but it presently reached a comparatively mature form. Thenceforth it was marked more by the change of men, in the succession of appointments, than in the introduction of new methods.

Meanwhile, the subjects and variety of the lecture courses increased. In 1880 there were only three courses, by Rayleigh, Stokes and Schuster. During the next thirty years courses multiplied, and a complex system of physics teaching was evolved. With the new wings the teaching was nearly completely housed in the Laboratory. This removed the lack of contact between lecture and practical work of the earlier days, when much teaching was still done in the colleges, and little or no apparatus was available except that provided from the lecturers' private stores.

The chief function of the college tutors was now to supplement the Cavendish courses, and help students who needed personal instruction.

With the concentration of the physics teaching and research the problems of organisation were simplified. Those working in physics were all in touch with each other, developing a feeling of comradeship. This stimulated the growth of the Cavendish tradition, which was transmitted to future generations in Cambridge and elsewhere.

12

New Physics, and a New Professor

After the discovery of the electron J. J. naturally searched for the fundamental unit of positive electricity, corresponding to the negative unit. This proved elusive. As the investigation of the negatively charged rays had led to the discovery of the electron, it seemed probable that the positive unit might be discovered in positively charged rays in discharge tubes. J. J.'s apparatus for investigating positive rays is shown in figure 11.

J. J. analysed the rays by submitting them to combined electrostatic and magnetic fields, as in his determination of the ratio of charge to mass of the electron, but in this case applied the two fields simultaneously. A fluorescent screen was placed in their track, particles being focused by the two fields according to the magnitude of their charge and velocity.

Those particles with the same ratio of charge to mass impinged on the screen at points lying on a curve, their place on the curve depending on their velocity. If the ratio of charge to mass were constant, then the flashing points produced by these particles would lie along a parabolic curve. This was what J. J. observed.

The particles in the rays with different ratios of charge to mass consequently produced two or more parabolas on the screen. The problem now was to identify the particles producing any particular parabola. One parabola, which appeared in all photographs, was identified as being due to hydrogen atoms carrying a single positive charge. Other parabolas could then be identified by comparison with that of hydrogen. J. J. obtained parabolas due to atoms of carbon and oxygen, and also molecules of carbon monoxide and carbon dioxide.

He recognised parabolas produced by particles that consisted of

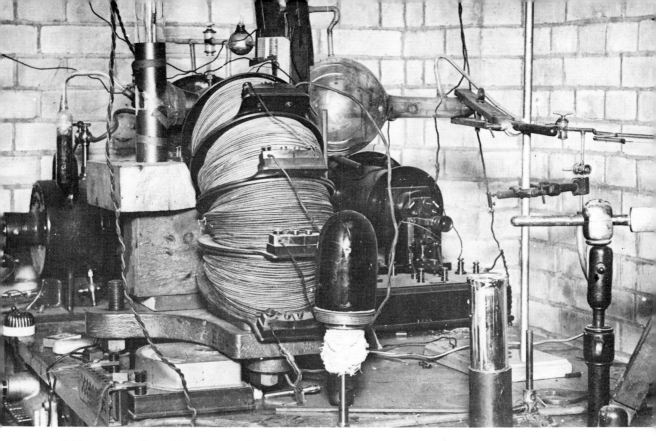

11   *J.J.'s apparatus for positive-ray research*

carbon atoms combined with one, two or three hydrogen atoms; these were new to chemistry, because, while they were individually recognisable, they could not be prepared in quantity. He also obtained parabolas produced by the hydroxyl particle, the combination of a single oxygen with a single hydrogen atom. He found that the mercury atom could produce at least seven different parabolas, indicating that the mercury atom could have at least seven different electric charges or masses.

Most remarkable of all, he discovered that the inert gas neon could produce two parabolas, corresponding to atoms of atomic weights 20 and 22 (see figure 12).

Researches in radioactivity had already revealed the existence of atoms of different mass but indistinguishable chemical properties. Soddy, who was the first to recognise such atoms, named them isotopes. J. J. had consequently discovered a method of separating isotopes.

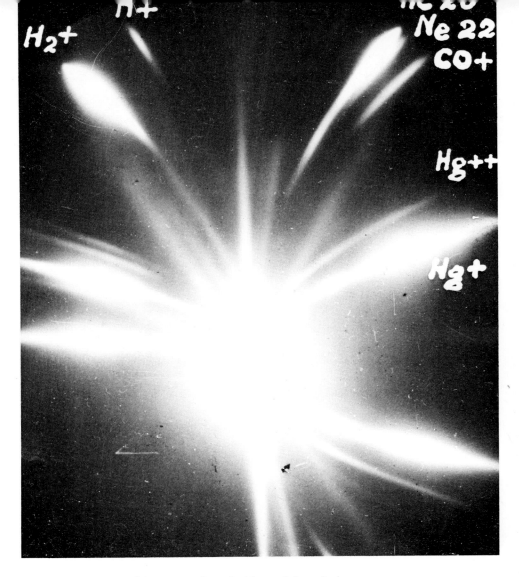

12   *J. J.'s separation of the isotopes of neon by his parabola method*

J. J.'s demonstration of the possibility of recording such particles as hydroxyl was of great significance for analytical chemistry, and his demonstration of the possibility of separating isotopes marked the beginning of the analytical technique that was used in separating isotopes of uranium for use in the first atomic bombs.

Both theoretically and practically this research was splendid, but J. J. did not develop it with quite the same power as he had shown in clinching the discovery of the electron. He was now in his middle fifties, and increasingly involved in the public affairs of science, besides directing the ever-growing Laboratory. The lead in British science passed to Rutherford at Manchester.

The transference was signified by Niels Bohr's move from Cambridge to Manchester in 1912 when he found that J. J. did not appreciate his new notions on a quantum theory of the atom. At Manchester he received a cautious but sympathetic hearing from Rutherford, who always listened carefully to any ideas that seemed to support his own work, however odd they might appear at first. Bohr decided to stay at Manchester.

In the Cavendish other researches of deep significance were started and developed during the last few years before the outbreak of war in 1914.

In 1910 J. J. required a personal experimental assistant, and F. W. Aston of Birmingham was recommended to him by J. H. Poynting. Aston had received an engineering training, and was a very skilful manipulator. He carried out for J. J. most of the experimental researches on positive rays in the period 1910–13. Continuing this line of research on his own initiative, he devised his focusing method for separating isotopes and the determination of the mass numbers of the isotopes of the majority of the elements, which was to become one of the major future contributions of the Cavendish.

Aston was particularly well qualified to express an opinion on J. J. as an experimentalist. In 1940 he wrote that working under him never lacked thrills.

When results were coming out well his boundless, indeed childlike, enthusiasm was contagious and occasionally embarrassing. Negatives just developed had actually to be hidden away for fear he would handle them while they were still wet. Yet when hitches occurred, and the exasperating vagaries of an apparatus had reduced the man who had designed, built, and worked with it to baffled despair, along would shuffle this remarkable being, who, after cogitating in a characteristic attitude over his funny old desk in the corner, and jotting down a few figures and formulae in his tidy handwriting, on the back of somebody's Fellowship thesis, or on an old envelope, or even the laboratory check book, would produce a luminous suggestion, like a rabbit out of a hat, not only revealing the cause of trouble, but also the means of cure. This intuitive ability to comprehend the inner working of intricate apparatus without the trouble of handling it appeared to me then, and still appears to me now, as something verging on the miraculous, the hallmark of a great genius.

The last decade of J. J.'s professorship, from 1909 until 1919, was distinguished by the researches of C. T. R. Wilson, W. L. Bragg, G. I. Taylor and E. Appleton, besides those of Aston.

163

Wilson's research led to the development of his cloud chamber, and its application to the photographing of the tracks of ions, the device that Rutherford described as 'the most original apparatus in the whole history of physics'.

In 1900 Wilson had been appointed a University Lecturer, at a salary of £100 a year, and Demonstrator at £50 a year; he was also elected a fellow of his college. Feeling his living secure, Wilson wondered about applications of his cloud chamber. 'For years I used to think and dream of what the cloud method could be made to reveal.' His first cloud-chamber arrangement had been devised to reproduce atmospheric cloud phenomena, and his second to assist in research on ions (see figure 13).

13   *Wilson's cloud chamber*

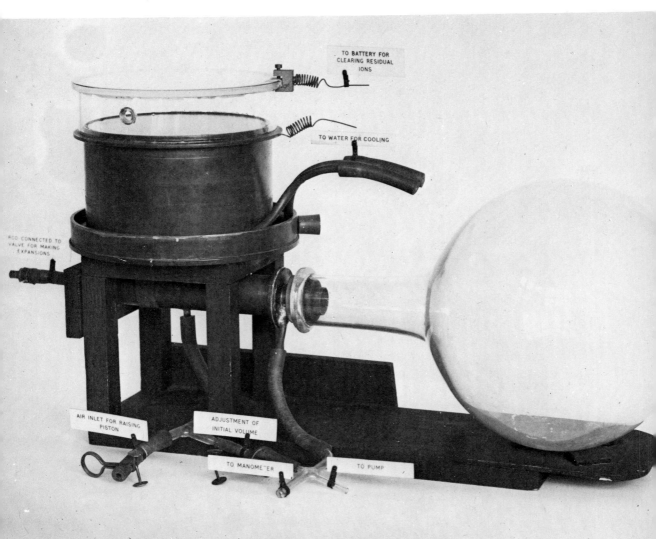

After ten years of thinking and dreaming he began, about 1910, to design a third arrangement, to measure the electric charge on an atom, by condensing drops on ions, thus making them visible. By measuring the total charge and counting the number of drops, the charge per atom could be calculated. He dropped this research, because R. A. Millikan had achieved the same end by a different method, but he had also conceived of 'the possibility that the track of an ionising particle might be made visible and photographed by condensing water on the ions which it liberated'.

He completed an apparatus for this purpose in 1911. He ionised the air in the chamber by X-rays, and made an expansion. He had 'little expectation of success', but was 'delighted to see the cloud chamber filled with little wisps and threads of clouds'. These were the tracks of the electrons ejected by the X-rays. Then he placed some radium in the chamber, and 'the very beautiful sight of the clouds condensed along the tracks of the α-particles was seen for the first time...' Typical cloud-chamber photographs are shown in figure 14.

He showed W. H. Bragg one of his first good pictures of α-ray tracks shortly after it had been obtained, and Bragg immediately produced a diagram in which he had forecast what it would be like. 'The similarity between the actual photograph and Bragg's ideal picture was astonishing.' His pictures revealed examples of single and compound scattering, the existence of which Rutherford had already deduced.

Rutherford wrote to W. H. Bragg in 1911 that he had worked on the same subject in 1906, but got nothing definite because his apparatus was too contaminated with radioactivity. He was sure the experiments could not have been easy. Wilson's results were a 'really splendid piece of work', and it was 'really very fine to see the things one has seen in imagination visibly demonstrated'. In 1913 Wilson introduced stereoscopic photography of the cloud chamber, so that the tracks could be seen in three dimensions.

With the outbreak of war in 1914 he put his research aside, and did not resume it until 1921. His apparatus became one of the prime instruments of nuclear physics. When W. L. Bragg was assembling historic apparatus for the museum in the Cavendish, he asked Wilson whether the cloud chamber for photographing atomic tracks, which he presented, was the original. Wilson replied in his strong Scottish accent: 'Therr was neverr but the one.' Its construction had cost only about £5.

165

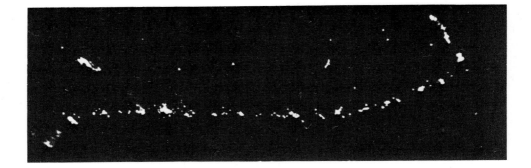

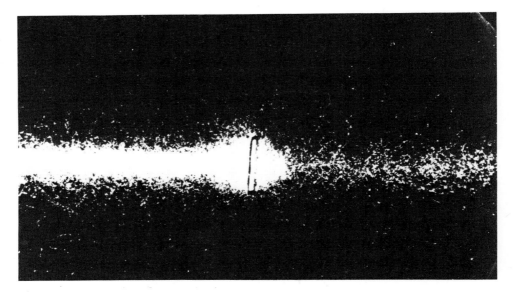

14 *Wilson cloud-chamber photographs of tracks of electrons released by X-rays. (A) Electron track showing how the ions occur partly in pairs and partly in groups. The groups were interpreted as indicating that in many cases the electron ejected by the primary β-ray may itself have energy enough to ionise. (B) Passage of an X-ray beam through a silver screen showing the absorption of the primary radiation.*

This almost superhuman effort of experimental genius was by no means the extent of Wilson's achievement. He was an extremely slow worker and had very few pupils. This arose from shyness and artistic perfectionism. Wilson did not like discussing things with others, and wrote notes of his thoughts which were virtually dialogues in soliloquy. These were highly original, and very stimulating to generations of subsequent readers; many young research men found inspiration in them for their first researches.

Another reason why Wilson's researches were slow was that he took his lecturing and demonstrating very earnestly, and laboured at them very thoroughly. His lectures were on light, and completely thought-out *a priori*, and consequently quite original. Their effect was great.

W. L. Bragg entered Cambridge as a student in 1909. He recollected that

For two years I was in his Part II Practical Class at the Cavendish. C. T. R. Wilson had organised a series of experiments for this class. There were not many of them, perhaps some 15 or 20, but they covered the whole of physics. There was one set of gear for each experiment and we took them on in turn (there were about 15 in my class). What I well remember is C. T. R.'s curbing of our impatience to finish an experiment and count it as 'done'. He forced us to regard it as a small research and look for strange unexplained effects. In his very hesitant gentle way he was quite firm in not letting us give up an experiment until we got everything out of it we probably could.

His lab boy was Crowe, then in his teens. I well remember that Crowe, when our finals approached, made out a list of the classes which we would all get and to the best of my recollection he was almost 100% right. There was really no need for the external examiners.

C. T. R. Wilson lectured on optics. The lectures were the best, and the delivery was the worst, of any lectures to which I have ever been. He mumbled facing the board, he was very hesitant and jerky in his delivery, and yet the way he presented the subject was quite brilliant. I think his lectures on optics set the standard for similar lectures all over the country when later his pupils got chairs.

I remember so well the start of the cloud chamber. He came down to the laboratory to tell me about the first photographs of α-rays and I remember his saying about the tracks that they are as fine as little 'heers' (i.e. 'hairs'). He had expected to see broad and woolly clouds and was astonished that the tracks were so narrow. He chose me, a great honour, to go with him to the Royal Society Soirée and demonstrate the cloud chamber for the first time. A highlight of this occasion was my explaining an electron to Larmor; in my gauche student way I had not realised who he was.

Later when Hilger had asked me to organise a series of stereoscopic pictures of crystal structures, I showed the Hilger letter to Wilson. I have a vivid recollection of his saying: 'Oh yes, about a year ago Hilger wrote to me about stereoscopic pictures of cloud tracks. I 'reely' must reply to that letter.'

I loved him dearly, he was one of the kindest, most considerate and modest of people. I think he had a somewhat kind regard for me because he usually confided in me about his work sometimes at quite an early stage.

On one occasion a suggestion from him was wonderfully fruitful. I had worked out theoretically that one could consider X-ray diffraction as due to reflection of the X-rays in the crystal plane. It was Wilson who suggested I should try specular reflection from cleavage sheets of mica. The experiment came off and I published it in *Nature* and this caused quite a furore of similar experiments... It was C. T. R.'s treatment of a grating diffracting white light, which he had given us in his lectures, which set me thinking on the right lines when I gave a simpler treatment in November 1912 of Laue's diffraction experiments. You will see how much I owe to him.

In 1911 E. V. Appleton went to Cambridge as a student. J. J.'s lectures interested him in electromagnetic waves, as did C. T. R. Wilson's in the phenomena of atmospheric electricity. As a student he conducted research on thunderstorms under his guidance.

Like W. L. Bragg he was fascinated by C. T. R.'s lectures. He found that he treated the subject with an entirely fresh originality, which could not be found elsewhere; his manner of lecturing was absolutely his own. He lectured with both hands, writing on the blackboard with one hand, and wiping off what he had written, a few seconds later, with the other. His hearers had to be pretty intent to follow what was going on, which they were most anxious to do, knowing that what was being said was of unique quality. They used to speculate why material was removed so rapidly from the board, and they thought that possibly, as a cautious Scot, he was careful not to leave anything there for fear it might be wrong. However, they came to the conclusion that this would not be in keeping with his personality. Ultimately they concluded that he didn't want another of the great men of the Cavendish, G. F. C. Searle, to acquire his material for one of his textbooks! Searle used to contend that J. J., Rutherford and the rest sacrificed teaching for research, and battled for what he regarded as the undergraduates' and teachers' rights. He resented Rutherford's tendency to raid the small stock of teaching apparatus in order to turn it to research purposes. As so many were worshipping the idol of research, Searle, who was a vegetarian and had contrary views on many topics, devoted himself to the idol of teaching. If Rutherford raided apparatus for research, Searle was justified in raiding material for teaching.

It was this difference of approach that convinced Wilson's students that he was deliberately trying to conceal his material from Searle. Wilson described his lectures as on 'Light', while Searle described his on the same subject as 'Optics'. The students noticed that Wilson used to leave on the board what he wrote on atmospheric electricity, in which Searle was not interested, and they claimed that this was a proof of the correctness of their theory. Appleton found the most conspicuous qualities in Wilson's researches to be their elegance and perception.

Wilson inverted Voltaire's principle that the best should never be allowed to be the enemy of the good, by never allowing the good to be the enemy of the best. For him there was always a best way of asking Nature questions: the simplest and the most elegant. All his experiments had a peculiar distinction, born of his depth of conception, aesthetic sense, skill and patience. He seemed to proceed on a kind of exclusion principle, by which the complications of all other questions, except the one being asked, were excluded. He was a scientist's scientist, and the more one knew about his subject, the more profound was one's appreciation of his work. Those associated with him scientifically were invited not merely to his house, but *into his home*, which was a very different thing. His pursuit of scientific perfection prevented him from having more than a few pupils. These acquired from him a unique sense of quality in conception, execution and description of experiments. Unique as that quality was, all of his pupils esteemed his personal friendship even more. Unlike other supreme artists in science, such as Leonardo or Henry Cavendish, he was entirely humane in his relationships.

In the 1920s Wilson allowed Appleton to use his special aerial and hut on elevated ground on the Madingley Road in Cambridge to study the change in the electric field due to thunderstorms. C. T. R. asked him whether he had ever felt nervous when the thundercloud approached the station. Appleton replied that he had been too busy observing to worry about possible danger, and Wilson commented that when he had felt a bit apprehensive in the hut, he used to comfort himself with the reflection that it had never yet been struck.

Appleton went on to his fundamental researches on the conducting layers in the upper atmosphere, and the measurement of their heights. In the Department of Science and Industrial Research R. Watson-Watt followed and extended Wilson's researches on thunderstorms: C. T. R.'s influence extended beyond his own creations and discoveries.

When the First World War started in August 1914 the Cavendish Laboratory had completed a stage in its history. J. J. was fifty-seven years old, not a great age, but his major personal researches had been accomplished. He was the most eminent British scientist, and was drawn into the advisory and presidential duties appertaining to such a position. The Cavendish Laboratory was now a comparatively large research and teaching institution, according to the conceptions of the period. It had a well-matured system, and deep foundations for future lines of fundamental research had been laid, but the precise shape that this would take naturally could not as yet be foreseen. The visible spearhead of British physical research had clearly gone to Manchester.

The war suddenly brought the regime of the Laboratory, now forty years old, to a halt. The professor and the staff had never had anything to do with the applications of science, in industry or in war. They had pursued physics as part of the cultural equipment of nineteenth and twentieth century educated society. When the war started they acted in the main just like the other sections of the British educated classes. They stopped doing their usual work and joined the Army, or some other part of the national forces. The idea that science might be important in war was not unknown, but it was kept in the background, or was secret. J. J.'s predecessor, Rayleigh, had, in fact, already been involved through Haldane in important military scientific work as chairman of the Government's committees for research on explosives and on aeronautics. In the latter committee, other former Cavendish men, R. T. Glazebrook and W. N. Shaw, were also members.

In 1914, however, the life of the Cavendish was almost innocent of the activities, and to a large extent even of the existence, of such committees. Science and warfare were generally thought to have little to do with each other, and the notion of science as a major military weapon occurred to few scientists.

J. J.'s son G. P. Thomson was in the Army as a Second Lieutenant, and was at the front before the end of 1914. J. J. immediately saw the war from the personal as well as the national point of view, and was very much concerned about his son's conditions in the field. Nearly all the research students remaining in the Laboratory enlisted in the officer's training corps, and J. J. made arrangements for them to work in the laboratory in the evenings, after receiving military training in the afternoons. At the suggestion of Lefroy of the Admiralty, some research on a hot-wire radio receiver was undertaken, and C. S. Wright was sent to work on it.

The idea of mobilising the Cavendish Laboratory for military scientific research was so far from the Government mind that in December 1914, 125 men were billeted in the big room of the Laboratory used for teaching medical students.

But as military research forced the mobilisation of resources on military problems, the Laboratory's workshops were assigned to the making of gauges, in intense demand for the armaments industry. Then, as the importance of science became more evident to government and public through the course of the war, some of the young research workers who had joined the military forces at the outbreak were directed to scientific research of military importance. G. P. Thomson worked on military aeronautics, and W. L. Bragg on the sound-ranging of enemy guns. Aston, Whiddington and many others joined the research staffs of military scientific establishments.

The scientific talent of the Cavendish Laboratory was dispersed, but many of the research men were able to make useful contributions in the organisations to which they were presently assigned. The early defeats in the war led to the recognition of the absence of any proper mobilisation of science for military and national ends. Scientists themselves were presently shocked by the deaths of some of the most talented young scientists, such as H. G. J. Moseley in 1915, who were killed in routine military operations in which no use at all was being made of their special scientific gifts.

As the military significance of submarines, aircraft and radio became clear to many from the damage they were causing, public and political pressure for coping with them increased. The military establishment had propagated the belief, as a means of supporting morale, that the British forces were so well armed and prepared that nothing more was needed. The accumulation of military reverses destroyed this belief, and the public now began to believe that the authorities were obstructing military science and invention, and that the national talent was being neglected.

In 1915 A. J. Balfour, then First Lord of the Admiralty, formed a Board of Invention and Research. The Chairman was Lord Fisher, and the other members were J. J. Thomson, Charles Parsons, and J. T. Beilby. Among the Board's consultants were W. H. Bragg and Rutherford. The Board was to consider all scientific ideas and inventions that might be of use in the war.

Thousands of suggestions came in from the public, especially after any spectacular event, such as the early raids by Zeppelin airships. For a time the Board received letters at the rate of five hundred a day on how they might be coped with.

J. J. acted as Chairman of the Board when Fisher was absent. He and his colleagues spent hours listening to the various types of inventors. J. J.'s shrewdness was an apt quality for dealing with many of them. Altogether, about one hundred thousand inventions were considered, of which about thirty proved to be of practical use.

It was an example of the decline of laissez-faire as the principal motive in the advance of science and invention. The mere attention of ingenious minds was no longer sufficient to solve scientific and technical problems. These required prior recognition, and sustained and organised study. The most successful part of the Board's work was on anti-submarine devices, carried out by W. H. Bragg.

J. J. apparently did not invent any military device that came into use, but he designed a mine that would explode without contact, and a device for recording the pressures caused by explosions. This consisted of a piezo-electric crystal connected to a cathode-ray oscillograph. His most valuable work was in recommending suitable men for the various kinds of physical research that had to be undertaken. For this, he was able to draw on his unrivalled Cavendish experience.

The experience of the Cavendish Laboratory in the First World War indicated that the conception of science as a pursuit of a cultivated leisured class was being replaced by a new one, in which it was regarded as a more closely-knit activity, with more consciousness of its relations to the other aspects of human activity.

Towards the end of the war, in 1918, the mastership of Trinity College had become vacant with the death of Montagu Butler. J. J., the most eminent British scientist, now sixty-one, his main research life completed, and increasingly involved in affairs, was clearly a very strong candidate for the position.

The appointment is made by the Crown (Trinity College was founded by Henry VIII), acting on the advice of the Prime Minister, who was then Lloyd George. When he was asked twenty-three years later what considerations weighed with him in recommending J. J., he replied that 'his super-eminence as a scientist was known, even to a barbarian like myself who never had the advantage of any university training. As one of the War Directorate I knew what invaluable services Thomson had rendered in the conduct of the war...'

J. J. expected at first that the mastership would not diminish the time he could give to science. He intended to relinquish the salary of the Cavendish professorship, but to retain the control of the Laboratory and research, at any rate until the end of the war, after which a more permanent arrangement would be necessary. He seemed to think that the new professor might run the laboratory activity, while he might remain in general direction and control. J. J.'s tenacity, which was such an important factor in his achievement, came out in his first impulse to retain influence in the Cavendish, even after his resignation from the chair. Most fortunately he mastered this tendency, though only after some delicate and anxious exchanges.

J. J. resigned from the Cavendish professorship in March 1919, and was appointed to a special professorship of physics, without salary.

It was obvious that Rutherford should be the next Cavendish professor, but the way was not quite clear. It was equally obvious that Rutherford could not accept the position that J. J. at first had in mind. Rutherford knew his old professor and benefactor, and was aware both of his conscious and unconscious desire to continue in some position of control or influence in the Laboratory. Rutherford was convinced that, if he were to become a candidate, it was essential that it should be made absolutely clear that the Cavendish professor was completely in charge of the Laboratory. He wrote to J. J.

My dear Professor,
I have been thinking over the Cavendish matter but of course there are a number of factors that enter into the question. Before coming to any decision there are several important points on which I would like your views and frank opinion.

Suppose I stood and were elected, I feel that no advantages of the post could possibly compensate for any further disturbance of our long continued friendship or for any possible friction, whether open or latent, that might possibly arise if we did not have a clear mutual understanding with regard to the laboratory and research arrangements. It is for these reasons that I feel it very desirable to discuss the position in its various bearings. In the first place, I should say that if elected I would welcome your presence in the Laboratory and would be only too happy if you would help us as far as you feel able in helping research and researchers in the Cavendish. . .

Rutherford felt confident that there would soon be more advanced students than the Laboratory could provide for, or that the two of them could look after properly without carefully thought out arrangements. 'Under such conditions it might prove awkward for both of us to place intending researchers in the position of making a decision with whom they wished to work.'

He believed there were two possible solutions: that J. J. should work by himself, with his own laboratory assistants; or that the Director should have charge of all advanced students, and assign their line of work and supervisor. Under the latter arrangement the Director would place under J. J.'s supervision those students who wished to work on the lines in which he was interested. 'If we kept closely in touch on projected lines of research, I would know your wishes and would try to fill them by turning over, as occasion arose, men to work under your supervision, but of course students would be students of the laboratory first, and of their supervisors second.'

Rutherford also raised the question of changes in organisation of teaching and research, and of personnel. 'It would be a disaster if

173

any trouble should arise on that score, for it seems to me that the Director must take sole personal responsibility for the efficiency of the teaching and research in the lab.'

He also asked for information on the finances of the Laboratory, covering the workshop, upkeep, apparatus and salaries of teachers. He felt that he was troubling him a good deal on 'a number of delicate points', but his views would be most welcome, since he had to make a decision very soon as to whether he would stand.

One of the most important of the electors to the Cavendish chair was Larmor. He corresponded with Rutherford and J. J., in order to clarify the situation. Rutherford's Manchester colleagues were trying hard to persuade him to stay there, and dwelt on the personal and organisational difficulties that might arise in the Cavendish.

Larmor rather pointedly wrote to J. J. that there seemed to be a strong feeling in the University that Rutherford should be attracted if possible, on the ground that J. J.'s activities would be more and more withdrawn to other duties, and in any case could last only a limited time. If he remained in charge, and there was no adequate successor, 'the repute of the school now one of the essential assets of the University might fail'. He said that the matter must be settled in a few days, and he left it to his judgement whether he 'could do anything further' by way of communicating with Rutherford.

J. J. wrote to Rutherford, saying he wished to make the position as clear as possible, even at the risk of repeating what he had said before. 'The intention is to make the two Professorships as independent as if their laboratories were in separate buildings.'

He was sure that the University would put up a second laboratory for the new professor as soon as funds were available. Meanwhile, a few rooms in the present building had been assigned for the new professor, but these would be all that would be under his control. As for the rest of the Laboratory, he would treat it as if it were 'a mile away'. He regarded the new professorship to which he had been appointed as in no way a part of the Cavendish professorship.

When he had sent in his resignation of the Cavendish chair, he had thought out the question, and had determined to sever himself entirely 'from any connection with the management and policy of the Cavendish laboratory', so that his successor would have a 'perfectly free hand to carry out any policy he might see fit to adopt'. He would confine himself to his own rooms, and do all he could to get a new laboratory built.

He added a postscript that everyone in the University was doing everything in their power to meet Rutherford's view, and there was 'a very keen hope' that he would see his way to come to Cambridge.

Rutherford stood for the chair at the latest possible moment, making the application by telegram on the day of election: 2 April 1919.

Rutherford was forty-seven years old, with a career of extreme achievement and distinction behind him. He had discovered the nature of radioactivity, had established the nuclear theory of the atom and founded nuclear physics. In 1917, at Manchester, during the First World War, he had first succeeded in disintegrating the atom.

It was splendid that he was to come to direct the Cavendish Laboratory, but was it conceivable, as an experimental physicist approaching the age of fifty, that he would be able to sustain his own achievement, and carry the Laboratory to the heights of 1897 again? What new qualities could he bring to the Laboratory, as its Director?

13

Rutherford's Background

Clerk Maxwell was a Scottish laird, Rayleigh a landed English aristo-crat and J. J. Thomson a middle-class Lancashire man. Ernest Ruther-ford (see figure 15) came from a very different social background: the agrarian democracy of New Zealand. He was born near Nelson, in the South Island, in 1871, only twenty-nine years after that town had been founded, and in the same year as the foundation of the Caven-dish Professorship of Experimental Physics in Cambridge.

The settlers were energetic and able craftsmen, nonconformist in religion, and intent on escaping from the appalling industrial condi-tions in Britain in the 1840s. They sailed in ships only about 40 metres long. One of the early parties included George Rutherford, a wheel-wright from Perth. He had been specifically engaged to build a saw-mill near Nelson. He sailed from Dundee in 1842 with his family, including his son James, then aged three. James was to become the father of Ernest Rutherford.

The voyage to New Zealand took more than six months. There, life had to be established in virgin forest. The trees were laboriously cut, and a water-driven sawmill was built to saw the timber. The tree-stumps were cleared from the ground, which was then set with wheat. Each grain of seed was wiped by hand to remove rust disease. The corn was reaped with sickles, thrashed by hand-flails, and carried on the back for six kilometres to a flour mill.

James Rutherford was given the rudiments of education by his family, and did not attend any school until he was ten. He grew up in the pioneer life of saw-milling and farming. When he was a boy of sixteen in 1855, a twelve-year old girl named Martha Thompson was brought from England by her widowed mother. Her father had had scientific interests, and she grew up to become one of the first school

15 *Rutherford*

mistresses in New Zealand. She and James Rutherford met, and in 1866 they married. They lived in a little wooden cottage, and produced a family of seven sons and five daughters. James lived to the age of eighty-nine, and Martha to ninety-two. Ernest was their fourth child and second son.

James flourished modestly; besides practising his craft of wheelwright, learned from his father, and running his small farm, he built a mill for preparing the wild flax gathered from the neighbouring swamps, and undertook some contracting for bridge-building and railway construction. He secured a contract for supplying 40 000 railway sleepers at 2s.8d. (13.3p) each, which were cut in his sawmill. Rutherford said that his father had natural engineering ability, and might, in other circumstances, have become a notable engineer.

His parents had a hard but fundamentally happy life. Ernest came of physically and psychologically healthy stock; his father understood craftsmanship and his mother education. He grew up in an energetic, inventive environment, and in addition to all this he belonged to a small closely knit community of pioneers, to whom united action in attacking problems was a primary condition of life.

This factor was of profound importance when Rutherford became a director of laboratories; it was natural for him to become a personal leader of a communal group. In this respect he was different from J. J. Rutherford became an Achilles of science, brandishing the sword of discovery in his own hand, as he rushed forward into the unknown, at the head of his followers. J. J., in contrast, was an Agamemnon who commanded the hosts, devising a grand strategy and directing his troops, but winning his battles through the hands of others.

Besides his fortunate endowments of family happiness, physique, and health, and an inventive, self-confident and successful social environment, Rutherford was also fortunate in receiving an excellent higher education, remarkable at such an early stage in a pioneer country.

He went to the village school at five, and at the age of ten acquired a primer on science written by Balfour Stewart, the teacher of J. J. at Manchester. At the age of eleven he went to another primary school, run by an excellent teacher, who gave free Latin lessons to promising pupils before regular school hours.

At the age of fifteen Rutherford competed for and won a scholarship to Nelson College. The head was no less a person than the famous cricketer W. J. Ford, who had been educated at Repton and Cambridge, and had been a housemaster at Marlborough College. His chief teacher was W. S. Littlejohn, yet another excellent teacher, and noted author of school textbooks.

Nelson College was a distinctly unusual institution. It had many of the features of an English public school, but conducted in an agrarian democracy instead of an aristocratic and upper-middle class society. Rutherford acquired some of the technical advantages of the English school system without its social prejudices. Almost his last words before he died were: 'Remember, a hundred [pounds] to Nelson [College].'

His all-round qualities won him prizes for reading, languages and mathematics, and he played football. He emerged as a leader among younger boys, who nicknamed him 'Grandpa'. He ended his school life as head boy, and competed for a scholarship at the University College, Canterbury. He came fourth out of ten. In 1890 he entered Canterbury College and in 1892 was awarded the Senior University Scholarship for mathematics.

Again, at Canterbury College, he was fortunate. The professor of mathematics, C. H. H. Cook, was a former fellow of St John's College, Cambridge, and a very capable teacher. The professor of chemistry and physics was A. W. Bickerton, who had original ideas on the equilibrium of gaseous cosmic spheres, and on chlorine as a cure for tuberculosis. He speculated on the collisions of novae, and, as Marsden remarked, this idea may have influenced the formation of Rutherford's thought on the planetary theory of the atom, and on the importance of collision phenomena in nature, including collisions between atoms.

He completed his university course by securing first-class honours in both mathematics and physics, which had been achieved only once before in the history of the university. His first researches in experimental physics were made during the last two years of his work for his honours degree. He had to take a course in practical physics, which included experiments on secondary circuits, and the passage of transient currents in conductors.

In his first research notes he recorded that it had occurred to him that the thickness of the wire in the secondary circuit would affect the current at make and break. He recorded that he had 'not observed until later that the same idea had occurred to Sir W. Thomson'. Rutherford's characteristic self-confidence had already appeared. Sir William Thomson (Lord Kelvin) was then about the most famous scientist in the world, and here was this youth in New Zealand arriving independently at the same ideas.

He carried out his experiments in a cold draughty cellar, used by the students as a cloak room. He reported that the use of magnetised steel needles as 'detectors' and 'galvanometers' had been explained, and 'their possible use also for measurement of the intensity of electro-magnetic waves'.

He used his 'magnetic detector' for the detection of radio waves passing through the brick walls of the laboratory, over a distance of sixty feet. He already showed his capacity for leadership, for he had no difficulty in getting fellow students and even young demonstrators to assist in the experiments.

Rutherford sat in 1894 for a scholarship offered to New Zealand by the Commissioners of the Exhibition of 1851. His mother received the telegram announcing the award. She went into the garden where Rutherford was digging potatoes, and called down the path that he had won the scholarship. Rutherford flung his spade aside, and ejaculated: 'These are the last potatoes that I shall dig.'

Before he left for England he became unofficially engaged to Mary Newton, a daughter of the landlady with whom he had lodged as a student in Canterbury. They were married in 1900.

He arrived at Cambridge in 1895. Besides demonstrating his radio discoveries, he soon collaborated fruitfully with J. J., and then started research in radioactivity, distinguishing α- and β-rays.

He went to Montreal in 1898, discovered the 'emanation' from thorium, which gave the clue to the explanation of radioactivity as due to spontaneous disintegration, and secured the collaboration of Soddy in working out the theory. His fame grew so rapidly that he attracted research students from Europe, the most eminent being Otto Hahn.

At Manchester Schuster built a model new physics laboratory largely at his own expense, and retired early in order to secure Rutherford as director. Arriving at Manchester in 1907 Rutherford found that he had inherited Schuster's personal assistant, Hans Geiger.

He drew up a list of thirty research topics, and was able to plan a general campaign of research on a larger scale than had been possible in Montreal.

Rutherford had noticed in 1906 that a beam of α-particles was scattered when passing through air. Geiger now examined the scattering quantitatively, and found that the average deflection increased with the atomic weight of the scattering medium. He was joined in his scattering observations by an eighteen-year-old student, Ernest Marsden. They discovered that there were far more very large deflections of particles, than were to be statistically expected. Some were being bounced straight back. Rutherford saw that the phenomenon implied that the atom must contain a nucleus.

Then, in April 1912, Rutherford was joined by Niels Bohr; this was the team who developed the quantum theory of the atom.

180

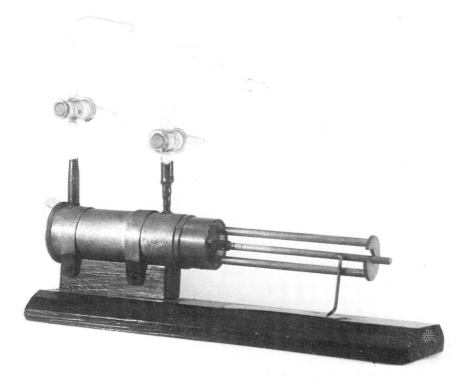

16   *The tube with which Rutherford first disintegrated an atom*

Marsden noticed that some hydrogen particles were knocked on by impinging α-particles, and had a range four times that of the impinging particle. Following this up, Rutherford, assisted only by his lab. boy, Kay, had evidence by November 1917 that atoms could be disintegrated by α-particles. Rutherford's apparatus is shown in figure 16.

At about this time Rutherford failed to attend one of the meetings of a war research committee. When asked why, he replied: 'I have been engaged in experiments which suggest that the atom can be artificially disintegrated. If it is true, it is of far greater importance than a war!'

By the early months of 1918 he had conclusive proof that he had indeed disintegrated the atom. He wrote up the research in four classical papers, published in 1919.

He had achieved the artificial disintegration of the atom with no assistance apart from Kay, using methods demanding the utmost physical attention only to be expected of a young man, when he was already middle-aged. What more could he do? What more could he expect to achieve? After a life such as this, was it not inevitable that the future must be an anticlimax?

Such was the situation when Rutherford was appointed Cavendish Professor on 2 April 1918, in succession to J. J. Thomson.

14

The Apotheosis of the Cavendish Laboratory

The prospects of the Cavendish Laboratory were not very clear at the conclusion of the First World War. Rutherford agreed to stand for the Cavendish professorship only at the last moment, after much deliberation and delicate negotiation. J. J. had retired after thirty-four years of direction, during which he had left the impressions of his personality, methods and ways of thought on the Laboratory. These had begun to age, and the four years of the war, when the Laboratory's characteristic lines of research were suspended, had attenuated them further.

In fact, while continuing to do magnificent work, J. J. had not, after the discovery of the electron in 1897, led the development of modern physics. W. L. Bragg defined this as 'the study of the fundamental particles of which matter is made, as contrasted with the study of matter in bulk and of large scale forces'. He commented that J. J. 'was a curious link between the old and the new physics. It has been said of him that he opened the door to the new physics, but never went through himself. . .'

Rutherford had gone through the door and was leading the investigators who were exploring the new universe within the atom.

In spite of his extraordinary personal researches at Manchester in the autumn of 1917, Rutherford did not feel that he was moving to Cambridge merely to continue a line of research in a flourishing condition and situation.

In May 1919 he received a letter from Hans Geiger in Berlin, reporting that he had survived after passing through severe struggles, and hoping that Rutherford still took 'a little interest in your old pupil who keeps his years in Manchester always in pleased memory'. Geiger remarked that on reading the literature he had concluded that 'no very striking discoveries' seemed to have been made in radioactivity.

A few days later, in June 1919, Rutherford replied that he was 'very glad to know that you have come through the struggle all right and are back at the old work again'. He said that he had 'kept in touch as far as possible with my old researchers during the War and am glad to know most of them are safe and sound'. He mentioned the latest news of most of his Manchester colleagues, and he remarked: 'We are all feeling very rusty scientifically after the war, and it will be some years before we can get going properly, for apparatus is very dear and difficult to get... I hope you will take up the number of alpha particles again... Well, I retain my old friendly feelings for my old researchers and hope that we may meet again when things have settled down to a more normal footing, but it will obviously be some time before this will be possible. I shall be glad to hear from you.'

So, Rutherford was feeling 'rusty' when he entered into the Cavendish professorship. The Laboratory was also pretty 'rusty', and he submitted a memorandum on its needs for presentation to the University. He attached to it an account of the *History of the Cavendish Laboratory*. This document, which was quite brief, placed great emphasis on the need to develop applied physics for industrial, state and political reasons.

He instanced Maxwell's electromagnetic theory as the precursor of radio communication, and an example of the kind of work that the Cavendish could contribute towards future practical interests. He added that, 'It is safe to say that Clerk Maxwell will take a place in history as one of the greatest men of science of all ages.'

He said that the output of important work had grown rapidly since 1895, and 'the Laboratory was soon recognised as the chief centre of research activity in physics'.

They needed more space for teaching and research. 600 students and 50 naval officers were under instruction in laboratories equipped to deal with half the number. It was difficult to overrate the importance of research, 'if the Laboratory is to play an active part in advanced study in research for the Empire'.

The First World War had shown the importance of scientists to the State, in military and industrial science. 'The rising prestige of this country and the eclipse of Germany will lead to an increase in the number of research students from neutral and allied countries who wish to work in the Laboratory. The international importance of attracting highly-qualified research workers not only from our own Empire but from friendly States, needs no emphasis...'

New laboratories were needed especially in applied physics, optics and general properties of matter. 'The need of a laboratory specially devoted to training in research in Applied Physics is of pressing importance if we are to play our part in the researches required by the State, and in providing well-qualified research men for various branches of industry and for the scientific departments of the State.'

The Laboratory was already equipped for radio research, but more provision was needed for the study of elasticity, and the motion of gas and fluids which 'lie at the bottom of many problems in Engineering'.

In applied physics it was intended to work alongside the School of Engineering, the physical aspects of problems of common interest to be studied in the Cavendish and the more technical in the Engineering Laboratories.

He summarised his main conclusions:

1 Increased laboratory and lecture space for the teaching of Physics.
2. Provision of new, well-equipped laboratories for Applied Physics, Optics and Properties of Matter.
3 Provision of three additional lecturers of high standing, competent to direct advanced study in research in the new departments mentioned above.
4 The endowment of another Chair of Physics in the University.

Nuclear physics was not mentioned in the document. The man who had just disintegrated the atom did not refer to the subject in his outline of future needs for his Laboratory. Perhaps he thought it was too obvious, but, for whatever reason, he did not mention it.

Rutherford's attention to, and conception of, financial needs, were no less remarkable than his emphasis on industrial, national and political considerations in the appeal for the extension of the Cavendish. He estimated that the new buildings required would cost not more than £75000, and that a further sum of £125000 would be necessary as an additional endowment.

In 1919 £200000 was a very large sum for a university science laboratory. The University was not entirely convinced of the necessity for the requirements enumerated by Rutherford, and in any case, the money was not available.

These plans and proposals were in striking contrast with Rutherford's personal methods of research. The apparatus with which he had disintegrated the atom was fantastically simple, and could be operated by one man alone.

Rutherford did not think only in terms of the 'string-and-sealing-wax' conception of scientific equipment. He clearly conceived the necessity for large-scale research and equipment. The reason why the Cavendish Laboratory did not at once embark on large-scale physics was due to the opposition of the University and the British social system, which had not yet come to understand the necessity for it, and the large expenditure it entailed.

Rutherford refused to solicit and wheedle for funds. He could easily have received large sums from America, but remarked that 'if the British want research, they can pay for it'.

Seventeen years passed before funds of this magnitude were provided for the Cavendish Laboratory. In 1936 the motor-manufacturer Sir Herbert Austin gave a quarter of a million pounds for its development. Shortly afterwards he was raised to the House of Lords. But in the intervening years research continued with the traditional small apparatus. Rutherford brought his personal apparatus with him from Manchester, and the 250 milligrams of radium that had been lent to him by the Vienna Institute of Radium in 1908, and with which a large part of his radioactive researches had been accomplished. He would have liked William Kay to have come with him, but Kay's wife did not want to lose the company of her Manchester friends, so Rutherford advised him to remain in Manchester.

Rutherford swiftly reorganised the Laboratory research on radio-activity. After the war young men of talent flocked to the University. They were of varied backgrounds and ages, wider than the range of undergraduates and research students arriving in pre-war times. There were several years, not merely one year, of new talent entering the Laboratory in 1919.

A development started which reached its apotheosis in 1932. It will be outlined in this chapter, and more will be said about it later on, according to the views of the principal participators.

Rutherford appointed Chadwick assistant director of research, which relieved him of much administration. Chadwick was a Lancashire man, who had gained scholarships that took him to Manchester University, which he entered in 1910. He became one of Rutherford's pupils at the time when the nuclear theory of the atom was being established. He undertook research under Rutherford's direction, and in 1912, at the age of twenty-one, he published experiments of a style similar to those by which he proved the existence of the neutron twenty years later. He was awarded a studentship in 1913 to work under Geiger in Berlin. When the First World War started in the following year he was promptly interned in Ruhleben. Instead of being broken by such an experience at such an age, Chadwick started scientific researches in the camp.

With him in Ruhleben was the British Army officer C. D. Ellis, who had been sent to Germany before the war for military studies, and consequently also had been interned. Chadwick and Ellis organised a little research laboratory. They received generous help from German scientific colleagues and friends, including Rubens, Nernst and War- burg.

Chadwick was a pale, thin, dark man, who did not look physically strong, but he showed character and determination. If he could run a lab. in Ruhleben he could run one anywhere. When he returned to England after the war he was weak in health, and had only £11. Rutherford gave him a job as a temporary demonstrator at Manches- ter, and then brought him to Cambridge.

Chadwick felt that he never quite recovered physically from Ruhle- ben. He had a pessimistic and critical outlook; this no doubt made him appear to himself and others less well than he was. However, his long life and great achievement belie basic constitutional weakness. No one could have done what he did without having a fundamental strength, whatever symptoms of illness were manifested.

Ellis followed Chadwick to Cambridge, determined to become a physicist.

The Admiralty had arranged for officers to attend the University on 'holiday courses'. P. M. S. Blackett, who had been in the Navy for ten years, and had commanded a destroyer in the war, was one of these. Younger research men, such as F. W. Aston and E. V. Apple- ton, returned from military scientific research or active service. W. L. Bragg, now twenty-nine years old, and already for four years the youngest of Nobel Laureates, also returned but soon left to succeed Rutherford at Manchester.

Rutherford resumed the disintegration experiments he had started at Manchester, in order to confirm his results by his original method, and also by other methods, in a carefully designed system of control experiments. He proved conclusively that the long-range particles produced when α-particles were fired into hydrogen were hydrogen nuclei, and he showed by magnetic deflection that the particles ejected from nitrogen were of the same nature.

He was usually assisted by new research students, of whom Ellis was one. Ellis was surprised and mildly shocked by the simplicity of the apparatus. This consisted of a small brass box; the scintillations produced by the particles on a zinc sulphide screen were viewed through a microscope. The apparently simple experiments required the highest experimental skill to furnish dependable results.

Rutherford was free from the social subtleties common to educated Englishmen. In this he was different from his three predecessors in the Cavendish professorship. He belonged to the democratic element in the British Imperial development, and said that 'the British Empire is all right'.

Rutherford was asked to deliver the Bakerian Lecture of the Royal Society in 1920. In it he reviewed his researches on atomic disintegration, and discussed their implications. It was evident that the nuclei of atoms contained electrons and hydrogen nuclei. How might these be put together to form the nuclei of other atoms?

It seems very likely that one electron can also bind two hydrogen nuclei, and possibly also one hydrogen nucleus. In the one case, this entails the possible existence of an atom of mass nearly 2, carrying one charge, which is to be regarded as an isotope of hydrogen. In the other case, it involves the idea of the possible existence of an atom of mass 1 which has zero nuclear charge. Such an atomic structure seems by no means impossible. On present views the neutral hydrogen atom is regarded as a nucleus of unit charge with an electron attached at a distance, and the spectrum of hydrogen is ascribed to the movements of this distant electron. Under some conditions, however, it may be possible for an electron to combine much more closely with the hydrogen nucleus, and in consequence it should be able to move freely through matter. Its presence would probably be difficult to detect by the spectroscope, and it may be impossible to contain it in a sealed vessel. On the other hand, it should enter readily the structure of atoms, and may either unite with the nucleus or be disintegrated by its intense field, resulting possibly in the escape of a charged hydrogen atom or an electron or both.

An atom of mass 1, with zero nuclear charge, would 'have very novel properties. Its external fields would be practically zero, except very close to the nucleus, and in consequence it would be able to move freely through matter'. Rutherford continued:

If the existence of such atoms be possible it is to be expected that they may be produced, but probably only in very small numbers, in the electric discharge through hydrogen where both electrons and hydrogen nuclei are present in considerable numbers. It is the intention of the writer to make experiments to test whether any indication of the production of such atoms can be obtained under these conditions. The existence of such nuclei may not be confined to mass 1, but may be possible for masses 2, 3, 4 or more, depending on the possibility of combination between the doublets. The existence of such atoms seems almost necessary to explain the building up of the nuclei of heavy elements, for, unless we suppose the production of charged particles of very high velocities, it is difficult to see how any positively charged particle can reach the nucleus of a heavy atom against its repulsive field.

He forecast the existence and properties of the neutron and the existence of deuterium, and discussed the existence of hydrogen and helium atoms of mass 3. This proved to be one of the most significant inspirations in the history of science. It was not that atoms of a hitherto unknown mass had not been imagined, for Harkins and Orme Masson speculated on the possible existence of neutrons, and J. J. Thomson had found evidence for the existence of hydrogen atoms of mass 3 in his parabola experiments on the separation of atomic particles by magnetic and electric deflections. The peculiar feature of Rutherford's deductions was that they were handled with the complete atmosphere of physical reality. Their particular importance was that, from 1920, the neutron was regarded in the Cavendish Laboratory as a real, if concealed, thing, which it was up to them to find. Glasson and Roberts made a search for this neutral close combination of a proton and an electron, but without success.

In 1920 Rutherford named the nucleus of the hydrogen atom the proton. He and Chadwick embarked on a systematic study of the artificial disintegration of the lighter elements. They soon succeeded in disintegrating aluminium, phosphorus and fluorine, and found that in general it was easier to disintegrate atoms with odd than with even atomic numbers.

Direct attempts were started in 1921 to discover the neutron, and artificial radioactivity. One idea was that the electron and the proton in a hydrogen atom might be made to combine by an intensive electric field, such as that produced by a violent spark discharge. If such a combination occurred, it might be accompanied by an emission of heat. The experiments were unsuccessful. Three years later, in 1924, they were still thinking about it. Chadwick wrote to Rutherford that 'we shall have to make a real search' for it.

The detailed analysis of radiations from radioactive substances, which was already a classic feature of Rutherford's researches, was continued at Cambridge. Ellis was asked to pursue the work on β and γ radiations, which Rutherford, Robinson and Rawlinson had carried out at Manchester.

Rutherford was as vigorous and enthusiastic as ever, but he had less opportunity for personal research, owing to the growth in his administrative work, and the calls on him as a national and international figure. He served on many committees, became Chairman of the Council for Scientific and Industrial Research, and President of the Royal Society. He delivered many scientific and general lectures and addresses. He was now old enough to be the father-figure of his younger research colleagues, and became known privately among them as 'Papa'.

Yet the passion for research and discovery continued unabated. When invited to give a course of lectures in America in 1923, he wrote to Boltwood: 'It is a pity but I cannot go. I am bound to the wheel in my laboratory... I should very much like to be free to spend a few weeks in Yale again ... but life for me is very busy in these days and I have to drive the boys along.'

As Rutherford's memorandum of 1919 on the needs of the Cavendish showed, he was quite aware of the desirability of expensive equipment. As Cambridge and British society did not then provide him with the necessary finance, he and his students continued with his classical type of small-scale experiments. But he encouraged such technical development of apparatus as could be carried out within the available means. His Japanese pupil Shimizu devised a type of Wilson cloud chamber in which expansions could be repeated rapidly. This reduced the labour of making large numbers of cloud tracks of atomic particles. Shimizu also arranged that the cloud tracks should be photographed simultaneously from two directions, thus stereoscopically revealing their direction in space.

In 1921 P. L. Kapitza arrived in Cambridge. He had been trained as an electrical engineer, and after illness had ravaged his family in the post-war and post-revolutionary privations, he had been sent to England as a technical attaché of a Soviet purchasing mission. The Soviet authorities approved of his desire to work in the Cavendish, and he was accepted by Rutherford as a research student.

He started research, at Rutherford's suggestion, on the loss of energy when an α-particle penetrates matter by measuring the heating effect. For this he devised a modified form of Boys's radiomicrometer, and with it succeeded in measuring the variation of the energy of the α-particle along its path. With Shimizu he collaborated in cloud-chamber researches. He conceived the idea of producing very powerful magnetic fields by short-circuiting accumulator batteries, which would be strong enough to deflect the comparatively heavy and very energetic α-particles. This offered prospects of increasing the refinement of the analysis of the properties of α-rays, and possibly of producing changes that might throw further light on the nucleus.

Blackett set to work on the development of the cloud chamber, and by 1924 had succeeded in photographing the disintegration of an atom.

The stream of experimental discoveries stimulated the development of theoretical atomic physics. This was led by Bohr, whose institute in Copenhagen attracted many of the ablest young theorists.

Important contributions also came from Paris, where Louis de Broglie had in 1924 propounded the wave theory of matter. In 1925 Heisenberg proposed the new quantum mechanics, based on the mathematical description of observable quantities, and the exclusion of all hypothetical entities. He joined in the Copenhagen researches, distinguished, as he put it, by 'der Kopenhagener Geist der Quantentheorie'. In 1926 Schrödinger announced his new wave mechanics, which was later shown to be mathematically equivalent to Heisenberg's, though the approach was very different.

In general, the leadership of theoretical physics remained in continental Europe, following the original discovery of the quantum theory by Planck, and the theory of relativity by Einstein. The concentration of theoretical research on the Continent emphasised the concentration of the British contribution in experimental physics.

Rutherford represented this concentration in the fullest degree. He said in 1925 that the theorists 'play games with their symbols, but we, in the Cavendish, turn out the real solid facts of Nature'. His attitude was a corrective to the mathematical *spielerei* which was the vice of the theorists, but it did not stimulate the creation of a school of theoretical physics in Cambridge comparable in weight with the Cavendish experimental tradition. Rutherford tended to look on theorists as mathematical handymen. His own son-in-law, R. H. Fowler, did, however, make the Cambridge theorists more useful to the experimental investigators, and his pupil P. A. M. Dirac was the first British theorist of the twentieth century to emerge in the same rank as the great continental theorists.

In 1928 Dirac showed how the quantum and relativity theories could be combined in his relativistic quantum mechanics. This relativistic form of the wave equation, besides describing the positive energy of the electron, also had roots that apparently referred to particles with a 'negative energy'. What could this be? Dirac sought for a physical interpretation of it. He conceived that it might refer to what he called a 'hole' in the universe, consisting of the absence of a negative electron. He showed that such a 'hole' would move about and behave like a positively charged particle, and he tried hard to identify it with the proton. But the latter was nearly two thousand times heavier than the electron. In 1932 C. D. Anderson discovered the positron experimentally, and it was evident that this was the particle whose existence Dirac had deduced. By deduction Dirac had discovered 'anti-matter', and presently hinted at the notion of the anti-universe.

Rutherford was considerate to great theorists such as Bohr and Dirac, however impatient he may have been with lesser men, and he thoroughly appreciated theoretical help with his own particular problems.

Experimental research had shown that α-particles of five million volts energy could escape from the nucleus of the uranium atom, while α-particles of ten million volts energy, when fired at it, failed to penetrate. Evidently, the nucleus contained some kind of ten-million-volt potential barrier. Rutherford was deeply impressed by this paradoxical phenomenon, and puzzled over the nature of the nuclear structure which could have such an extraordinary property. In 1927 and 1928 Gurney and Condon, and Gamow, showed that the paradox could be explained by the new wave-mechanical theory of matter.

According to this, the position of a particle was governed by a principle of probability. The chance that it would be inside or out-side the potential barrier at any moment was defined statistically. The chance could be calculated precisely. If the α-particle happened to be outside the barrier, then it escaped.

Rutherford tried to conceive the radioactive nucleus as a vibrating system of particles, from which α and β particles escaped according to statistical laws. Intranuclear rearrangements of particles changing from one excited state to another were accompanied by the emission of γ-rays.

This view of the nucleus did not have the simplicity of the earlier Rutherfordian conceptions. He had for some time shown a certain irritation with the absence of simple, clear-cut discoveries of funda-mental importance which had characterised his earlier work. One or two of his younger collaborators even left the subject, uncertain of its future. In this atmosphere Rutherford particularly welcomed the discovery in 1929 by Rosenblum in Paris of the fine structure of α-rays emitted from radioactive substances, accomplished with the aid of the great electromagnet designed by Aimé Cotton for the French Academy of Sciences. This furnished the kind of detailed information required to make the conception of nuclear structure more precise.

A special magnet was constructed for Ellis to extend the fine-structure researches started by Rosenblum. Cockcroft assisted in its design, after he had worked with Kapitza on the design of his dynamo for producing very strong magnetic fields. Classical experimental atomic physics had lost its earlier simplicity by 1929–30.

It seemed that the era of Rutherfordian atomic physics must be ending. Rutherford himself was nearly sixty. Major discovery in physics beyond that age is extremely rare. Though not satisfied with the situation, he pressed on. Since the first days when he perceived the potential of energetic α-particles as missiles with which the interior of the atom might be explored, he had begun to think of producing artificially accelerated atomic missiles, even more energetic than those shot out of radium. In 1913 he was already saying that 'it is a matter of pressing importance at the present time to devise electrical machines of the highest possible voltages'.

This was still on his programme at Cambridge. He encouraged T. E. Allibone and E. T. S. Walton in the design and construction of high-voltage machines, and Cockcroft began the construction of a high-voltage direct-current machine, while he was still assisting Kapitza.

In 1930 there was an intense atmosphere of activity in the Cavendish Laboratory, and also some frustration and unrest. The effects of the First World War had passed out of sight, and the number of research workers had grown. They came from many countries, especially the British Dominions. Unfortunately the laboratory space was very limited. For example, Kapitza was erecting his dynamo in a shed that had formerly been in the department of physical chemistry, and Cockcroft was assembling his accelerator in a corner of a chemistry store room, out of glass cylinders of the type used in petrol pumps.

Never were so many gifted experimental physicists packed in so narrow a space. The competition for room and apparatus was severe, and strong repulsive forces were sometimes generated between the rabid investigators. They formed a kind of nucleus of experimental talent, which needed all the binding force of Rutherford's genius to hold them together.

There were differences of opinion as to whether particular research workers were getting their fair share of available resources. One young man who boldly complained to Rutherford that he was being starved of apparatus was met with the observation: 'Why, I could do research at the North Pole!' Rutherford had a tendency to fix a lot of attention on particular men, especially those from the British Dominions. It was perhaps partly due to the fact that he had no son. His only child, a daughter, died in 1930.

Some considered that it was not always those who made the most discoveries who received the lion's share of the resources. When they expostulated, Rutherford told them that it was not necessary to have expensive apparatus to make discoveries.

Rutherford served as President of the Royal Society from 1925–30, and in 1930 he became Chairman of the Advisory Council of the Department of Scientific and Industrial Research. With so many duties and preoccupations, and experimental nuclear physics becoming so complicated and obscure, it seemed impossible for Rutherford to preside once more over major discoveries, even if he were very active.

Then, in 1930, came the announcement by the German physicists W. Bothe and H. Becker of their discovery of a peculiar radiation that was very penetrating. They had been engaged in studies of the effects of the bombardment of atomic nuclei, using a strong radioactive source, which produced very fast and energetic α-particles. Such particles usually disintegrated the nuclei of light elements, but Bothe and Becker found that this did not happen with the light element beryllium. Instead of being disintegrated and emitting particles when bombarded, the beryllium nucleus emitted a radiation of great penetrating power, far greater than that of the protons and other particles produced in the familiar disintegrations of atoms.

The beryllium radiation attracted the attention of the investigators of cosmic rays, for they seemed to offer a radiation of a power intermediate between that of radioactivity and cosmic rays. The American physicist R. A. Millikan in a lecture in Paris referred to the similarity of Bothe's rays to cosmic rays, and suggested that this should be investigated experimentally. Irène and Frédéric Joliot-Curie started such an investigation, and soon discovered that the rays could produce a very remarkable effect. If paraffin wax, or another substance containing hydrogen, was placed in the path of the rays, numerous particles were knocked forward, out of them. These were examined, and found to be protons. When submitted to a powerful magnetic field, they were only slightly deflected, showing that they were moving with a very high energy. As the beryllium rays were not like the well-known particle radiations from radioactive substances, the Joliot-Curies proceeded on the assumption that they were wave radiations.

They supposed that this very energetic wave radiation was knocking the protons forward, and since the energy of the protons was known, the energy of the supposed new wave radiation could be calculated from the known laws governing the exchange of energy and momentum between waves and particles. The calculation showed that the beryllium radiation, if it consisted of waves, must have the extraordinary energy of 50 million volts. This was ten times as energetic as the particles emitted in the spontaneous disintegration of radioactive atoms.

Now if the beryllium radiation did indeed consist of waves of this energy, the energy of the bombarding particles that produced them could be calculated. When this calculation was made, the result did not agree with the known energy of the particles used in the bombardment of the beryllium. The source used by the Joliot-Curies for providing the bombarding particles was polonium, which emits only α-particles and no electrons or wave radiations, so there was no danger of confusion with electrons or waves emitted by the source.

The Joliot-Curies concluded that the known law of interaction between waves and particles did not apply to the interaction between the beryllium rays and protons. They suggested that they had discovered a new type of interaction between wave radiation and matter, to which the conservation of energy law did not apply.

As soon as the Cavendish investigators heard of this work, they suspected the truth. The idea of a neutron had been in their thoughts for more than a decade. As Pasteur said, chance favours the prepared mind. Chadwick and Webster examined the beryllium radiation, and found anomalies in its intensity in various directions. Its intensity along the direction of the α-particles that had produced them was much greater than in the reverse direction. The difference could not be explained on the assumption that the beryllium radiation consisted of waves.

Chadwick was inspired to consider whether the anomalies could be explained on the assumption that the beryllium radiation consisted of neutrons. His prepared mind recognised the implications of the

195

facts, and twenty years' experience in radioactive research provided the skill with which they could be quickly and decisively confirmed.

Chadwick had been a young research worker under Rutherford at Manchester when he discovered that atomic nuclei, when bombarded by particles emitted from radioactive substances, might be made to emit wave radiations of high frequency. He discovered the neutron twenty years later by experiments similar in style to those of his early researches.

17 *The chamber with which Chadwick discovered the neutron*

There was also another necessary factor: the provision of the appropriate experimental means. The Joliot-Curies had been assisted in their experiments by the possession of a powerful source of polonium, the radioactive element originally discovered by Marie Curie, and named after her native country (Poland).

The refinement of the Cavendish radioactive technique was reinforced by a fortunate acquisition of polonium. N. Feather, Chadwick's colleague, had been working in America. He was given a number of old radon tubes by G. F. Burnam and F. West, of the Kelly Hospital, Baltimore. These provided a more powerful source of polonium than had previously been available in the Cavendish, and facilitated the decisive experiments proving the existence of the neutron. Chadwick's apparatus is shown in figure 17.

Although inheriting the Curie tradition in radioactive research, and having access to the powerful polonium sources which led to their pregnant experiment, the Joliot-Curies lacked the preparatory idea of the neutron foreseen by Rutherford and Chadwick. They later explained that they had not read Rutherford's Bakerian Lecture of 1920, as such lectures were generally summaries of past work, and rarely contained original ideas; if they had read and digested it, they believed they would not have missed perceiving the true implications of their experiment. Shortly afterwards they discovered artificial radioactivity. Rutherford, in congratulating them, said that he had sought for the phenomenon since the early years of his research life, but they had found it, and that was well.

Chadwick gave his first account of the discovery of the neutron to the Kapitza Club, a few days before he published the discovery in *Nature*, on 27 February 1932. Kapitza had founded his club to promote more informal discussion of scientific problems and researches in Cambridge. As a Russian, Kapitza possessed a cultural tradition distinctly different from the British; he was accustomed to the much more uninhibited Russian tradition of discussion. Unlike most foreigners he thought the university education system in Cambridge was, on the whole, better than that in Continental Europe. Students and teaching appeared to him to be on a higher average level, and more disciplined. But he thought the Cambridge scientific life, research and teaching too formalised. Young British students refrained from expressing opinions in the presence of their elders, for fear of exposing their ignorance and making fools of themselves. Kapitza opened the discussions in his club with apparently random

197

remarks. He sometimes made intentional howlers, which even the youngest and least experienced would not be afraid of correcting. In this way he promoted a freer and livelier standard of scientific discussion.

Some consider that Kapitza's contribution to the Cavendish and Cambridge from another cultural tradition was even more important than his inventions of apparatus and stimulus of research. The uniformity of the Cambridge scientific tradition is conspicuous when Cambridge scientists, so diverse in personality among themselves, are seen against a distinctly different social background, such as the Russian.

Chadwick's account stimulated a discussion which, within an hour or so, led to the proposal of important new experiments and lines of research for investigating the properties of neutrons.

The discovery of the neutron and the realisation of one of Rutherford's visions, created an unparalleled atmosphere in the Cavendish Laboratory. There had been other great moments in British physics, as when Newton published the *Principia* and Faraday announced his discovery of electromagnetic induction. But the situation in 1932 differed from those occasions in being the triumph of a community of investigators as well as that of a genius. Rutherford was also the leader of a closely knit group of human personalities. His 'boys' were growing up into major investigators and leaders themselves.

By the early months of 1932, after several years of work by Allibone, Cockcroft and Walton, the construction of a high-voltage apparatus for accelerating atomic particles had been completed. Rutherford had been somewhat impatient of the rate of progress on this project, and in the exultation and confidence created by the neutron discovery, he became still more pressing for experimental results.

At last, in February 1932, Cockcroft and Walton had indubitable evidence that they had succeeded in disintegrating atoms by electrically accelerated particles. They observed that when a target of the light metal lithium was bombarded by accelerated protons, scintillations appeared on a zinc sulphide screen, which seemed to be of the kind made when α-particles impinged on it. These massive, very energetic helium nuclei made characteristic scintillations. When Rutherford was brought to inspect the scintillations he recognised them at once, for, as he afterwards said, he had been present at the birth of the α-particle, and he instantly recognised that these were legitimate children. The apparatus used by Cockcroft and Walton is shown in figure 18.

Cockcroft and Walton published their discovery of the disintegration of atoms by highly accelerated protons in *Nature* on 30 April 1932. On this day modern large-scale physics was born.

The excitement of these events was prodigious. Here was the enormously energetic α-particle, discovered and named by Rutherford thirty years before, being ejected or released from the nucleus of a lithium atom by a proton of only about one-seventh of its mass, and

18 *The apparatus with which Cockcroft and Walton first disintegrated atoms by electrically accelerated particles*

with an energy excited by a field of a mere 125 000 volts. It seemed that a subtle skeleton key had been found to the armour-plated and belted atomic nucleus. This key had been fashioned with electrical engineering machinery.

The prospect of the release of atomic energy, with its million-fold increase in the scale of energy reactions, had been brought nearer. Engineering would henceforth be a major instrument in atomic research. Rutherford's mastery of the craftsman's approach had not prevented his encouragement of the engineering approach, with its considerable organisational requirements. His insight into the need for large-scale physics has not been sufficiently appreciated, because his mastery of craftsman's physics was so outstanding.

It seemed likely that when an accelerated proton struck a lithium atom, the two particles momentarily combined to form a new particle of atomic mass 8, which immediately broke up into two α-particles

$$^7\text{Li} + \text{p} \rightarrow 2\alpha$$

Cockcroft and Walton devised an experiment to see whether the α-particles were being emitted in pairs, and found evidence of this. P. I. Dee and Walton then secured magnificent cloud-chamber photographs of the twin disintegrations.

Measurements showed that the α-particles had an energy of 8 MeV, so that 16 MeV was released in each atomic reaction. The amount of energy released corresponded to the slight contraction of mass that occurred when a lithium atom combined with a proton to produce two helium atoms.

This was the first direct experimental proof of the conversion of mass into energy, and the truth of Einstein's celebrated formula for the equivalence of mass and energy.

As a great deal more energy was released than the amount of energy of the impinging proton, it was evident that the release of atomic energy was now a problem for consideration. The Cockcroft and Walton apparatus was not, however, a practical machine for releasing atomic energy, because only a few of the thousands of millions of protons accelerated caused a disintegration. The total energy required to operate the whole machine was much greater than the amount released in the few lithium atoms disintegrated.

The Cockcroft and Walton experiment suggested, however, that the release of atomic energy on a practicable scale for social purposes could not be very far away. Cockcroft himself was rather optimistic, but Rutherford adopted an extremely cautious line. He was pursued by the world's press, who tried to persuade him to make some sensational pronouncement on the release of atomic energy, but he rather crushed such approaches. He even tried to produce physical arguments indicating that atomic energy would never be released.

Whether Rutherford really believed in these arguments, or was using them as a fence to prevent his own people from making premature predictions, which he hated, is a matter of opinion. He was a particularly severe critic of scientists who claimed more than their work justified. He was also not subtle in his use of words. It was amazing that his simple ideas and simple phrases were so often right.

In 1936 he said in a lecture that while there seemed to be little hope of releasing nuclear energy for industrial purposes by accelerated particle bombardment, 'the recent discovery of the neutron and the proof of its extraordinary effectiveness in producing transmutations at very low velocities opens up new possibilities, if only a method could be found of producing slow neutrons in quantity with little expenditure of energy'. He died in the following year, and in the year after that, his old pupil Otto Hahn discovered fission. Six years later Fermi had released nuclear energy on a potentially unlimited scale by slow neutrons. One might justly regard those late words of Rutherford as a final example of his almost superhuman insight.

The discovery of the neutron, and disintegration of atoms by machinery, brought the Cavendish achievement to its apex. The discoveries themselves created the conditions that necessarily led to the Laboratory's relative decline in comparison with other laboratories. The kind of physics inaugurated by the neutron and the big accelerators needed, required a new scale of resources beyond those of Britain.

The Cavendish Laboratory reached the supreme heights of its achievement when it was an imperial institution, directed by a Dominion citizen from the other end of the world. Under Rutherford, research workers had come in numbers from all quarters of the globe. Massey and Oliphant came from Australia; Ahmad and Bhabha from India; Schonland from South Africa; Shenstone and Terroux from Canada; Kara Michailova from Bulgaria; Chao from China; Bjerge and Jacobsen from Denmark; Goldhaber, Kuhn and Riezler from Germany; Occhialini from Italy; Shimizu from Japan; Niewodniczanski, Sosnowski and Wertenstein from Poland; Bretscher from Switzerland; Bainbridge and Oppenheimer from the United States; Chariton, Gamow, Kapitza and Leipunski from the

U.S.S.R., among many others. (The research workers in 1932 are shown in figure 19.)

With the dissolution of the British Empire, the Cavendish could no longer draw upon the same wide resources of men. Neither was Britain prepared to finance research of the magnitude required for the exploration of the possibilities opened by the use of big machines, on the same scale as that which could be pursued by the new superpowers.

With Rutherford's death in 1937, and the advent of the Second World War, the situation for Britain, and for the Cavendish Laboratory, changed fundamentally.

Before considering the succeeding period, the appearance of the Cavendish Laboratory in its apotheosis to some of those who were inside it, will be examined.

19   *Research men in the Cavendish in 1932*

15

The Rutherford Era

Chadwick, who was Rutherford's chief colleague, recalled that when Rutherford moved to Cambridge one of the first things he suggested to him was that he should think of some method by which the nuclear charge of an atom could be measured by α-particle scattering, to confirm the nuclear charge given by the atomic number.

Chadwick was hard-up, but Caius College helped him along with a studentship of £120 a year. When he began work in the Cavendish, he was involved in the delicate relationship between J. J. and Rutherford, and the interviews between them were among his most intense memories.

Rutherford had to ask J. J.'s permission to make use of any part of J. J.'s rooms that he was not using. J. J. never came in until 1.30 p.m. Rutherford always insisted that Chadwick should be present when they met for discussion; he was determined that there should be no misunderstandings. Personally, Chadwick always found J. J. very kind.

The veneration that Rutherford showed J. J. in these private interviews was very impressive. He did not show the same veneration in public.

At some date between 1932 and 1935 Chadwick had a discussion with Rutherford on the conditions that had to be satisfied to get more energy out of a nuclear reaction than was required to sustain it. They concluded that at that time they could see no means by which it could be done. Chadwick felt that no one except himself realised how cautious Rutherford was.

As Rutherford's closest colleague, and in effect his deputy, Chadwick expressed opinions on many of the matters that were left to him. Rutherford often adopted these as a matter of course, in making his decisions. An example of this was Kapitza's entry to the Cavendish

Laboratory, although Kapitza himself probably did not know that it was due to Chadwick. When Joffe and Kapitza visited the Cavendish, Rutherford took Joffe away to talk with him, and left Kapitza with Chadwick. A few days later Kapitza asked for permission to enter the Cavendish. Rutherford asked Chadwick what he should do. Chadwick said that Kapitza was very bright, and would be an asset. Kapitza was accepted, on the condition, suggested by Chadwick, that he should not introduce politics.

Rutherford and Chadwick kept notebooks in which they jotted down during the year ideas for experiments. This facilitated the allocation of experiments at the beginning of the year, when they knew how many research men there would be, and what qualifications they had. It was found that two people working together was not always a good arrangement. Difficulties used to arise, for example, when both competed for the same college fellowship.

In later years Chadwick reported to Rutherford daily at 11 a.m. on what was going on in the Laboratory.

Chadwick argued with Rutherford about administration, and, especially, money; he thought that Rutherford could have done much more with more money, but, in Chadwick's experience, he had a profound distaste or inhibition in asking for it. One of Chadwick's jobs was to assign working space. He assigned a large room for Rutherford, and a small part of it for himself. Rutherford objected to Chadwick having any of it, but Chadwick stood his ground. Rutherford commented: 'You think I'm being unreasonable. Well, perhaps you're right!'

The Cavendish in Rutherford's time was particularly happy because he never exercised his eminence. He once remarked that he had done enough for one man. It was the business of the younger men to get on with finding out about the nucleus.

Up to 1925 Chadwick was almost part of the Rutherford family. Then he married, and the situation changed greatly. Lady Rutherford and Chadwick's wife had been brought up differently, and had different ideas.

Oliphant presently became particularly intimate with the Rutherford family; like them, he came from the Antipodes, and fitted into the social situation more closely. He did not have much to do with the running of the Laboratory until 1935, when Chadwick left for Liverpool.

With regard to the organisation of the Cavendish, Chadwick

wanted to set aside two or three rooms with special equipment, such as a cloud chamber, all ready and working properly, so that when new and obscure effects appeared, they could be at once sensitively examined. Rutherford did not agree to this. He preferred apparatus to be set up to deal with the experiment immediately in hand—what might almost be called a makeshift approach. In Chadwick's view, if his proposal had been adopted, they might well have discovered the Compton effect, the positron and other things. In their earlier experiments with γ-rays they discovered that there was a 'something else'. They put a layer of lead across the beam, but positrons were not found, though there must have been many.

Apparatus and finance were inadequate to provide the required refinement of experimental observation. For years there was no professional glass-blower in the Cavendish, and there was a shortage of expert technical assistance. In the period 1919–37 the average expenditure on apparatus for research and teaching was about £2000 a year for 400–500 students. (In the 1970s the annual Cavendish budget was of the order of £1 million). You could not buy apparatus because there was no money available; it was heart-breaking: Rutherford would not ask for money. The General Electric Company and the British Thomson Houston Company sometimes gave a little.

When Dr Anderson, the Master of Caius, tried to help Rutherford's Cavendish researches financially, he would not accept it. Anderson was personally very fond of him as 'a great big lovable bear'. As an example of the financing of research, the first job G. I. Taylor was paid for was personally financed by Schuster.

In 1919 J. J. supervised four students, and Rutherford fourteen. One or two were allocated to Appleton, and none to Ellis. In 1971 there were so many that the exact number was difficult to remember. Students paid £5 a term, of which £3 went to the University and £2 to the supervisor.

Change was inevitable. There had been no possibility in 1932 to follow up the discovery of the neutron. There was no suitable apparatus and Rutherford would not spend the necessary money. Chadwick felt he had to leave the Cavendish; Rutherford could not really help him, nor was he willing. Chadwick had wanted him to build a high-tension laboratory, but Rutherford would not; Chadwick thought that perhaps he was right. The Jacksonian chair became vacant, but Chadwick was not appointed to it because he was a nuclear physicist; this would have brought him into conflict with Rutherford.

Rutherford was extraordinarily kind and generous about this incident. He wrote a letter which almost brought tears to Chadwick's eyes. Chadwick destroyed it since he did not want anyone to see it.

Chadwick told Rutherford that the Cavendish must change. When he had gone to Liverpool and started to build a cyclotron, he visited Rutherford at Cambridge. After breakfast on a beautiful sunny morning at Newnham Cottage, Rutherford criticised him for encouraging Cockcroft to build a cyclotron.

There was a dreadful occasion when Rutherford attacked Chadwick because Oliphant had decided to leave Cambridge and go to Birmingham. Chadwick had said that there was no future for nuclear physics in Cambridge. There was no space, finance or other facilities. Molecular physics was going to be the interesting field of research, but Rutherford would have nothing to do with it.

The occasion of the row over Oliphant was the last time that Chadwick saw Rutherford. Only a few days before he died, Rutherford wrote a letter to Chadwick, which indicated that he had got over it.

During the Second World War Chadwick and Oliphant went to Washington to explain the experiments and conclusions arrived at in Britain that provided the first concrete evidence for the practical possibility of an atomic bomb. Chadwick was attacked by oyster poisoning just before a very important meeting. Oliphant nursed him like a brother, and got him through a very difficult moment.

At one stage, they had to share a room, and Oliphant listened in the early hours of the morning to agonised groans. Oliphant felt that Chadwick's misery was psychological as well as physiological.

Chadwick became the head of the British scientific team sent to the U.S. to collaborate with the Americans on the development and production of the atomic bomb. He was one of the few scientists able to deal with the redoubtable General L. R. Groves, the military man at the head of the American organisation for producing the bomb. Chadwick showed notable diplomatic and managerial gifts in this very responsible position. No doubt his long experience, as well as his natural talents, in dealing with one formidable personality assisted him in dealing with others. Rutherford had not overawed him, nor did Groves, nor Cherwell.

Chadwick remarked that he had got to know and like Lord Cherwell, in spite of his faults and mistakes. Cherwell prevented Tizard from knowing anything about the bomb and nuclear energy; it was a wicked action. He was fond of doing lightning calculations, but many of them turned out to be wrong. Cherwell was unable to stand up to Churchill. Bohr had described to Chadwick how Churchill had shut Cherwell up by saying the subject under discussion was not science, but politics. Chadwick believed that he would not have shut up in the same situation. He had a poor opinion of the social common sense of many scientific people. He thought that Bush and Conant were all right, but most scientists just collapsed in a difficult situation of a general rather than scientific character. In Chadwick's opinion Groves in his disputes with the scientists was generally right.

In his war work Chadwick saw something of British Intelligence. He noted that, like the Cavendish, it was short of money.

He ascribed the virtue and special quality of the Cavendish to the tradition created by the first three professors, Maxwell, Rayleigh and J. J. Thomson. Maxwell in particular he considered to be a very great man. Cambridge had always attracted good people, carefully selected, and these then came under their influence.

One of the merits of Cambridge University was that it was flexible. For example, when research students, of whom Rutherford was the first, were admitted in 1895, the changes in the University which sanctioned the innovation, were not ratified by the Privy Council until 1896. Strictly, Rutherford's admission to the Laboratory at the time was illegal, but, Cambridge being what it was, nobody bothered.

Rutherford and Chadwick influenced the policy of admitting people from the British Dominions, attracting people like Massey. Chadwick thought this a very valuable development.

The closeness and intimacy in which people worked in the Cavendish was an outstanding feature, in contrast with many other laboratories. For example, when Chadwick visited Siegbahn's laboratory in Sweden, he was astonished by the quiet; one had, as it were, to walk a mile before meeting anyone doing anything. After the Second World War he often used to visit F. P. Bowden's laboratory in Cambridge, which was incorporated into the Cavendish in 1957. It was very crowded; all were close, talking and discussing together, which was very stimulating.

He recollected that the details of organisation in the Cavendish were very odd. In 1919 no proper professional accounts were kept; Thirkill did them as a voluntary effort.

The material limitations of Cambridge were remarkable. For example, the University Library was formerly so small that there was no place for keeping Rayleigh's papers. (Now they are in America.)

Chadwick started on the preparation of the publication of Rutherford's collected papers. All of Rutherford's popular articles as well as his scientific papers were collected. When Chadwick left his large house in North Wales to move into a flat in Cambridge, he burned the huge collection of popular papers. He regretted this when he came to prepare notes for the Rutherford Centenary in 1971. He was particularly sorry not to be able to find Rutherford's notes for his idea of a Committee for Prevision, in which the trends of science and its application in industry and business were to be forecast. Rutherford had collected a lot of material on which his views were based.

Chadwick expressed the opinion that the electors to the Cavendish chair had made the correct selection, when they chose W. L. Bragg as Rutherford's successor. However, it must have been a severe disappointment to Chadwick. After Bragg left the Cavendish, Chadwick himself became one of the electors.

With regard to the judgement of scientists, opinion can change. He himself in his earlier days had not esteemed Sir Oliver Lodge. He came to know him very well in his last years, and this completely changed his opinion. Among other things, Lodge was one of twenty-four children, and had had eight himself; he knew a lot about people. Chadwick said that Rutherford had, in a greater degree than any other man he had known, 'the ability to fix the whole power of his mind on a matter for a long time without getting tired or losing interest'. One might add to Chadwick's observation that Isaac Newton and Thomas Young, when asked to what quality they ascribed their achievements, both mentioned precisely this. For Rutherford, 'familiarity with new facts and problems did not diminish his wonder at them'.

To Ellis it appeared that for Rutherford 'the laboratory was his greatest experiment, and like the apparatus in an experiment it grew organically with but little planning'.

They worked on the disintegration of light elements by α-particles, detecting the products of disintegration by the scintillation method. Rutherford usually had two or three students to help him. The experiments started about four in the afternoon, and all went into the dark laboratory for a preliminary half-hour to sensitise their eyes. They sat drinking tea, while Rutherford talked of all things under the sun, in a curiously intimate yet impersonal way, coloured by his characteristic of considering statements independently of the person who made them. Presently Crowe brought in the radioactive source, and the researchers closed their eyes while the lights were turned up, so that the source could be fixed.

It was very striking how Rutherford kept entire control of the experiment. In the comparative darkness, working quickly because of the decaying source, Rutherford would sum up instantly the implications of the results, and give further directions. Ellis believed that he could do this because he had thought deeply about the experiments before they were performed, and anticipated the results. He had calculated the relevant quantities in advance, and he instantly detected whether anything unexpected was happening. 'He was the complete antithesis of the man who observes first and then goes home to work up the measurements to find out what has happened.' If anything went wrong, such as contamination, he expressed his dissatisfaction in no uncertain way; there was no question of philosophic calm; he was far too human and personal for that.

He had a genius for working in the right direction. Any competent experimenter could have done most of his work, but anyone except Rutherford would have strayed from the path.

Ellis thought that perhaps the most characteristic example of his personal genius in this period was in 1927, when he proposed a theory of the origin of α-particles. This theory was superseded very soon afterwards by the wave-mechanical theory, but in his paper he marshalled the experimental facts, which showed how completely he had appreciated the difficulties in explaining them in terms of classical physics. It exhibited his characteristic ability to seize on a definite experimental fact, and realise that, independently of all other arguments, it showed that the accepted classical picture was fundamentally wrong. In this case others found the answer, but in the earlier case of the scattering of α-particles, he had also found the answer: the nuclear theory of the atom.

In both cases his characteristic genius was at work, in the recognition of the fundamental contradiction between the experimental facts and classical theory.

His grasp of the implications of the experiments going on in the laboratory was extraordinary. But this was by no means always immediately apparent. In his discussions with the researchers as he came round, he often spoke in a way so confused that there was a temptation to correct him. After due experience, this was not lightly undertaken. Afterwards, when one's own conclusions had become clearer, one would go to Rutherford and enthusiastically expound them to him, only to find that he had seen it all far more precisely, and the whole idea had really arisen through his inspiration. He was unbiased by current theoretical ideas. He did not think logically, but had an artistic feeling of how nature works.

It was always so clear to him what was the next thing to do that he neither would give reasons for it nor felt the need to do so. He did not seem to be greatly interested whether current theory was for or against his ideas. He was a natural leader not because he wanted to dominate, but because he was sure that what he wanted to do was right, and intended to go on doing it.

He bore his share of administration, but he did not like committee work. 'He never really appreciated the essential democracy of Cambridge, although he of all men was, in some senses, the most democratic.' He was just as likely to tell a distinguished person what some unimportant person had said, as vice versa. Such distinctions did not exist for him. In moments of relaxation he would discuss anything except physics. He was at his best in talking about his early days; and, with his interest in people, he could recount events and conversations in a vivid manner which would hold the interest of any group of people.

The style of work pursued by Chadwick and Ellis was continued particularly by Feather. He worked, as he put it, within the sound of Rutherford's voice from 1926 until 1937, except for two years, one at Baltimore and the other at Liverpool. He grew up scientifically under Rutherford's wing, and was at first too innocent to be afraid when his voice boomed. Later, when he became a fellow of Trinity and met Rutherford at high table, he realised that it boomed more often in simple boyish enjoyment than in anger.

In 1936, less than three months after Feather had been appointed a college lecturer, Rutherford took him as one of a delegation of three to call on the Master, J. J. Thomson, to deliver a congratulatory address on his eightieth birthday. Less than a year later Rutherford was dead, and J. J. contributed to a composite obituary of him. J. J. himself died on 30 August 1940, the sixty-ninth anniversary of Rutherford's birthday.

After the First World War, when Rutherford succeeded J. J., only the senior teaching staff remained substantially the same. Even as late as 1926 a few students still worked in J. J.'s large room on problems he had suggested or approved. Beyond this, and content in his isolation, was Aston, who had two rooms, one containing his mass-spectrograph, and the other his measuring equipment. Few ever penetrated there, and as the years passed, the isotopic analyses of the elements, and the measurements of their mass numbers, were quietly carried out. Feather was of the opinion that as early as 1927, Aston's measurements of isotopic masses indicated that the phenomenon of atomic fission might exist, and it might easily have been discovered then. The information that there would be a big release of energy if the uranium nucleus split was evident.

Aston's isolation was not typical of the Cavendish. Throughout the rambling building, with bits and pieces added to it from time to time, there was bustling activity, and a great feeling of community of purpose, which was a very strong feature of Rutherford's laboratory. When Feather went for a year to work in Rowland's laboratory in America, he was struck by the difference. Rowland was a very good physicist, but there was a complete absence of this sense of community.

Rutherford's 'boys', irrespective of the quarter of the globe whence they came, soon found themselves members of the family. They drank the tea that he provided, but supplied their own buns, which were bought by the man who was acting, at the time, as secretary of the Bun Club. The quarters were cramped, there were no chairs or tables and smoking was taboo. Feather aptly described it in the words of the injunction: 'They shall eat it standing, with their loins girded and their staff in their hand.'

In the summer the 'family' was photographed, forming a unique series of pictorial records of the great scientists, and the young men who profited inestimably from association with them.

The Cavendish Physical Society met on alternate Wednesdays throughout the year. Tea was served in the space between the lecture bench and the blackboard. Lady Rutherford usually presided. Rutherford used to give a general talk at the first meeting each year, in October, on the work in progress in the laboratory. There was usually a large audience, not confined entirely to his own department. The young research worker sometimes felt slightly hurt if Rutherford did not refer to his particular bit of work, but this was soon forgotten under the influence of the enthusiasm, humanity and greatness of the speaker.

Later on the student would appreciate Rutherford's handling of eminent visiting lecturers; the warm welcome, and the equally warm assurance conveyed to his 'boys' in a very audible aside at the end, that what was going on in the Cavendish was at least as good as the information to which they had been listening, and that he and they knew as much about it as the distinguished lecturer.

Alternating with the Cavendish Society meetings were the Research Colloquia, which met at 5 p.m., and were of a smaller and more specialist character. They were arranged by Chadwick, who became responsible for the general organisation of research.

Graduates who were accepted for research were put through a 'nursery course', learning the art of making such items as gold-leaf electroscopes, and testing their eyes for scintillation observations. This was done in an attic above Rutherford's room, just under the roof; it could be oppressively hot or draughtily cold there.

After they had passed through the 'nursery', they started on their own research, or paired with someone who was about as inexperienced as themselves. Chadwick was watchful, cryptic and helpful. Now and again Rutherford would look in, announced only by his great voice. In their second year, if they were lucky, they might begin to get results. Rutherford would then be less impatient. If they had something interesting to report he would call for a stool, perch himself on it, and bring out one of those old pencil stubs, with which, as Feather put it, he painted rather than wrote, on the first reasonably clean scrap of paper they could find. He would begin to show that their result had some significance, and then they felt lucky indeed.

After three years a thesis would be submitted, and Rutherford might be one of the examiners. He would be friendly, and put the examinee at his ease by asking entertaining questions, such as his views on the self-inductance of a wedding ring. On one occasion, when a co-examiner asked a particularly tricky question, Rutherford exclaimed to him: 'Why, I don't know the answer to that myself; do you?', and it turned out that the co-examiner didn't, either!

16

Cloud Chambers

In January 1919 Blackett was sent to Cambridge by the Admiralty, as one of four hundred other young naval officers, to take a six months' course to absorb some general culture, and enjoy a change after serving four years at sea during the First World War.

Three weeks after arriving at Cambridge he resigned from the Navy, sacrificed a substantial gratuity, and became an undergraduate in the University, graduating in physics in 1921. He became a research student under Rutherford, who gave him a research problem ideally suited to his interests and aptitudes. This was to investigate the disruption of nitrogen nuclei by fast α-particles, by means of the Wilson cloud chamber. Rutherford had set Shimizu this problem in 1919. After taking a few thousand photographs with his improved apparatus, Shimizu returned to Japan for family reasons.

Rutherford asked Blackett to take over the experiments, which he gladly did. By 1924 Blackett had succeeded in photographing the phenomenon, and obtained the first photograph of the transmutation of an atom. Rutherford had set him an almost perfect problem, on which a good deal of ground work had already been done, the solution of which was bound to lead to important results.

It had been alleged that Rutherford had once boasted that he had never set a research student a dud problem. This reminded Blackett of Napoleon's saying that there are no bad soldiers, only bad generals. It would have been in keeping with Rutherford's personality for him to have said that there are no bad research students, only bad professors.

When Blackett began research in the Cavendish, experiments were usually done by one person, who built most of the apparatus himself. There were about thirty research students and, as Blackett put it, '$1\frac{1}{2}$ mechanics'.

The technical means in use were very simple and mostly home-made. The counting of scintillations caused by α-particles on zinc sulphide screens observed under a low-power microscope was the most common technique. Rutherford had founded nuclear physics with it. It depended on the accidental properties of zinc sulphide containing slight impurities.

C. D. Ellis and others were measuring β-ray spectra with photographic plates, and Wilson's cloud chamber was developing as a tool in nuclear research. No Geiger counters were used in the Cavendish until about 1930.

20   *The transmutation of an atom photographed for the first time by Blackett*

BLACKETT'S PHOTOGRAPHS
1 AND 2 H PARTICLE EMITTED IN FORWARD DIRECTION.   3 BACKWARD DIRECTION.

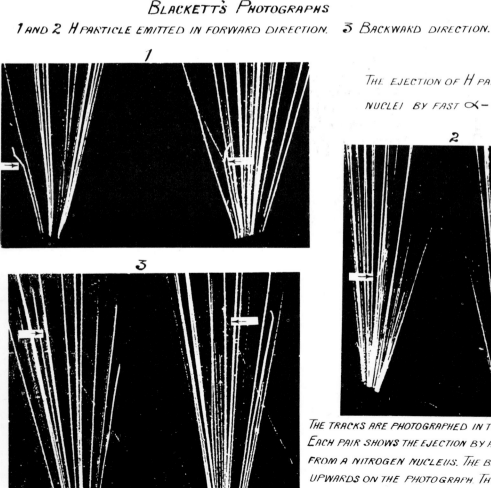

THE EJECTION OF H PARTICLES FROM NITROGEN NUCLEI BY FAST α-PARTICLES.

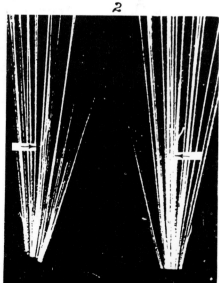

THE TRACKS ARE PHOTOGRAPHED IN TWO DIRECTIONS AT RIGHT ANGL
EACH PAIR SHOWS THE EJECTION BY AN α-PARTICLE OF A H PARTI
FROM A NITROGEN NUCLEUS. THE BUNDLE OF α-PARTICLES PAS
UPWARDS ON THE PHOTOGRAPH. THE STRAIGHT FINE ARM IS TH
TRACK OF THE H PARTICLE, AND THE SHORT SOMEWHAT BENT A
IS THE TRACK OF THE RESIDUAL NUCLEUS. NO THIRD ARM FOR T
TRACK OF THE α-PARTICLE AFTER THE COLLISION CAN BE SE

Blackett found himself in a research room, empty apart from a few bits of Shimizu's apparatus, and was simply told to get on with it. Rutherford was not much interested in the building of the apparatus, but when results began to come, he became extremely enthusiastic and helpful. After two or three years of carpentry, machine work and electric wiring, Blackett got an automatic cloud chamber working, and took some 25 000 photographs in a few weeks. They contained 400 000 tracks, including six of the anomalous events that Rutherford had asked him to look for. The photographs showed that the transmutation was an integration rather than a disintegration; the nitrogen nucleus absorbed the impinging α-particle, and was transmuted into a then unknown new isotope of oxygen (see figure 20).

For thirty years Blackett built and used cloud chambers for studying nuclear collisions, except for the year 1924–5, when he went to Göttingen to work under James Franck, then laying down the experimental foundation of the quantum theory of the excitation of spectra and the physics of stationary states. He vividly remembered Rutherford's grudging permission for him to leave the Cavendish for a year (his first sin), and to study the outside of the atom rather than the nucleus (his second sin). Blackett found his Göttingen experience extremely valuable, both through the widening of his knowledge of physics, and by becoming familiar with the continental way of life.

Besides making spectacular discoveries Blackett converted the cloud-chamber technique into a systematic tool of nuclear research. He worked out reliable methods of interpreting atom-chamber photographs. These aspects of Blackett's achievement were particularly esteemed by Joliot-Curie.

In 1932 his research underwent a change, when Occhialini arrived from Italy to work with him. He brought the technique of the coincidence counting of cosmic rays using Geiger counters, which had been developed by Rossi in Florence. Soon after Occhialini arrived, he and Blackett thought of applying the coincidence-counter technique for detecting cosmic rays passing through a cloud chamber. Occhialini had come for three weeks, but he stayed for three years. In 1933 they confirmed K. Anderson's discovery of the positron, and discovered the existence of showers of positive and negative electrons, which confirmed Dirac's relativistic theory of the electron published in 1928, and his conception of the existence of anti-matter.

By 1933 Blackett felt rather cramped in the Cavendish; the very limited room for research students made it difficult to keep them. This was an important reason why he left to take a chair in London. He went with the determination to get money for apparatus, and get on with research. He succeeded in getting money, which became much more freely available.

Men wanted to get on with their own research, and nuclear research in the old style was running into the sand. A new style was required, such as has nowadays been developed in CERN at Geneva, where a man of normal ability can join a team, and in it do useful work.

One of the qualities that Rutherford conveyed to his pupils was the importance of alertness in noticing and exploring the unexpected. At Montreal he had noticed that the opening and shutting of a door in his laboratory produced effects that led him to the discovery of the thorium emanation, of crucial importance in developing his theory of radioactivity as due to the spontaneous disintegration of atoms.

It was in his spirit to study cloud-chamber photographs and emulsion photographs of cosmic rays in the expectation that important new knowledge would emerge, though no one had any idea what they were going to find. Starting with the positron, a wealth of new particles was found. Then came various sorts of particles unpredicted by theory. By looking at photographs of cosmic-ray showers, the charged and uncharged 'strange' particles were discovered, though theory had not previously given any hint of their existence.

Speaking in his later years Blackett felt that though theory was becoming extremely sophisticated, discovery by accident was still very important. It was as vital then as in Rutherford's time to keep one's eyes open for something new, whenever doing an experiment. The scale of work had changed since his youth, though not so much as was commonly thought. There were fields where big instruments were necessary, but many where small-scale work could be very profitable.

It has often been discussed whether Rutherford would have been as successful in research today as he was in his own day. Blackett's remarks suggest that Rutherford would have discovered significant new fields of research; in any case, he would probably have revelled in the flood of new particles revealed by the giant accelerators, and might have inspired faster progress in their interpretation.

Rutherford's prejudices also had their effects. His pre-wave-theory conception of the nucleus in the 1920s was unfruitful. He and his school failed to discover artificial radioactivity, though it was easy to produce and observe. This was due to his slowness in introducing Geiger counters. Like Chadwick, Blackett believed that the positron could also have been easily produced and observed before 1932.

In relation to Blackett's remarks, Rutherford's late and cautious attitude to the quantum theory was a handicap in the Cavendish, though much less than might have been expected. He had a peculiar gift for getting round awkward theoretical preconceptions. When they got in the way of interpreting any significant experimental results, he was inclined to jettison them. He used theoretical ideas as a kind of scaffolding for making experiments, which could subsequently be abandoned without compunction.

Following the Cavendish tradition, the theoretical education of research men in experimental physics in British universities took a rather haphazard course, without many formal lectures or direction. In the United States a much higher level of theoretical instruction for experimental physicists has developed. As long as fields can be found, as in Rutherford's day, when great discoveries could be made with little deep theoretical knowledge by relying mainly on a mastery of experimental method and an eye for significant phenomena, the British tradition will continue to be fruitful, as for example in the field of the unstable elementary particles. But when deeper theoretical knowledge is required to design and interpret an experiment, the older Cavendish approach will be insufficient.

Blackett thought that the practical solution is probably to give the experimentalist an intense education in that part of theoretical physics immediately bearing on his experimental field. This requires much reading, special lectures and colloquia, much personal discussion, and attending specialist conferences often of an international character. A mixed group of senior and junior experimentalists is necessary; it should contain about half a dozen members, for fewer than that cannot keep up with the many different branches of literature, lectures and conferences, besides doing their experiments.

In the 1920s the Cavendish was such a group, studying different aspects of the nucleus. With the growth of the scale of research in the 1930s, groups were no longer sufficient, and teams became necessary.

The necessity for groups and teams is now evident. In Rutherford's Cavendish the organisation of the appropriate kind evolved largely without conscious conception; now its necessity can easily be grasped.

But, observed Blackett, the odd experimental genius may turn up who works best on his own. How easily and delightedly the director of a laboratory will then make the appropriate special arrangements! No one is administratively more easily provided for than the lone genius, and few things are more certain than the disappointment of research men of ordinary calibre, who do not receive the appropriate help of being directed to a narrow field, and supported by a compact group of workers.

As to how experimental physics was actually done in the Cavendish in the 1930s, Blackett's essay on 'The Craft of Experimental Physics', in *Cambridge University Studies* published in 1933, is classical.

The experimental physicist is a Jack-of-All-Trades, a versatile but amateur craftsman. He must blow glass and turn metal, though he could not earn his living as a glass-blower nor even be classed as a skilled mechanic; he must carpenter, photograph, wire electric circuits and be a master of gadgets of all kinds; he may find invaluable a training as an engineer and can profit always by utilising his gifts as a mathematician. In such activities will he be engaged for three-quarters of his working day. During the rest he must be a physicist, that is, he must cultivate an intimacy with the behaviour of the physical world. But in none of these activities, taken alone, need he be pre-eminent, certainly not as a craftsman, for he will seldom achieve more than an amateur's skill; and not even in his knowledge of his own special field of physics need he, or indeed perhaps can he, surpass the knowledge of some theoretician. For a theoretical physicist has no long laboratory hours to keep him from study, and he must in general be accredited with at least an equal physical intuition and certainly a greater mathematical skill. The experimental physicist must be enough of a theorist to know what experiments are worth doing and enough of a craftsman to be able to do them. He is only pre-eminent in being able to do both.

Such was Blackett's account, straight from the fire of the Cavendish research furnace. He noted, however, that a rapid change in the technique of experimental physics was taking place. Improvements in technique made possible experiments that hitherto had been impossible. To an important extent this was due to improvements in industrial technique. The relation was reciprocal. A laboratory discovery in one decade might lead to a new industry in the next, while the purely commercial products of industry may provide instruments that extend the field of the technically possible. The recent introduction of the wireless valve and the photographic plate were examples of indispensable aids to modern research, which had been carried by industrial demand to a pitch of refinement far beyond the resources of a research laboratory.

The organisation and tradition of most English laboratories required that the experimenter should rely mainly on his own resources. It was perhaps not fanciful to relate this tradition to the vogue for practical hobbies. 'So perhaps English experimental physics has derived strength from the social tradition and moral principles which led the growing middle class to spend the leisure of its prosperity in the home rather than the café.'

The good experimenter must design his apparatus and make it work. He must spot faults quickly, and judge between the various possible causes of failure. It is hard to acquire this facility without handling apparatus, but once it is acquired, the handling may often be delegated.

Optical apparatus made with cardboard and wax was often preferable both for ease of alteration and practical insight gained from its use and manipulation.

But experiments were requiring more and more complex apparatus. This taxed the single experimenter. 'To some extent the difficulty is being met by team work. The collaboration of an electrical engineer, a radioactive chemist, and an expert in valve circuits, may make possible an experiment which would be impossible for one alone.'

The rapidity with which an alteration to an apparatus can be carried out was a matter of primary and not secondary importance. If it took a long time, the experiment was frequently abandoned. One of the strong points of self-made apparatus, even if amateurish, was that it could often be altered very much quicker than that requiring professional assistance.

Experimenters rarely have the aptitude or energy to master abstract physical theory after spending an eight-hour day in the laboratory looking for elusive leaks in their apparatus. But their general knowledge of the behaviour of the objects of every-day life could be a help in predicting the behaviour of simple mechanisms. 'No one, in this mechanical and ball-playing age' could fail to have a practical knowledge of the solutions of many differential equations in classical dynamics. 'The parabolic path of a projected tennis-ball or the interchange between rotational and translationary forms of the energy of a yo-yo are matters of everyday knowledge.'

Thus the British ball-playing tradition was of help to the Cavendish experimenter. The Laboratory formed a cricket team from its staff and research workers. In the 1930s, when J. D. Cockcroft and H. Massey were its opening batsmen, they were quite formidable in their games with teams from other laboratories. They also formed a mixed hockey team (see figure 21).

As the advance of physical knowledge compelled the abandonment of the description of atomic phenomena in terms of the continuous motion of particles in space, it was no longer possible to conceive a purely mechanical model of the atom; but fortunately for the un-mathematical, the semi-mechanical model conceived by Bohr enabled nearly a generation of physicists to think 'atomically' without advanced mathematics. It became part of the laboratory equipment of the experimental physicist.

The experimental physicist goes about his work in the laboratory with varied manual and mental skills, an amateur in each alone, but unique in commanding them all. 'It is his job both to make and to think, and he can divide his time as he thinks fit between both these pleasurable occupations.'

21   *Cavendish hockey players including Cockcroft, Blackett and Massey*

Blackett's essay threw much light on the nature of the activities in the Cavendish in the 1930s.

Dee has described how he was drawn into the atomic researches at the time when Chadwick was establishing the existence of the neutron. Dee, who was working with C. T. R. Wilson at the Solar Physics Laboratory, happened to call on Feather in the Cavendish. Feather was engaged in photographing the recoil tracks of nitrogen nuclei projected by neutrons. Shortly after this Chadwick gave the preliminary account of his experiment to the Kapitza Club.

The essential question was whether the radiation from beryllium consisted of photons or neutrons, waves or particles. At the Kapitza Club meeting it seemed to Dee that the study of electrons recoiling from the beryllium radiation would very clearly distinguish between photons and neutrons because of the large differences in the recoil energies, according to the two hypotheses.

Dee and Wilson were studying the condensation of water on gaseous ions. They had perfected a technique for operating the cloud chamber under very clean conditions, which would be very suitable for looking for electrons recoiling from the impact of neutrons, since these would lie within the energy range that they were studying.

Dee therefore approached Chadwick, who immediately arranged for him to have the loan of the radioactive source for overnight experiments, while he and Feather were resting from their daytime labours. The expected short tracks of recoiling electrons were not, however, observed at once. Wilson, who had at first been a little disturbed by the interruption of their experiments on condensation, now became interested. It appeared desirable that more thorough shielding from unwanted γ-radiation was necessary, so he turned up at the Laboratory with all his gold medals, including the one for his Nobel Prize, to replace the lead absorbers they had been using. Dee suggested that the absorbing powers of the gold medals would be improved if the embossed heads were beaten flat, but Wilson demurred to that.

These experiments were the occasion of Dee's first personal contacts with Rutherford. Almost daily he had to report his initial failures to observe the desired recoils. He felt Rutherford's comments to be devastatingly scathing. Chadwick was unshaken by these failures; he took them as evidence that the neutron–electron collision cross-section was very small.

In spite of these difficulties it was abundantly clear to Dee that

Rutherford was completely convinced that the neutron he had predicted had been discovered. The way of its discovery, by the straightforward application of the laws of conservation of momentum and energy to the collision processes involved in the explanation of the relative sizes of the pulses observed by Chadwick, convinced him beyond all other considerations. He referred to the naturalness of the neutron concept in relation to isotopes, and with regard to the electron recoils concluded: 'Well, they have to be there; we have to have it'; that is, they had to have the neutron.

It gradually emerged that Dee's difficulty had arisen from the small neutron–electron cross-section. He felt that he would have been saved many painful hours if only he could have received some comfort from theorists, but none was forthcoming, despite his attempts to find it.

A few weeks later, when Dee exhibited beautiful pictures of the desired recoils at a colloquium, Rutherford interjected in a thunderous voice: 'and what is that spot in the top right-hand corner?' Dee admitted in mortification that it must be due to contamination. 'You will have to be more careful in future', Rutherford commented severely. A very long time afterwards Dee overheard Rutherford telling this story; he had not realised that he had been teasing him.

Rutherford arranged for Dee to work in the Cavendish to obtain cloud-track pictures of the disintegrations of atoms, which Cockcroft and Walton had obtained with their particle accelerator. Rutherford used to call in at the end of the day to look at any new pictures. He derived intense pleasure from examining them, and above all from hypothesising about new processes that might explain the observed tracks. With one of his pencil stubs he would rapidly calculate the expected ranges of various kinds of particles, according to momentum and energy conservation. He exercised his grasp and easy facility in small calculations of this kind with exactly the same pleasure and assurance as are shown by a master chess player demonstrating combinations on a chessboard. It was the gift that had enabled him to discover the nucleus from α-particle scattering. He had the same process of thought even in playing golf; he once hit the man in front of him with his ball. When the struck man strode back in anger to protest, Rutherford said quite naturally: 'Now, for heaven's sake, be *reasonable*.' It was natural for flying particles to go off in odd directions sometimes; after all, had it not been odd scattering that had led to the foundation of nuclear physics?

He once told Dee that, when playing golf with Aston in the regular Sunday foursome, he liked to put a ball in a bunker to hear Aston complain about having to get it out.

Rutherford always seemed to listen intently to what people said; he had an exceptional power of concentration. Then, when he spoke, it was with certainty and emphasis. Either his mind had been made up, or the time for decision had not come.

His pride in the Cavendish Laboratory and his interest in the welfare of his collaborators knew no bounds. Nothing that could help the work in the Cavendish was beneath his notice. Shortly after Dee was married, he and his wife attended one of the Rutherford Sunday tea parties. At a convenient moment he walked Dee's wife to the bottom of the garden, and explained to her that physicist husbands must be freed from all household tasks, so that they would be free to devote themselves wholly to their work. Dee thought that Rutherford would always be remembered by those who knew him for his boisterous and friendly personality, and for his happy and total absorption in his own work, and that of his collaborators.

Dee worked in the Cavendish from 1930–7. He stayed on after Blackett and others had left to take up chairs. Rutherford used to come almost every night to his laboratory, sit on a stool and talk. He had become lonely. Nuclear research had been a 'very blue-eyed kind of activity', a very cut-throat type of existence.

There was a broad interplay between professor and undergraduates. By 1972 the whole social situation had changed. In Rutherford's time the regime in the Laboratory was extraordinarily Victorian, autocratic and paternalistic. Research students were not independent. Today, Dee felt, students have their own ways. It is not as in Rutherford's time, when professors and seniors gave up an evening to look after students—a kind of intellectual slumming. Dee thought it very sad that the old regime had gone.

Students now have their spokesmen in the manner of shop stewards. Today there are able students who don't want research grants from the Science Research Council. That would have been inconceivable in the 1930s. The research student now wants to know what job the grant will lead to, before he accepts it.

Rutherford was a great marshaller of his forces. In 1932 everything was funnelled into the vital line. Within a week, he brought Dee from the Observatory and put him, with his cloud-chamber technique, onto neutron research.

Nowadays young people are confident in their right to choose; in Rutherford's day research students did not argue about their line of work. Rutherford ran the Laboratory almost like a school, of which he was headmaster.

17

Engineers and Electricians

Cockcroft was once asked in his later years what he most liked doing. He pondered for a while, and then said: 'I like making things.'

He was born in 1897 at Todmorden in the Yorkshire moors, near the border with Lancashire. His family built a textile mill in 1846, and in 1922 the mill went over from steam to electric power, utilising a water turbine. Cockcroft was keenly interested in the mill machinery, and he used to go on long walks on the moors, which developed both his physique and his quiet, reflective habits.

He decided to qualify as an electrical engineer, and entered Manchester University in 1914. He attended the lectures of the eminent mathematician Horace Lamb, and the first-year lectures in physics, more for amusement than instruction. The lecturer was unable to preserve order, and after hilarious chaos the head of the department, Rutherford, took them over. So at the age of seventeen, Cockcroft gained a first and lasting impression of Rutherford.

After the outbreak of the First World War Cockcroft served in the Army from 1915 to 1918 as a signaller. He went through some of the bitterest fighting, and on one occasion was the sole survivor at a forward post. His experiences affected him deeply; he rarely spoke of them, but he acquired a firm contempt for war.

After the war he studied electrical engineering under Miles Walker, and became a student apprentice at Metro-Vick in 1920. Miles Walker gave personal instruction to two outstanding students, of whom Cockcroft was one. He was asked to nominate a candidate for a scholarship that had been founded in memory of the victory in the war. He first approached the other man, who seemed to be the abler, but he turned it down. Walker then approached Cockcroft, who accepted it with enthusiasm. Walker advised him to go to Cambridge to take the Mathematical Tripos, and gave him a letter of introduction to Rutherford.

Cockcroft accordingly called on Rutherford, and found him sitting on a stool in the old Maxwell Wing of the Laboratory. Rutherford received him very kindly, and gave him permission to devote such time as he could spare from mathematics to work in the advanced practical class under Thirkill and Appleton. He graduated in mathematics in 1924 with distinction, and was accepted as a member of the Laboratory, taking the introductory training course under Chadwick. He collaborated with a fellow student in making the favourite tool of the Cavendish, a gold-leaf electroscope. Then they made a Macleod gauge, which blew up just as Rutherford walked in at the door; he was sympathetic, but explained that he did not keep students who had too many accidents.

During the first decade of his professorship at Cambridge Rutherford pursued research along three main lines, the transmutation of the lighter elements, the investigation of the fields of force around atomic nuclei and the study of the radiations from radioactive substances. In the later years of his first decade he encouraged the invention of new techniques in several directions.

Cockcroft's early papers, dating from 1925, dealt with such topics as applications of electronic valve oscillators and the effect of curved boundaries on electrostatic fields around conductors. In 1924 he started a research on depositions from molecular beams on low-temperature surfaces, which he published in 1928.

After his introductory training Rutherford asked Cockcroft to work with Kapitza. He helped with the installation of the short-circuit alternator, made by Metro-Vick to Kapitza's specification. In this and in subsequent work with Kapitza, Cockcroft played an important part. He helped with Kapitza's designs for hydrogen and helium low-temperature liquefiers, and with the procurement of compressors and ancillary electrical and mechanical equipment and special machines.

Cockcroft was present at a remarkable interview with Rutherford, when Kapitza persuaded him to raise money to build the Mond Laboratory for the development of this work. In due course Rutherford secured £15000 from the Royal Society to build the Royal Society Mond Laboratory, a large sum for fundamental research in those days. Cockcroft was closely associated with the design of the Mond Laboratory, and when Kapitza did not return from the U.S.S.R. in 1935, undertook its direction at Rutherford's request.

At Kapitza's invitation Cockcroft joined his club, and found it very stimulating.

During the last decade of Rutherford's professorship much attention was given to the development of technique. Wynn-Williams and his collaborators devised the electrical recording of particles. Feather and Nimmo, E. J. Williams, Terroux and others were building cloud chambers, and spreading the influence of C. T. R. Wilson in the Laboratory.

Rutherford had called in his Presidential Address to the Royal Society in 1927 for artificially accelerated particles whose energy exceeded that of the particles emitted from radioactive substances. The first step in response to this demand was made by T. E. Allibone, who was seconded by Metro-Vick to the Cavendish to work on high-voltage accelerators for charged particles. Allibone brought with him a 500 kV Tesla coil, the cracks from the sparks of which annoyed the dons of Corpus Christi College, opposite the doorway of the Cavendish. With it, he produced accelerated electrons of 300 kV energy.

E. T. S. Walton arrived from Trinity College, Dublin. At Rutherford's suggestion, he worked on the acceleration of electrons by spinning them in the electric field produced by a changing magnetic field. This was the principle subsequently successfully applied by Kerst in the betatron. Walton worked out the theory of the stability of the orbits of the revolving electrons. He attacked the experimental and engineering problem of designing and making the machine with inadequate resources, but was unable to solve it successfully.

Allibone, Cockcroft and Walton worked in the same room, the overcrowding of which was not relieved by the violent screeches from Allibone's Tesla coil. A good deal was known about transformers, for they had been developed for dealing with electrical power, but they were not easily adapted to atomic experiments. A high, steady voltage under control was required.

Cockcroft thought about the problem of accelerating charged particles, and spoke on it to the Kapitza Club. Then came news of Gamow's theory of the emission of α-particles from radioactive substances, as due to a tunnelling process through potential barriers in the nucleus, arising from the wave properties of fundamental particles. This explained why the emitted particles had much lower energies than the potential barriers they had penetrated.

In November 1928 the typescript of a second paper by Gamow

arrived in the Cavendish, in which he showed that the disintegration of light atomic nuclei by bombardment with α-particles was probably due to the same wave-mechanical property that enabled particles to penetrate potential barriers at much higher energies.

Cockcroft immediately saw the bearing of Gamow's work on the problem of disintegration by artificially accelerated particles. It was not necessary for the particles all to have much higher energies than the barrier protecting the nucleus. He calculated the probability of particles with energies in the range from 200 kV to 3 MV, and concluded that one in every thousand of the protons in a beam of 200 kV energy should penetrate the nucleus of a boron atom. He wrote a memorandum suggesting that an accelerator capable of producing protons of 300 kV energy should be built, and submitted it to Rutherford.

In December 1928 Cockcroft began to experiment with components for a high-voltage vacuum accelerator. He persuaded Rutherford to secure £1000 from the University to buy a 300 kV transformer, and the components for constructing rectifying and accelerating tubes.

In January 1929 Gamow spoke on his theory to the Kapitza Club, and members could remember standing around, while Cockcroft calculated the probability with which a proton beam would disintegrate lithium.

At this stage Cockcroft was joined by Walton, whose attempts to make an electron accelerator on the betatron principle had not succeeded. They attempted to utilise Allibone's technique in applying high voltages to vacuum tubes. Voltages up to 280 kV were applied, but without result.

They then moved to a much higher and more convenient room, an old lecture theatre from which the internal fittings had been removed. They redesigned the apparatus, making use of the Greinacher circuit for doubling voltages. By an ingenious doubling of switches consisting of rectifiers, alternating current was used to charge transformers during one half-cycle. Then, during the other half-cycle, the charge was transferred, so that a steady direct high-voltage current was obtained. One of the technical details on which the success depended was the introduction of non-volatile plastics and oils by C. R. Burch, which enabled a high and manageable vacuum to be produced in the large volume of the tube in which the electric discharge occurred.

Protons were fed into the tube, and accelerated by the high-voltage electric field. Targets of various materials were placed at the bottom of the tube, and a zinc sulphide screen was arranged, so that any particles ejected from the target under the bombardment of the fast protons could be viewed through a microscope. 'It was all in the prevailing Cavendish tradition, but in place of string and sealing wax, it used glass tubes from Bowser petrol pumps, and Plasticine.'

After many difficulties had been overcome the apparatus worked, and steady beams of protons of up to 600 kV energy were obtained. 'At last we obtained a high energy proton beam and brought it out into the air through a window to check on its energy and range in air. After wasting a certain amount of time in this way, until prodded by Chadwick and Rutherford, we directed it onto a lithium target and at once observed, with a zinc sulphide screen, the bright scintillations which were obviously due to particle emission from the lithium.'

Rutherford was called in to see them, and described them as the most beautiful sight in the world, but he would not allow the discovery to be published until it had been conclusively proved.

The proof was obtained by rough coincidence observations carried out by two observers, and the discovery was announced by Cockcroft and Walton in April 1932. They remarked that

> It seems not unlikely that the lithium isotope of mass 7 occasionally captures a proton and the resulting nucleus of mass 8 breaks into two α-particles, each of mass four and each with an energy of about eight million electron volts. The evolution of energy on this view is about sixteen million electron volts per disintegration, agreeing approximately with that to be expected from the decrease of atomic mass involved in such a disintegration.

If a single 600 kV proton could release 16 MeV of energy, it was obvious that the release of atomic energy was probably not very far away, even if, as in this experiment, so few protons produced disintegrations that the atomic energy released was very small compared with the total energy required to accelerate the whole beam of protons, including the vast number that did not produce disintegrations. Cockcroft was reasonably optimistic that their experiment heralded the future release of atomic energy.

Rutherford persuaded Oliphant to join him in the accurate measurement of the energies of the particles released in disintegrations of light elements by the accelerated protons. Cockcroft helped them, with his characteristic kindness and competence, in the technical problems of setting up the large-scale electrical apparatus.

When E. O. Lawrence began to obtain disintegrations with his cyclotron in California, he believed he had evidence that when deuterium atoms were projected onto a target, their nuclei were disrupted into a neutron and a proton. Cockcroft and Walton suspected that the conclusion was incorrect, and were able to prove that it was due to contamination. Cockcroft and Lawrence had a long correspondence on the problem, which led to a permanent mutual regard and respect.

After a tour of the American laboratories Cockcroft tried to persuade Rutherford to obtain a cyclotron, but did not at first succeed. In 1936, after the Austin gift of £250 000, Rutherford agreed. Cockcroft visited Harvard in 1937 to deliver some lectures, and after this went on to California, where he secured drawings of Lawrence's 36-inch cyclotron.

He supervised the building of the cyclotron in the Cavendish, and kept in touch with Chadwick, who was building one in Liverpool. Cockcroft, together with Oliphant and Dee, succeeded in securing the purchase of a Philips high-voltage accelerator on the Cockcroft and Walton system.

Cockcroft said that Rutherford took no particular pleasure in the extension of the Cavendish, because it was building for his successor. But when the University asked Rutherford to raise money for the Laboratory, he did it with characteristic ability. Though he often made scathing remarks about Eddington, he asked him to write the brilliant pamphlet he composed on the Laboratory. He requested Cockcroft to prepare an official paper on what should be done, and he asked the Chancellor of the University to look for a large sum of money. After receiving the Austin benefaction he sent his 'boys' round the Continent to collect ideas on how it should be spent, and gave them a pretty free hand in designing the big new extension. He enjoyed spending some of the money on a cyclotron, jingling the money, and saying, 'There won't be very much left for my successor.'

After 1935 Cockcroft ceased making experiments in nuclear physics, though he remained influential in questions of equipment and policy. Though he participated in one of the greatest discoveries in the history of physical science, he published only sixteen papers.

Cockcroft was essentially more of an engineer and mathematician than a physical thinker. His greatness was in the organisation of large-scale physical research, of which he was effectively the founder.

In 1935, after Kapitza's settlement in Moscow, Cockcroft supervised the duplication of the equipment, and the transference of the original to Moscow. In the Mond Laboratory he directed research on liquid helium, with J. F. Allen. Within a short time important low-temperature work issued from the Mond.

Cockcroft's ability to drop personal research in nuclear physics, in which he had achieved world fame, and move into another branch of physics with immediate success, was remarkable and characteristic. When the Austin gift arrived Rutherford passed the design and construction of the new Austin Wing over to him.

His capacity for organisation and construction was exceptional. In 1933, while he was carrying out these projects in physics, his college, St John's, had appointed him their Junior Bursar. He held this post for six years, and in the period modernised and extended a large part of the College buildings.

As a teacher he was clear, but generally too concise, except for the ablest students. He was more gifted in telling people what to do.

He was elected to the Jacksonian chair in 1939, which he held until 1946, when he was appointed Director of the new Atomic Energy Establishment at Harwell. During his whole tenure of the chair, he was away from Cambridge, but the effects of his activities during the Second World War on the Cavendish were great.

With the threat of war Cockcroft had been one of those approached confidentially by Sir Henry Tizard for collaboration in secret research on radar. He had a leading part in recruiting about eighty good physicists from Cambridge and elsewhere. His efforts were extremely successful. He was a member of the Tizard Commission which went to the United States, to establish confidence and co-operation in the military scientific field. In December 1940 he was appointed Chief Superintendent of the Air Defence Research and Development Establishment at Christchurch in Hampshire.

Beneath his quiet, calm exterior he was prepared to support rebellious and even provocative actions, when he thought them necessary to get things done.

In 1946 Cockcroft said that

Although Rutherford died in 1937, his influence was a major factor in the scientific supremacy of Britain in the war. The Senior Staff and research of the laboratory, together with members of other physics schools, were mobilized in the first days of the war and developed for Britain and the allied cause the centimetre radar which turned the tide of the U-boat battle, directed the bombing of Germany and helped decisively to sink the Japanese fleet. The Liverpool branch of the Rutherford School, with Frisch and Peierls from Birmingham, initiated the work on the atomic bomb which ended the war with Japan. One can only wish that Rutherford had been alive to deal with its consequences.

231

The spirit in which Rutherford might have approached this task may, perhaps, be deduced from his posthumous address read to the joint session of the British Association for the Advancement of Science and the Indian Science Congress, in Calcutta in January 1938. He said:

This is in a sense a scientific age where there is an ever-increasing recognition throughout the world of the importance of science to national development. A number of great nations are now expending large sums in financing scientific and industrial research with a view to using their natural resources to the best advantage. Much attention is also paid to the improvement of industrial processes and also to conducting research in pure science which it is hoped may ultimately lead to the rise of new industries.

With regard to the Need of National Planning in India, he said:

While I cannot lay claim to have any first-hand knowledge of Indian industries and conditions yet I may be allowed to make some general observations on the importance in the national interest of a planned scheme of research in applied science. If India is determined to do all she can to raise the standard of life and health of her peoples, and to hold her own in the markets of the world, more and more use must be made of the help that science can give. Science can help her to make the best use of her material resources of all kinds, and to ensure that her industries are run on the most efficient lines. National research requires national planning. If research is to be directed in the most useful direction, it is just as important for a nation as for a private firm to decide what it wishes to make and sell. It is clear that any system of organised research must have regard to the economic structure of the country.

These last words of Rutherford written in 1937 preceded and virtually forecast the attitude to science usually attributed to its use on a national scale in the Second World War.

In this final statement Rutherford spoke with the words and vision of a world statesman of science. Perhaps, if he had survived, he might have had a very great influence on the world situation after the Second World War and the invention of the atomic bomb.

Cockcroft's diplomatic qualities were of immense value in the civil development of atomic energy. When this was started the U.S. Government put restrictions on the transmission of secret atomic information, which created quite unnecessary and provoking difficulties. Cockcroft handled the situation with boldness and discretion.

His part in creating the scientific side of the British state atomic-energy development belongs more to the state than to the Cavendish.

W. B. Lewis, who became head of the scientific side of the Canadian atomic-energy establishment at Chalk River, Ontario, entered the

Cavendish Laboratory in 1930, when the development and application of electronic instruments in nuclear research had become pressing. From his childhood he had wanted to be an engineer. He was sent to Haileybury School, where his physics master, who had worked with J. J., stimulated his interest in physics. After this, Lewis went to Cambridge.

In his experience he did not find Rutherford shy of engineering or of money matters. He accepted gifts for the Laboratory from the Metropolitan-Vickers electrical firm without aversion.

Rutherford's competitive spirit had been roused by Rosenblum's experiments in Paris; he did not want to be outdone. He demanded a smaller electromagnet than the Cotton giant, which would nevertheless produce more intense lines. 'Can't we use our brains, and get the same result from a smaller machine?'

Cockcroft, and McKerrow of Metro-Vick, accordingly designed the 'bun-magnet'. Lewis was uncertain who did most of the designing. The magnet cost £250; 'A lot of money', commented Rutherford, 'as much as a grant for a research student, but we would get much more from it'.

One of J. J.'s services, which has perhaps not been sufficiently appreciated, was his defence of pure research at the end of the First World War. This helped to create the situation in which so much was done in the period between the wars.

Lewis remarked that the close working relations and mutual discussion in the Cavendish developed a healthy capacity to take criticism. It was understood that work was being criticised, not people. It might have been thought that American physicists, brought up in the practical tradition of their country, would have been far more thick-skinned than the products of Cambridge, Trinity College and the Cavendish, but this was not so. American scientists were apt to regard criticism of their work as personal. Perhaps this was connected with the greater importance of work as a factor in social status in America.

In the Cavendish every one knew what they wanted to do. In other laboratories people were more the product of a certain uniformity of teaching. American and German physicists tended to regard Cavendish men as only half physicists: they did not sufficiently relate their experimental to theoretical, or theoretical to experimental work.

In his childhood Lewis had been given a Meccano set, and he had made up his mind then, never to make a model that was not of his own design. Later on he found that this independent attitude fitted very neatly into Cavendish requirements, when the invention of new types of electronic instruments became pressing. The independent attitude was also specially attractive to students from the Commonwealth. They tended to be self-standing people, who found the regime congenial.

Possessiveness of ideas is not compatible with good research work. So many people are possessive, and wrapped up in commercial security. Proprietary commercial information can be very frustrating, when one wants to see ideas tested for the sake of the advance of knowledge.

When scientists make much use of others to carry out their ideas, they tend to believe that *they* have thought of everything. J. J. made so much use of other people that he tended to think he had done everything himself.

The strength of the Cavendish tradition for free publication was illustrated in the early 1930s, when Cockcroft and Walton were shocked by Rutherford, who swore them to secrecy for a few days on their discovery of the artificially induced disintegration of atoms, because he wished to ensure that the publication should be made correctly, with no exaggerated claims.

The discoveries in the early 1930s were closely associated with advances in technique, especially in vacuum pumps and electronic circuits. Lewis thought it odd at the time that in 1932 the scintillation method was still being used. But scintillations were not so subject to interference as electronic devices in an electrically unprotected laboratory.

Cockcroft owed his illumination as a physicist to Gamow. Gamow went to Bohr; Bohr sent him to Rutherford; Rutherford put him on to Cockcroft. The combination of Cockcroft and Gamow turned disintegration from a dream into an engineering project.

In Germany and the United States men tried to be 'complete' physicists. In the Cavendish people with the right specialisms came together and collaborated.

Chadwick was very much an experimental physicist. He read the literature very widely, and was Rutherford's 'intelligence man'. Lewis was very much surprised when Chadwick decided to accept the chair at Liverpool; he was converted to the merits of the cyclotron before Rutherford.

Lewis shared Rutherford's philosophy of getting more accuracy into measurement.

The rapid extension of science since the 1930s has produced a plateau of discovery which disguises the high peaks that formerly used to be so prominent.

When the threat of war arose Lewis was admitted, in March 1939, into radar secrets. He was working on the duplex telephone, which fitted in with radar. Blackett and Cockcroft were already involved in confidential war work. In June 1939 the Air Ministry asked Cambridge to release Ratcliffe and Lewis for radar research and teaching. One of Lewis's young men was Martin Ryle, at that time very intense and temperamental, with the strong impulses that are a factor in genius.

Ratcliffe and Lewis took over Searle's job of teaching practical physics. Lewis had gained Searle's confidence. He had discovered a mistake in Searle's textbook, which went back to a detail from Seebeck in 1840. Rayleigh had copied this, and Searle had copied Rayleigh. Searle was very much impressed by this detection.

Ratcliffe excelled in clarity of exposition. Blackett was a very good teacher. In Lewis's opinion Appleton was not technically a radio man; he lectured on the subject, but without depth. He gave the impression that he did not welcome questions. After the war he preferred to talk on radioastronomy.

Kapitza and his club were a very important influence. Kapitza believed in fun in research, and in enjoying life. In Cambridge he was a unique figure, though in the U.S.S.R. there have been other scientists with a comparable outlook, such as Flerov. He was good at what Rutherford called 'fool-experiments', which turned out to be important. He entertained unusual ideas on the nature of ball-lightning, which horrified Schonland. His work in low-temperature physics, and in the background of research in magnetism was very important.

The Kapitza Club, with its free-and-easy criticism, contributed greatly to the valuable 'thick-skinned' Cavendish style.

18

Russian and Australian Views

Kapitza regarded Blackett, Chadwick, Cockcroft, Ellis and Oliphant as Rutherford's chief collaborators in his Cavendish period. He said that the thirteen years he spent in the Cavendish were the happiest in his life, and he found Rutherford a 'very kindly and sympathetic man'.

When Cambridge was celebrating the centenary of Clerk Maxwell's birth, Rutherford asked Kapitza what he thought of the reminiscences of some of Maxwell's pupils during his period as Cavendish professor. Kapitza replied that they were 'sugary'. Rutherford laughed, and told him that after his own death it would be up to him to tell future generations what he was really like. When Kapitza found himself in 1966 having to do this very thing, he remarked how well he now understood the feelings of Maxwell's pupils.

By 1966 about one-fifth of all the enormous output of papers on physical research in the world related to the investigation of nuclear phenomena.

Conceiving science as understanding the laws of nature, as the basis on which we may use nature for our cultural development, the significance of Rutherford's work had been prodigious.

The basic characteristics of his thinking were great independence and hence great daring. He was not interested in technical details. When shown blueprints, he said he merely wished that the principle on which the machine worked should be stated to him simply.

Kapitza never heard Rutherford argue about science. If anybody contradicted him, he listened with interest, but did not answer. Rutherford's thinking was inductive, while J.J.'s was deductive. Though not particularly manipulative, he could deal quite skilfully with such things as fine-walled glass tubes filled with radon.

Rutherford was very strong on appreciating students, and said it

236

was important not to be jealous of one's pupils. He was generous, which was one of the reasons why he had so many good pupils.

For all his outspokenness he had a deep fundamental sense of modesty. When Millikan visited the Cavendish, Rutherford introduced him to Kapitza, and said he should tell him about his experiments. He added, however, that he doubted whether he would be able to, for Millikan would immediately begin to tell him about his own experiments. Rutherford then laughed loudly, with Millikan joining in less enthusiastically. Kapitza soon found that Rutherford's prophecy was correct.

With regard to politics and Fascism in Europe, Rutherford had taken too optimistic a view. Rutherford thought that when the trouble arose it would be over soon.

He was struck by Rutherford's interest in Chekhov's play *Uncle Vanya*. He remembered how clearly, vividly and simply he related the plot, sympathising completely with Uncle Vanya.

In his discussions on atomic energy with Kapitza, Rutherford never expressed any interest in its release. He was not interested in technical problems, and those connected with business. 'You cannot serve God and Mammon at the same time' was one of his favourite phrases.

After 1934, when Kapitza did not return to the Cavendish, Rutherford wrote to him at least once every two months. He gave the news on science, Cambridge, jokes, good advice and cheering remarks.

When Kapitza entered the Cavendish in 1921 Rutherford had warned him not to engage in Communist propaganda in his laboratory, but it now appears that he and his pupils were laying the foundations for a scientific–technical revolution.

For Kapitza, the death of Rutherford marked the end of an epoch in science, and the beginning of a new period in the history of human culture, characterised by the scientific–technical revolution. It was to be hoped that humanity would in the end use it beneficially.

Oliphant, like Rutherford, came from Australasia. He became a close intimate of the Rutherford family, besides probably the closest of Rutherford's later scientific collaborators. Kapitza also had particularly close personal relations with Rutherford, but he saw Rutherford's Cambridge period in the perspective of his very different Russian cultural background.

Oliphant was born near Adelaide in 1901. After studying physics he became an assistant to Professor Kerr Grant at Adelaide University. His background and education were particularly valuable in understanding Rutherford.

237

He first saw the Rutherfords in 1925, when they visited Adelaide, and Rutherford spoke in the physics department at the University on the researches in progress in the Cavendish Laboratory. As a young assistant he was not introduced to him, but he was fascinated by his talk, and decided that he would try to go to the Cavendish Laboratory. In 1927 he successfully competed for an Exhibition of 1851 scholarship. Kerr Grant had recently visited the Cavendish; he told him that the atmosphere was more lively than in any other physical laboratory in England, and urged him to go there. Oliphant had already telegraphed to Rutherford and had been accepted by him, and Trinity College had awarded him a research studentship.

Oliphant has described his Cambridge days in *Rutherford: Recollections of the Cambridge Days*, which is one of the best sources on the Cavendish of that time, and is closely followed here. He arrived at the Cavendish in October 1927, a raw student admitted to what he then regarded as by far the greatest physical laboratory in the world.

He was told to wait outside Rutherford's office by W. H. Hayles, the lecture assistant who also acted as Rutherford's secretary. Hayles, who took down Rutherford's letters in long-hand, appeared to him a formidable little man.

Oliphant stood in the uncarpeted passage, and found himself in the presence of two other new research students waiting their turn; one was C. Eddy and the other E. T. S. Walton. All agreed on the unprepossessing appearance of their surroundings: dingy varnished pine doors, and stained plaster walls under dirty, dull skylights. They speculated mutually on what the Professor might be like. When Oliphant was called in, he entered a small room littered with papers, and saw a cluttered desk which, he had been taught at school, was the sign of an inefficient mind. It was raining, and water drained down the grime-covered panes of the uncurtained window.

He found himself genially greeted by a large florid man with thinning hair, who reminded him of the keeper of the general store and post office in a village in the hills near Adelaide. Rutherford at once made him welcome and at ease. He sputtered as he talked, holding a match to his pipe, which produced smoke and ash like a volcano. He was most interested to learn that he was married, and immediately invited both of them to tea on the following Sunday.

Oliphant told him of his project for research on the bombardment of metals with positive ions. Rutherford discussed it, and agreed that he should proceed, saying he would read his paper, and advise whether, and where, it should be published. He then told him to go round the Laboratory and make himself known to 'the boys, particularly Aston and J. J.', whom he would find working in the 'garage', or nearby, and who ought to be interested in what he wanted to do. Oliphant paused as he was about to leave the room, wondering where that 'garage' might be. Rutherford noticed his uncertainty, and boomed cheerfully that he should not be diffident, and should tell everyone he had sent him.

As he walked out he nearly collided with two tall young men, coming out of a door opposite. One of them, a handsome and impressive man, said to him with a charming smile: 'I'm Blackett. This is Dymond. Who are you?'

Oliphant told them of his instructions from Rutherford, and his ignorance of the place. They took him to the 'garage', the large basement laboratory where J. J. worked, and said that he would find Aston in a room beyond.

Oliphant regarded Aston and J. J. with as much awe as Rutherford, and had never imagined that he would meet them in the flesh. He was relieved of much embarrassment on finding that they were out, and he had an interesting chat with their assistants, Everett and Morley, on the pleasures of motor-cycling. He observed the mass of glass tubing built up in J. J.'s laboratory, and secretly observed to himself that he could do better than that. With reinforced self-confidence he now went to see his tutor at Trinity College.

Oliphant found that Rutherford was very fond of the College, enjoying especially dining with the company at High Table on Sunday evenings. Oliphant did not find it easy to fit in with the social atmosphere of Trinity.

Oliphant received a written invitation from Lady Rutherford for his first Sunday tea. He felt it was like an invitation to Buckingham Palace, and arrived in his best suit precisely at the official time of 4.30 p.m. He was met at the door by a maid in cap and apron, and was conducted into the handsome living room in Newnham Cottage, where he was greeted by the hosts, and placed on one of the chairs in a circle of seats. He found Lady Rutherford short and plump, with a rather abrupt manner.

She addressed everyone as 'Mister', whether he was a Nobel laureate or a research student. Kapitza, who had recently married Anna Kryloff, was present, and Rutherford recounted with gusto and affection how they had met in Paris, and been married there.

There was no distinction of status in the seating, and the Rutherfords drew everyone into the conversation. After about an hour, if it was fine, Lady Rutherford asked everybody whether they would like to see the garden. They were then conducted round, finishing at the exit gate, where they were firmly bidden goodbye.

Oliphant noted that Rutherford's hand-clasp was remarkably limp, in contrast with his boisterous salutation of 'Hello, my boy!' He thought he was shy of physical contact with other people.

As a young research student Oliphant attended various courses of lectures at the Cavendish. He found Aston's lectures on isotopes and mass spectroscopy dull, because he read them from his book. Eddington's on relativity were attractive, because they were not too mathematical, and interspersed with speculations, but he felt he did not learn much about relativity from them. Hartree's lectures on quantum theory were pedantic, but helpful and thorough; Mott's subsequent lectures on this subject were excellent. A crowded audience attended the first lecture in C. T. R. Wilson's course on atmospheric electricity, but by the third lecture the audience had dwindled to four. Oliphant did not attend any more, because Wilson appeared to be embarrassed by his inability to express himself. In contrast, J. A. Ratcliffe's lectures on the ionosphere were the most lucid and best-presented he ever heard; Ratcliffe was then about twenty-six, and working with Appleton.

Rutherford gave a course each year on general atomic physics. As he became increasingly busy with engagements in London, he had occasionally to depute these lectures. Presently, this task fell to Oliphant. When this happened Rutherford called him to his room, and went through the lecture he was to give. He had an extraordinary collection of notes, written on scraps of paper, all pinned together in the correct order. He had used these notes for years, altering and adding to them as knowledge and experience advanced. The emendations were written in pencil, often almost illegible. They went through these together for an hour or more, making more alterations, and muddling their order; Oliphant found it necessary to write out a set of notes of his own, and as he received only a day's notice, this was quite a heavy task.

The experience of going through Rutherford's notes with him was extremely illuminating. It became clear what he thought was important, and what could be omitted; this was often quite different from most textbook opinion. When he was questioned on any point, or asked to clarify it, his insight into atomic physics became apparent. Oliphant said that it was at these moments that he grew to appreciate 'his profound sense of history. He understood how knowledge grew and how spasmodically new ideas were born.'

Rutherford's historical sense has not often been commented on, and this observation by Oliphant offers a clue to yet one more of the factors, the combination of which constituted Rutherford's extra-ordinary power. At first sight, Rutherford appears as the essence of the non-historical, being able to see with a piercing glance the meaning of phenomena immediately before him, free from the burden of misconceptions acquired from his predecessors. In fact it was not like that. Rather, he understood other people's mode of thought, went through it consciously or unconsciously, and sensed where it was wrong; but in doing this, he came to understand others' modes of thought, and was able to sympathise with them. The consciousness of his research associates that their mental struggles towards dis-covery were sympathetically understood inspired their creative effort. This aspect of Rutherford's power of research inspiration was related to his interest in all human affairs and conditions of men.

His natural contact with people was illustrated by Oliphant's account of how, after getting entangled in calculations about the radius of the electron, he turned to the class, and told them they could complete it for themselves; in any case it had 'very little meaning'. On another occasion, getting into a contradiction about the apparent increase of mass of an electron with increase in its kinetic energy, he suddenly said to the class: 'You sit there like a lot of numbskulls, and not one of you can tell me where I've gone wrong.'

Technically, he was not a good lecturer, except when speaking on a recent discovery of his own. Ordinarily, he hum'd and ha'd too much.

Rutherford had an extraordinarily vigorous and active imaginative life. He often spoke as if he had difficulty in finding words to describe what he could see in his mind. However, clarity is not always a sign of rich thinking.

For the first five years of his research in the Cavendish Oliphant worked on the properties of positive ions, and the separation of iso-topes by electromagnetic methods. At various times P. B. Moon and R. M. Chaudhri collaborated with him. After Cockcroft and Walton's disintegration of atoms by accelerated particles, while he was staying with him in his cottage in North Wales, Rutherford told Oliphant that he wanted to exploit the Cockcroft and Walton technique as much as possible in the Cavendish, and invited him to collaborate in setting up a second accelerator system. Oliphant would be able to make use of his experience in positive-ion experiments. Their aim was to increase the accuracy of the physical measurements of the effects produced by accelerated particles.

A simplified form of Cockcroft–Walton apparatus was designed, operating on about a quarter of the voltage and provided with an improved form of tube, which gave a stronger stream of protons. It was set up in a low-ceilinged room (see figure 22) adjacent to the one where Rutherford and Chadwick had done most of their famous work on the disintegration of light atoms by α-particles. Because of the low ceiling the accelerating tube was placed horizontally. The end where the beam of particles emerged was separated by a brick wall from the part where the high voltage was applied, thus reducing the space penetrated by the by-product of unwanted X-rays. This left more room for equipment and instruments for observing and measuring the particles produced in the atomic transmutations.

The low ceiling, the compact arrangement and the working part of the room shut off by the brick wall produced a cosy effect. There were a number of tall four-legged laboratory stools, on which callers were cheerfully commanded to sit.

Collaborators at various times in this little room, all of whom were excellent physicists but very varied personalities, were A. E. Kempton, skilled in composing light verse; Miss Reinet Maasdorp from Rhodesia, and subsequently an able General Secretary of the Association of Scientific Workers; B. B. Kinsey, whom Oliphant described as a character straight from Wodehouse; and P. Harteck, a tall, fair, very polite Austrian who had been trained in Berlin.

With their magnetic-analysis technique they were able to identify and measure precisely the energy of the particles. They had the aid of Rutherford's personal assistant George Crowe in the highly skilled task of splitting sheets of mica of precise thickness, in order to measure the energies of emitted particles. At first they used the scintillation method for observing the particles emitted in nuclear transmutations, assisted by the skill of Rutherford and Crowe achieved through their long experience of interpreting the scintillations. This was soon superseded by the electronic methods developed by C. E. Wynn-Williams and W. B. Lewis.

Rutherford gave Oliphant and his collaborators a completely free

242

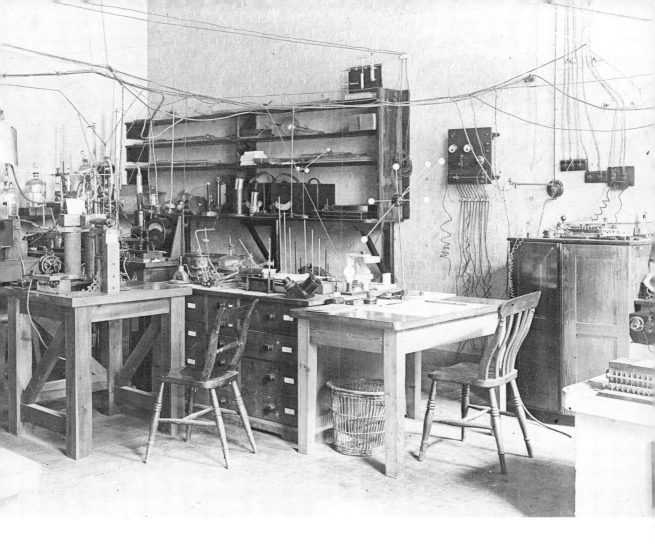

22 *Rutherford's research room*

hand in the design of the apparatus and the conduct of experiments. While not personally taking part in these, he followed the progress of the research with intense attention, discussing at every stage what to do next, showing deep interest in experimental results and pondering on their significance. This caused the young research workers to pursue the investigation with the sustained effort that helped to produce the exceptionally high Cavendish fertility in discovery.

The apparatus was largely constructed from bits and pieces of instruments that happened to be there. It consequently often gave a good deal of trouble in functioning. Rutherford got very annoyed by the consequent delays, but showed no interest whatever in raising the money required for purchasing more reliable equipment.

He usually inspected the experiments twice a day, late in the morning and just before six in the evening. If exciting effects were being observed, he would come in to see how they were progressing more frequently. The investigators generally found this embarrassing. With Rutherford breathing down their necks, they tended to mishandle the apparatus. In his impatient enthusiasm Rutherford would seize parts with his hands, which at this time in his life had begun to shake; at least twice he smashed the thin mica windows through which the products of transmutation emerged. Air rushed into the evacuated tubes, and the vacuum pumps began to race; panic ensued, until everything was shut off. Rutherford was humbly apologetic and went away for hours, even days, while the collaborators cleared up the mess.

If Rutherford came in at the end of an experiment, he could scarcely wait for the examination of the photographic records. He would sit at a table in the next room, studying the record still wet with fixing solution, which dripped onto the working papers and his clothes. He dribbled ash from his pipe over the sticky paper, which he damaged with a stump of pencil he took from his waistcoat pocket, in order to mark the soft material. In his anxiety to examine the most interesting parts he would pull out the long record of the experiment from the roll held by Crowe, so that it sometimes fell on the floor; then he trampled on it when he got up. The collaborators had to do their best to retrieve the situation, completing the fixing and drying of the bits of damaged paper. They usually tried to prepare and measure the record carefully when he was not there, but when he was this was impossible.

On one occasion, after a laborious day, the experimenters decided to develop their record the next day. Rutherford happened to come in just before he was leaving, and was extremely angry when he heard that they proposed to go home. He insisted that the record be developed at once, and roared that he could not understand how, with these exciting results, they were too damned lazy to look at them immediately.

The record was developed at once, but the materials were exhausted, and it came out badly. Nothing could be made of it, and Rutherford went off muttering that he did not know why he had been blessed with such a bunch of incompetent colleagues.

After dinner that night, he rang Oliphant, and hesitatingly apolo-

244

gised for having been so bad-tempered, asking him to call round at his home in the morning. When Oliphant arrived he found him even more contrite. He reported that his wife had said he had ruined his suit, and he asked whether they had been able to salvage the record.

Oliphant said that he drove them mercilessly, but they loved him for it.

In 1933 the eminent American physical chemist, G. N. Lewis, visited the Cavendish. He had developed the method of preparing heavy water (deuterium oxide, ^2H_2O) by electrolysis. He presented Rutherford with half a cubic centimetre of the liquid. Oliphant devised a procedure by which the deuterium oxide could be used over and over again, thus enabling significant experiments to be made with the small amount of material available.

He and his collaborators found that if a beam of deuterons (deuterium nuclei), were projected at any kind of target tried, a copious stream of protons was produced. E. O. Lawrence had observed such a stream, and had concluded that it consisted of protons and neutrons, arising from the disintegration of the deuterons on impact at the target. Oliphant suspected that this was not the correct explanation, for his group had observed that the emission of protons from impact at a steel target grew with time. Some of the deuterons from the impinging beam were sticking to the steel surface, and when these were struck by succeeding deuterons from the beam, the interaction of deuteron on deuteron was adding to the emission of protons.

Following this, Harteck prepared compounds containing heavy hydrogen. The first was heavy ammonium chloride. When this was bombarded with the beam of accelerated deuterons a very large emission of long-range protons was observed, even at accelerating voltages as low as 20 or 30 kV. Rutherford was excited by these results, and urged them on. They made certain that the emitted long-range particles were protons, and set up a chamber containing helium at high pressure, for detecting impinging neutrons. This promptly recorded the simultaneous emission of neutrons. Besides these, the emission of particles carrying a single charge but of shorter range was observed. They were investigated, and found to be a new isotope of hydrogen of mass number 3; they named it tritium.

It was produced by the fusion of two deuterons, which formed an unstable helium nucleus. This then disintegrated into an atom of tritium and a proton

$$^2H + {}^2H \longrightarrow {}^1H + {}^3H$$

This was the last of the major discoveries made under Rutherford's general direction. Tritium has since been of fundamental practical and political importance, as a constituent of hydrogen bombs.

As Rutherford pored over the records of the observed particles he made calculations of the range of tritium nuclei from the known ranges of α-particles and protons, to see whether the data agreed. He did this by rapid mental arithmetic and approximations, at which he was peculiarly skilled. He nearly always arrived at the correct result, which could not have been obtained with complete certainty without a complex calculation.

He was so excited when he believed he had got onto the identification of the particles that he kept making slips as he expounded the argument to his young colleagues. The measurements and analysis ultimately showed that the range and properties of the observed particle were very close to those now accepted for tritium. But this was not accomplished without one of Rutherford's strokes of genius.

In the course of the analysis a search was made by special experiment for the emission of very short-range particles from the target. When this was done, a group of particles that carried a double charge was detected. They were equal in numbers with the protons and tritium nuclei, and appeared to be α-particles. The result caused consternation, for the equality of numbers suggested that all three groups of particles were produced in the same process.

Rutherford worried over the figures all afternoon, trying hypothesis after hypothesis until everyone was tired, and all went home to think about it. Oliphant went over the figures again, and rang up Cockcroft to ask whether he could offer any suggestion, but he had none. Oliphant finally went to bed, quite tired out.

At 3 a.m. the telephone rang. The Oliphants feared bad news, and Mrs Oliphant, who always woke instantly, went to answer the call. She came back with the news that 'the Professor' was on the phone, and wanted to speak to him. Oliphant, still half asleep, heard Rutherford apologising for waking him up, and then say excitedly: 'I've got it. Those short-range particles are helium of mass three.'

This suddenly woke Oliphant up, and he asked how he could possibly have arrived at this conclusion, for there was no way in which two particles of mass two could combine to give two new particles, one of mass three, and the other of mass one.

Rutherford roared back on the phone: 'Reasons, reasons! I feel it in my water.' He then explained that the helium nucleus of mass three was the companion of a neutron, produced in an alternative reaction, which happened to occur with the same probability as the reaction that produced the protons and tritium nuclei.

Oliphant returned to bed, but did not sleep. After breakfast he called on Rutherford at Newnham Cottage, and they went through the approximate calculation together. They agreed on further measurements to confirm his conclusion. Oliphant and his colleagues in the Laboratory carried these out, and by the end of the morning had proved that: 'Of course, Rutherford was right.' There was an alternative reaction between two deuterons, which produced a neutron and a helium nucleus of mass number 3, and the mass of ^{3}He was slightly less than that of tritium.

$$^2H + {^2H} \rightarrow {^3He} + n$$

Everyone shared Rutherford's excitement. They had discovered two new isotopes and measured their masses, and they had explained the remarkable deuterium reactions. In the evening Oliphant drafted a note to *Nature* on the work. The next morning Rutherford emended it all over in pencil. It was retyped, and sent off for publication. Oliphant said that at no other time had he ever felt the same sense of accomplishment, nor such comradeship as Rutherford radiated on that day.

At this time Rutherford was sixty-three, and Oliphant and his group were around the age of thirty. Rutherford was able to work through the night, when able young men half his age had been tired out. If he could work with this enthusiasm, energy, and insight in his sixties, what must he have been like in his thirties? Oliphant's account of the discovery of tritium brilliantly illustrates the force of Rutherford's genius.

Oliphant moved to Birmingham in 1937. War-research of great significance was accomplished in his laboratory, with the invention of the magnetron by J. T. Randall and H. A. H. Boot. In 1950 he returned to Australia as professor of particle physics, and Director of the Research School of Physical Sciences, at the National University at Canberra. He was the leading Australian authority on atomic energy, and was also active in affairs, especially in scientific relations with other countries, including the U.S.S.R. and the Peoples' Republic of China. In 1971 he became Governor of South Australia.

Kapitza also noted Rutherford's mental processes. He said that intuition, in the sense of a subconscious process of thought was no

247

doubt a factor in it, but he stressed Rutherford's enormous capacity for work. He searched and worked incessantly for discoveries. He published only results, and never told his colleagues of the many ideas that had passed through his mind, but which he had found to be untenable. This could be gleaned only from casual remarks he made.

Sheer intensity of enthusiasm, energy and work was a very large factor in his achievement. Kapitza commented that it was rarely recognised how hard scientists worked, and what a small proportion of the work they do is actually published.

Joliot-Curie was once asked for his opinion on the remarkable contrast between the apparently small amount of evidence and the extreme modesty of the language, compared with the importance of the results, in Rutherford's four great papers in 1919 on the disintegration of the atom. He replied that it was not widely understood that a scientist may publish only a hundredth of his evidence, leaving out especially the mass of fairly, but not completely, conclusive observations, none of which are tidy or convincing enough for publication, but which together, in the experimenter's mind, are overwhelmingly convincing.

19

Individual Genius also Flourished

While the main stream of research in the Cavendish was determined by the genius of the director, this did not stifle the independent investigator, even in the Rutherfordian period of concentration on atomic physics. Among the gifted research students of J. J.'s later years was G. I. Taylor. He started research in the Laboratory in 1908, and was appointed an assistant demonstrator in physics. He became a fellow of Trinity College, and a devoted College man.

His early research was in optics, designed to develop the theory that J. J. had proposed for the explanation of the photoelectric effect, before the quantum theory had been invented. In 1909 he published a paper on 'Interference Fringes with Feeble Light'; subsequently, he showed that when a beam of light of such low intensity that only one quantum could be near the slit, was directed at a double slit, the interference pattern was nevertheless the same as with a beam of ordinary strength.

Taylor was the son of an artist and a grandson of George Boole, the founder of mathematical logic. He was endowed with the same kind of individuality and ability to work entirely by himself as his grandfather. He became a universal physicist in the manner of Rayleigh, making major contributions in a remarkable range of subjects, and arriving at profound conclusions by inspired approximations. He was interested in exploration, and joined the *Scotia* expedition of 1913 as meteorologist. He began to study the dynamics of the atmosphere, and applied his intuitive insight into the behaviour of fluids, as Rutherford had into that of atoms. In the First World War he engaged in meteorology and experimental aeronautical research, and in 1915 qualified as a pilot.

Taylor's genius was stimulated by a combination of work in the laboratory and the study together with life in the open. He had been delicate in his earlier years, and yachted for health as well as for pleasure. Out of this came the invention of a new form of anchor, together with its theory and practical demonstration.

As an expert in aerodynamics he was associated with the first flight across the Atlantic, from Newfoundland to Britain, and was consulted in the design of the airships R100 and R101. His researches in solid-state physics were concerned with X-ray crystallography. In 1934 he put forward the explanation of properties of solids in terms of dislocations of their orderly crystalline atomic structure, a theory independently propounded by E. Orowan and M. Polanyi.

He contributed to the theory of the weather on the earth and on other planets. It has been suggested that the famous red spot on Jupiter is a cosmic example of the phenomenon of the 'Taylor column', a rotating column of fluid that is stable. His work on turbulence in fluids has been essential for the design of aircraft, jet-engines and atomic bombs. He was a member of the team that made the first atomic explosion at Los Alamos. In a paper only a page and a half long he calculated the blast effects of an atomic explosion. From it, the main features could be forecast — a typical piece of Rayleighian virtuosity in investigatory calculation.

His advice on protection against bomb explosions, and on the best methods of attacking submarines with depth charges was of great importance in the Second World War.

Taylor used to work in a room on the ground floor of the Cavendish with one assistant. Rutherford and Chadwick worked in a room next to his, and in order to reach Rutherford, visitors had to go through Taylor's room. Rutherford had many visitors, and after welcoming them, he used to introduce them to Taylor, and leave him to show them round.

Taylor did not have pupils, and worked on his own. Rutherford seemed quite to like this; he did not disapprove of those who did not join his team. Taylor was very helpful to any who sought his advice. He could be very patient in explaining his research to raw student enquirers. His influence was exerted through his presence and conversation rather than by formal teaching.

Taylor was one of the electors to the Cavendish professorship in succession to Rutherford. He remarked that none of Rutherford's 'boys' was equal to Rutherford himself, and the selection of an appropriate successor was difficult. The electors had long discussions as to whether he should be a nuclear physicist. Taylor thought that they did well to choose W. L. Bragg: the molecular biology under him turned out well. Taylor also admired electron-microscope development under Bragg.

In aerodynamics Taylor had the same degree of eminence as Prandtl. When the International Congress of Applied Mechanics was held in Cambridge in the 1930s, Prandtl stayed with him. He was completely taken in by the Nazis, and told him what fine people they were. While he was with Taylor, the news of the execution of Röhm arrived. Prandtl refused to believe it, and said it was a newspaper lie. At the Congress in New York Prandtl did not dare be seen reading the newspapers; he got Taylor's wife to read them to him. At the time of Munich Prandtl sent the Taylors pictures of Hitler blessing little children.

Taylor thought that, in its way, the Cavendish was unequalled. The German physicists were more inclined to theory. In his opinion Rutherford owed a great deal to his son-in-law, R. H. Fowler, for mathematical and theoretical help. He remarked that the Fowlers belonged to the family of the famous inventor of agricultural machinery.

When Taylor visited Russia in 1970 he was surprised to find that some of the physical laboratories appeared to be run down, and requiring renewal. He was astonished to learn how many Russians had read the novels of his aunt E. L. Voynich, the fifth of George Boole's daughters, and the author of *The Gadfly*. Taylor's grandmother was a daughter of Sir George Everest, whose name was given to the highest mountain in the world.

Like other members of the Boole family, Taylor had artistic insight. An Italian student, the son of a mayor of Rome, came to study in the Cavendish. Taylor noticed at once that he was a Colonnan, having previously seen a picture of one of the Colonna popes. Taylor did not like the Trinity portrait of Clerk Maxwell, which was a bad one, and was put out of sight; the portrait of Lord Butler had been put in its place. Taylor admired the portrait of Montagu Butler.

Taylor manifested in himself an extraordinary range of scientific and cultural interests, typical of his family. His contribution was a very significant feature of the Cavendish tradition, widening and deepening it in other directions while the pursuit of the electron and the nucleus was in full cry. G. I. Taylor worked in regions of classical as distinct from nuclear and quantum physics, and yet he flourished within the strong Cavendish community.

Another of the outstanding workers was F. W. Aston, whose exact measurement of the masses of atoms was of fundamental importance for nuclear physics, and yet he was as much an individual investigator as Taylor.

Rutherford's capacity for getting on with these highly individual workers, whose approach was so different from his own, was illustrated by the golf foursome, which he, Taylor, Aston and R. H. Fowler usually played on a Sunday afternoon. While Rutherford at the head of his 'boys' penetrated into the unknown, waving the sword of discovery as they rushed forward, his relations with the privateers of discovery, operating entirely on their own, could be amiable and understanding.

F. W. Aston was born in 1877. His father was a metal merchant of Harborne, Birmingham, and his mother a daughter of Isaac Hollis, a gun maker whose firm became the basis of the Birmingham Small Arms Company. He was brought up on his father's small farm, and from his boyhood was interested in animals, and in mechanical toys and chemical experiments. He inherited the Birmingham tradition of metal-working and craftsmanship. He made his devices and experiments in a loft of the family house, in the spirit of the Birmingham men who had contributed so much to the development of the Industrial Revolution. James Watt had to migrate from Scotland to Birmingham to find the fine mechanical craftsmanship necessary to make his steam engine a success; J. J. had to go to Birmingham to find an assistant with the appropriate mechanical and inventive skill to make a success of the isotopic analysis, which he had begun.

Aston went to Birmingham after his schooling. He studied chemistry under Tilden and Frankland, and physics under J. H. Poynting, one of J. J.'s close friends. When the course on fermentation chemistry was started at Birmingham he took it to qualify as a brewer's chemist, and in 1900 secured a job with Butlers, the Wolverhampton brewers.

Like many others, he had been excited by the discovery of X-rays in 1895. In his spare time he made X-ray vacuum tubes. He invented a new form of Sprengel vacuum pump, and started to investigate the properties of the Crookes dark space in discharge tubes. He made other instrumental inventions and researches in a tiny workshop at home. In 1902 he built his own motor cycle, and became interested in motor-racing; his brother W. G. Aston became a noted motoring journalist.

He returned to Birmingham in 1903, with a scholarship that

enabled him to pursue his researches on the Crookes dark space. He discovered a special region within it, which came to be known as the 'Aston dark space'.

In 1908 Aston's father died, and he inherited a modest fortune. Thereafter, he was fundamentally an independent Birmingham gentleman. One of his first actions was to go on a tour round the world. He became a passionate and accomplished traveller, especially on sea voyages, planning every aspect in precise detail. He applied the manual dexterity he exhibited in the laboratory to deck games, tennis, golf, swimming, skiing and playing the cello. He won many tennis tournaments, climbed Alps, and went surf-riding in Honolulu. When he became delicate in his later years and had to give up strenuous physical sports, he started to collect Chinese porcelain.

Aston was a striking example of the British bias towards particle physics. Like many other British experts in particle physics, he was skilful in ball games. He was physically sensitive, especially to the sound of other people's voices at night, and he did not like anyone else to be in his laboratory when he was doing delicate experiments.

By 1909 Aston's gifts as an instrumentalist and experimenter had become clear; he was appointed a lecturer in physics at Birmingham. Meanwhile, J. J. had obtained suggestive but inconclusive evidence, by his parabola method of analysing positive rays, that atoms and molecules of some substances might have integral masses. The experiments were technically difficult, and he needed skilled experimental assistance. He asked his friend Poynting if he knew of any suitable person, and Poynting recommended Aston. So, after three months as a lecturer, Aston went to Cambridge to assist J. J. He entered the Cavendish, and Trinity College in 1910, when he was thirty-three.

He soon advanced J. J.'s positive-ray analysis. With his skill in high-vacuum technique he introduced large low-pressure discharge bulbs. He designed the kind of camera that became standard for photographing the parabolas, and obtained the first pictures of the twin parabolas of neon with it.

He invented a stop-cock switch for moving objects in high vacua, and a very sensitive quartz micro-balance for measuring the densities of separated gaseous constituents.

With his improved apparatus Aston obtained a twin parabola for neon, whose branches appeared to correspond to atoms of 20 and 22 units of mass, respectively. But the method was not yet accurate enough to give completely decisive results.

He tried a fractional-distillation method of separation, but it was not successful. Then he tried diffusing the gas through the fine pores of clay tobacco pipes, utilising the principle subsequently applied in the plant at Oak Ridge, U.S.A., for separating uranium 235 from uranium 238 for making atomic bombs. His results were positive, if not incontestable.

At the British Association meeting in 1913 he announced his preparation of specimens of two kinds of neon, and in that year he was appointed Clerk Maxwell Student at Cambridge.

When the First World War started, Aston went to the Royal Aircraft Factory at Farnborough as a technical assistant in chemistry. He investigated the pigments and processes in doping the canvas used for covering aeroplane wings, and he invented the neon tube for producing short flashes.

The scientists lived in a civilian mess known first as 'Arnold House' and then 'Chudleigh'. Aston thought about the problem of neon in his spare time, and discussed it with the other scientists who happened to be there. These included, from time to time, F. A. Lindemann (Lord Cherwell); G. I. Taylor; Melville Jones; F. M. Green, the chief engineer of the factory; E. D. Adrian; W. S. Farren, later Director of the R.A.E.; D. H. Pinsent, a brilliant young scientist killed in an experiment in the air; H. Glauert; and G. P. Thomson. Aston was himself involved in an air experiment ending in a crash; fortunately he was not injured.

Aston's period at Farnborough was very important. He did not know much physics, had no talent for teaching and verbal communication, and was shy about discussing theoretical problems; no doubt he was conscious of, and sensitive about, his ignorance. The conditions at Farnborough made personal discussions as easy as possible for him. The wartime informality, and the fact that most people were working on topics that were not their main speciality, facilitated discussion. The scientists concerned were men of outstanding ability and intelligence, who could help him without being intellectually blunt.

He discussed isotopes with Lindemann, and learned from him something of the then novel quantum theory. After the war they published two joint theoretical papers on the separation of isotopes. Lindemann believed that the neon parabolas were due not to neon atoms, but to compounds of neon, or of carbon. Though Aston was unequal to Lindemann's theoretical arguments, he persisted in believing that they must be due to neon atoms of differing mass.

The highly creative and sympathetic atmosphere at 'Chudleigh' was an excellent stimulus for Aston. He had difficulty in adapting to situations, but this exceptional one suited him.

By the end of the war his ideas for the separation of isotopes had advanced. He tried an improved method for the separation of neon isotopes by diffusion, in which the gas was circulated thousands of times mechanically through pipe-clay. It failed, giving less good results than his simpler pre-war method. He was not discouraged, and turned to the construction of an improved apparatus for separating isotopes by electromagnetic focusing, which he had thought out at Farnborough. He carried this out in the Cavendish in 1919. The new instrument, which he called the mass spectrograph, made great improvements in intensity, dispersion and resolving power.

J. J.'s parabola method depended on deflecting particles with the same velocity. In the mass spectrograph all particles of the same mass, whatever their velocity, were focused on the same line. By this method, analogous in principle to an optical refracting system, far more particles were focused, and the lines obtained were much stronger. He published a paper with R. H. Fowler on the theory of the instrument—another benefit of his Farnborough discussions.

He quickly proved that neon, chlorine and many other elements were mixtures of isotopes. He formulated the rule that atoms have masses that are very near to whole numbers. This provided evidence that atomic nuclei might be made up of combinations of numbers of protons and electrons, which was of major importance for nuclear physics.

His results were immediately appreciated. He was elected a fellow of Trinity in 1920, where he lived for the rest of his life. He was awarded the Nobel Prize for chemistry in 1922.

Aston now constructed a more sensitive mass spectrograph (see figure 23), which measured atomic masses to one part in 10000. In order to ensure the reliability of the current producing the field from the deflector plates, he made the five-hundred lead accumulators required himself. The results obtained with this instrument were of extreme importance. The slight divergencies from the whole-number rule in the masses of the atomic nuclei gave a measure of the amount of atomic energy that might be released in the disintegration of a nucleus.

Besides providing fundamental information required for the release of atomic energy, Aston's 'packing fraction' measurements had extraordinary accidental importance. His first results on the ratio of the masses of helium and oxygen indicated a large difference between the physical and chemical atomic weights of hydrogen. This suggested that the separation of isotopes of hydrogen would be easier than might have been expected. This was a stimulus to Urey and Brickwedde, who carried it out successfully, and discovered deuterium. In fact, later measurements showed that Aston's original difference was too large.

Aston embarked on the construction of a third mass spectrograph, accurate to one part in 100000. It did not quite achieve this, but it provided accurate measurements of the mass defects of a number of important isotopes.

23 *Aston and his mass-spectrograph*

Aston perceived the significance of his researches: in 1936 he said that knowledge of the exact masses of atoms would allow nuclear chemists to synthesise new atoms, and in some reactions sub-atomic energy would be liberated.

He disagreed with those who said that the release of atomic energy should be stopped by law. He thought that the elders among our ape-like ancestors had probably objected to the introduction of fire and cooked food. He believed there was no doubt that sub-atomic energy would be released, and man would control its infinite possibilities. In his opinion, man could not be prevented from releasing atomic energy, and we could only hope that he would use it not exclusively for blowing up his neighbours.

20

The Cavendish in Transition

J. J.'s son, George Paget Thomson, was, as it were, born in the scientific purple. He grew up at the heart of Cavendish physics, in his youth enjoying the society of his father's pupils, and presently starting research in the Cavendish under him. He continued working there into the Rutherford period. He saw both J. J. and Rutherford from a unique standpoint, across two generations.

As the son of J. J. and by his own researches, he acquired an impregnable position in the scientific world. Accustomed from his infancy to this unsurpassed environment it never occurred to him to say anything other than what he directly felt. In the atmosphere of his father and Rutherford he took outspokenness for granted, making comments that no one else could have uttered in the most unselfconscious way.

In his opinion there was very little difference between J. J.'s and Rutherford's ways of running the Laboratory, except that in the end Rutherford got a little more money. Rutherford was rather more severe in keeping the group of research men working as a group. He concentrated research more along a few lines, whereas J. J. (Thomson generally referred to his father as 'J. J.') concentrated especially on the investigation of the nature of ionisation.

G. P. Thomson worked on hydrocarbon free radicals, which had been discovered by J. J. From this he went on to study slow collisions of positive ions. With the outbreak of the First World War he served in France, and later in aeronautical research. After returning to Cambridge in 1919 he published a large work, *Applied Aerodynamics*.

In the Cavendish, E. G. Dymond had been investigating the scattering of beams of electrons by helium, and had obtained anomalous results; the electrons and helium atoms did not collide and recoil

258

strictly like billiard balls. G. P. decided to try to gather more information on the laws of impact or interaction between electrons and atoms. In 1922 he was appointed Professor of Natural Philosophy at Aberdeen. He continued experiments on the scattering of positive rays, and though these were unsuccessful in their immediate objective, they led to his independent discovery of electron diffraction.

He attained success in his demonstration of electron diffraction by firing a beam of electrons through celluloid. He had picked up this tip from C. T. R. Wilson's exposition in the Cavendish of Thomas Young's experiments. C. T. R. was very good in a practical class, although a very bad lecturer. G. P. Thomson and W. L. Bragg both gained crucial technical suggestions from C. T. R. Wilson's Cavendish course on light, one for the discovery of electron diffraction, and the other for X-ray crystal analysis.

G. P. thought Searle a superb arranger of the teaching of practical physics, though too keen on F. E. Smith's electrical measurements, which seemed to him rather boring. The teaching of practical physics, started by Maxwell and developed by Rayleigh, was flourishing in J. J.'s time. J. J. did not alter the course much, but he appointed Searle, which was to prove a great thing for the further development of Cavendish teaching. G. P. was an undergraduate in 1910, and it seemed to him that the course then was probably not very much different from what it had been in the 1880s.

He was of the opinion that the eminence of a laboratory is due to the genius of the people in it. The Royal Institution, like the Cavendish, has had a similar history. Nowadays, one never hears of Schenectady; laboratories rise and decline, depending on the people they can get. The eminence of the Cavendish owed a great deal to the attractiveness of Cambridge, and its eminence as a centre of mathematics. One might say of G. G. Stokes, who occupied the Lucasian chair for forty-four years, that in his time he did about half the physics in the country.

G. P. regarded Rutherford as a good man of business and affairs. He could have become a tycoon or prime minister of New Zealand; he could have made a fortune out of radio. He thought Callendar had been underestimated. Barkla and Townsend did not advance beyond their earlier achievements because they were no longer surrounded by their equals, who could have told them where they were wrong. Townsend opposed the quantum theory. He ought to have done the experiments of Franck and Hertz. The Oxford atmosphere that surrounded him was bad, and there is still a strong anti-scientific movement there in the 1970s.

259

When Jean Perrin visited the Cavendish, J. J. entertained him to dinner at his home. He invited two recently arrived research students to meet him: Rutherford and Townsend. When Rutherford was introduced to Perrin as coming from New Zealand, Perrin's eyes opened wide: 'New Zealand! But they are black and they eat people!', he exclaimed in French. Rutherford knew enough French to understand what he said, and remarked: 'We don't really eat people much now; and in any case, we don't talk about it.'

G. P. was friendly with Cherwell. He liked him, and considered that he had done a great deal to raise Oxford physics; his subordinates liked him.

Cambridge University made a great mistake in allowing organic crystallography to leave the physics department and go to medicine. They should have kept molecular biology on in the Laboratory after Bragg's departure.

G. P. was one of the electors to the Cavendish chair after Bragg had left. The thing that hit the Cavendish so badly was that many leading men had gone to new jobs. They thought it wrong to leave them to return to the Cavendish. For example, Cockcroft had gone to Harwell, which he had built up. All the good experimentalists had got good jobs. In the circumstances, the electors chose Mott, who, though a theorist, had created a very successful school of physics at Bristol. It should not be forgotten that Clerk Maxwell, the founder of Cavendish physics, was primarily a theorist, and the appointment of Mott could be regarded as in part a return to the first Cavendish style.

Two of the most influential figures in Cavendish transitions were E. V. Appleton and J. A. Ratcliffe. Appleton acted as Director of the Cavendish in the interregnum between Rutherford's death and Bragg's election. Appleton created the Cavendish school of radiophysics; Ratcliffe was his pupil, and suggested to Ryle that he should pursue radioastronomy. Both of them were men from the North, Appleton from Yorkshire and Ratcliffe from Lancashire.

Appleton was born in Bradford in 1892. His father was a warehouseman, separated from the working class by wearing a bowler hat instead of a cloth cap. His parents were strict Wesleyan Methodists. As in many such families in the industrial North, they were interested in religion, choirs and music. They were intensely respectable, and 'getting on' was for them one of the first principles of life. Appleton grew up a neat, tidy, conventional boy. He went to the local elementary school at seven, won a scholarship to the local secondary school at eleven and one to Bradford Grammar School at sixteen. His father felt unable to support him there, so he did not take it. At about this time he became a 'temporary laboratory assistant' at Bradford Technical College. He sat for the London Matriculation, which he passed with first-class honours, and thought of proceeding to one of the London University colleges. But he happened to meet his old science teacher, J. A. Verity, in the street. Verity congratulated him on his success, and told him of a local scholarship, founded in memory of a prominent Bradford citizen, which would enable its holder to go to Cambridge and enter the Cavendish Laboratory. Appleton sat for the scholarship and won it, finding himself at the age of eighteen entitled to the, then, handsome income of £150 a year. He went up to Cambridge and St John's College in 1911.

He was a rather typical Yorkshire boy, passionately interested in cricket, and had indeed considered becoming a professional cricketer. He resembled in figure, and perhaps also in temperament, the great Yorkshire cricketer, George Hirst. The Yorkshire cricketers, and particularly Hirst, were famous for pulling games out of the fire and winning against all odds, by dogged skill, caution and self-command. Hirst never allowed himself to be ruffled. In a way, Appleton played his life as Hirst played cricket; when in a tight corner he closed his mouth quietly but firmly, and steadily played himself out of it.

Following his father, he was a competent musician and singer, and used his command of voice production quite skilfully as a public speaker.

With his competence, sporting interests and natural acceptance of the desirability of joining the upper educated classes, he thoroughly enjoyed his undergraduate days. In the Natural Sciences Tripos he gained first-classes. His interests covered a wide range, and were not immediately specialised. He was awarded prizes for mineralogy and physics.

In 1914 he started research as W. L. Bragg's first research student. He was initiated into X-ray analysis, and his first task was to analyse the structure of copper crystals. Almost immediately afterwards, the First World War started. He had been in the Officers Training Corps, and applied for a commission. This was slow in arriving, so he entered the ranks. After serving in them for 116 days he was posted for a commission, and trained for the Army Signal Service.

He was sent to the first course on radio organised for British officers, conducted at Malvern. This was the beginning of his concentration on radio; except for the war he would probably have become an X-ray crystallographer. He remained in signals and radio for the rest of the war, joining in radio-intelligence research, and the examination of captured equipment.

He was deputed to give courses on radio to officers, during which he developed his powers of teaching and simple exposition. His first papers, published in 1918, were on thermionic valves.

He returned to Cambridge in 1919, and worked for a time under J. J. When Rutherford became Cavendish professor, he appointed Appleton an assistant demonstrator. He started research on the detection of α-particles, but he wished to pursue research on radio, in which he had been involved during the war years. He had become aware of the variety and interest of the problems to be solved, and of their practical importance. Among these was the transmission of radio waves through the atmosphere. An understanding of the factors affecting this had been of great military importance, and was of equally obvious importance for civilian communications.

Rutherford encouraged him. Appleton proposed three main lines of work: the investigation of the physics of radio valves, the problems of atmospherics and the propagation of radio waves. His former research student, Balthazar Van der Pol, who later became one of the leading physicists in the Philips Research Laboratories, joined him in research on the nonlinear characteristics of radio-valve oscillations.

He was stimulated by C. T. R. Wilson's research on thunderstorms to attempt to correlate them with atmospherics, using Wilson's special aerial and hut. This work brought him into contact with R. A. Watson-Watt. They collaborated at Cambridge and Aldershot, but presently sharply diverged.

In 1924 Appleton was joined by Miles Barnett, a research student from New Zealand. They started systematic recording of the strength of B.B.C. radio signals. It led to the discovery that signals were arriving by two routes, one along the ground and the other by reflection from the upper atmosphere. The nature of this reflection required thorough investigation.

Appleton was hampered by his heavy load of teaching; he also felt rather out of things as one of the few in the Cavendish Laboratory not engaged in atomic research. He decided to apply for the Wheatstone chair of physics at King's College, London, which, in the 1860s, had been occupied by Clerk Maxwell. As professor, and head of his own laboratory, he would have more scope for organising his researches. He was appointed to the chair in 1924.

Meanwhile, the research on radio reflections proceeded fruitfully. In 1925 he published his most important discovery: the determination of the height of the reflecting layer, or ionosphere, in the upper atmosphere, which came out at about 60 kilometres. His measurement of the distance of a conducting surface by radio reflection contained the principle of radar. Practical radar for military and civilian purposes however, required a series of developmental inventions of the first order. These were inspired by the threat of the Second World War: they were not in the line of Appleton's genius, which was for pure scientific research.

After his determination of the height of the ionosphere, which gave substance to the theories and speculations of Balfour Stewart, Heaviside, Kennelly and Marconi on conducting and reflecting layers in the upper atmosphere, Appleton became the world authority on the subject. He discovered many other structural features and mechanisms of the ionosphere, which were of high scientific importance and practical significance.

He was appointed Jacksonian Professor at Cambridge in 1936, in succession to C. T. R. Wilson. When Rutherford unexpectedly died in 1937, he became acting director of the Cavendish.

He was not elected to succeed Rutherford, though he had strong supporters. The ionosphere and the propagation of radio waves were then far removed from the nuclear physics that was so much the dominant subject under Rutherford: nor was Appleton attractive to all as a director. There were some who contended that the Cavendish had become too concentrated on nuclear physics, and a change of direction to research on the solid state would be of benefit to the country as well as to physics.

The electors ultimately chose W. L. Bragg, whose subject was related both to atoms and to the solid state. Appleton expressed the opinion that the electors had made the right decision.

Historically, Appleton's line of research might be regarded as a continuation of the researches on electricity and magnetism pursued by Clerk Maxwell and J. J. It was carried on by his pupil Ratcliffe, who in his turn inspired Ryle to engage in radioastronomy. In this way, Appleton's work was one of the chief links in Cavendish research. It has even become of importance in relation to nuclear and subnuclear research, owing to its bearing on plasma and the problem of controlled hydrogen fusion. The most striking Cavendish heritage from Appleton's work, however, is through radioastronomy, which is now one of the most fruitful lines of Cavendish research, and has latterly led to the attachment of the Astronomer-Royal to the Cavendish Laboratory.

Thus the Cavendish investigators keep returning, expectedly and unexpectedly, to fundamental themes of physical thought, characteristic of Clerk Maxwell and his successors.

When Bragg departed from the Cavendish in 1953 Ratcliffe acted as director, until Mott arrived in Cambridge to assume the duties of the chair in 1954. It was singular that Ratcliffe repeated his teacher Appleton's contribution to the Cavendish, both in the line of research and in administration. They had the practical sense of men from the northern industrial regions, like J. J. and Cockcroft.

Ratcliffe entered the Cavendish in 1921. He recollected that when he started research in 1924 the atmosphere was confident, and the Laboratory intensely admired. Americans were coming to it to study; later on this movement stopped, and the British began to go to America. In the 1920s the British rarely went to America to study physics.

A period of British national confidence led to confidence in questioning nature. People expected the Cavendish to do things and make discoveries. J. J. had settled the status of the Laboratory, and Rutherford reigned at the top of the scientific world.

How far did the regime depend on the man, and how far on the nature of science at the time? The latter helped to decide what kind of a professor to have, and what he could do.

After Rutherford, many had expected that Appleton would have been elected. He would have been good at research, and at internal and external politics. In Ratcliffe's view Bragg's strong point was his support for the lines of others. Appleton would probably have organised most of the research along his own line. Bragg had an admirable gift of supporting people without running them, for example Perutz and Kendrew, and he did the same thing with Ryle. Some people thought Bragg was ineffective. This was because he disliked academic and university politics. The nature of Cambridge University had its effect. The Cavendish professor was expected to do three things: he had to run the Laboratory, be a scientific figurehead in the country and an influential figure in academic politics. Bragg had no taste for the third of these. In Cambridge, academic politics is very important, as was evident in the actions that led to the foundation of the Laboratory. It was the result of extended academic debate and manoeuvre.

Ratcliffe thought that Rutherford was by far the best of the Cavendish professors in external relations outside Cambridge. During

Bragg's time, the weight in the University of chemistry under Todd and engineering under Baker, increased relative to the Cavendish.

Mott was very active in the University's science politics, and his mastership of a college was an important factor in this. He restored the position of the Cavendish inside Cambridge, relative to chemistry and engineering.

Ratcliffe thought that if Appleton had succeeded Rutherford, it would have been very good for the Cavendish, for the University and for relations in the external world.

Bragg was wise; he was also financially independent and did things because he enjoyed them. He was good with his hands. His wife was a great help to him. Bragg's review of *The Double Helix* by J. D. Watson in the *Times* revealed the greatness of his personality.

The administration and the archives of the Cavendish were for long of a makeshift character. Up to the Second World War there was only one telephone; one had to go to Lincoln and ask: 'May I use the telephone?' Wynn-Williams, the electronics expert, fitted up an automatic exchange, which he put onto the circuit illegally. Before the war there were no extensions.

During the war the Cavendish men who went into many branches of important war work and institutions became used to secretaries and facilities. When they returned after 1945 they immediately pressed for, and received them.

Bragg brought in James, and then Dibden, as administrative secretary of the Laboratory, a position that is now of great importance. Ratcliffe had recommended Dibden, an able naval officer who had served in the Second World War. He was very good with technical assistants and workshop staff, who had increased in numbers and required looking after with more attention.

One day, when Ratcliffe began research in the 1920s, he was about to lift a heavy piece of apparatus. The laboratory assistant, Burt, stopped him and said: 'You mustn't lift that up, you'll rupture yourself.' In those days gentlemen did not lift heavy weights, and there was a wide social difference between scientists and technicians. The technical staff of a large laboratory now requires expert management, and this is one part of the administrative secretary's job. Dibden, with his naval experience, was particularly good at it.

The unique attraction that has characterised Oxford and Cambridge for centuries, and existed long before the Cavendish Laboratory, is passing. Some consider that the Cambridge undergraduate who was educated in this tradition has done more for Cavendish science than the talented research students coming from outside. They produced a large body of very good scientists of the second level, which provided an excellent support for the front rank of genius.

Large-scale science started from Cockcroft's work on particle accelerators in the Cavendish, but it soon became too expensive to be pursued in a university laboratory. The expense produced a distortion in the balance of science in the university, and when Mott became Cavendish professor he stopped the project for a linear accelerator. This was one of the reasons why D. H. Wilkinson left Cambridge for Oxford.

Ratcliffe was of the opinion that if science has immediate utility, a university is not the place for its effective development.

The success of the Cavendish development of radioastronomy was due to Ryle. He was extremely able, both experimentally and theoretically, and rich in ideas; besides all this, he had scientific courage.

Radioastronomy developed from physics, not from engineering. In the United States the type of radio research pursued by Ratcliffe, under whom Ryle started, was done by electrical engineers. In spite of the original discoveries of Jansky, radioastronomy did not develop there because the electrical engineers were not interested in asking fundamental questions of nature. As Jansky's work did not appear to have any bearing on practical radiocommunication, it was more or less dropped.

Bragg was an X-ray crystallographer whose researches had been carried out with inorganic materials. Yet, during his professorship, both molecular biology and radioastronomy were advanced, because general questions were put to nature, and efforts were not restricted to carrying on an established line of investigation.

The teaching of physics was important and pre-eminent in the Cavendish. For example, Alexander Wood's first-year lectures were extremely good. His contribution in this direction has not been sufficiently appreciated. Searle had a profound influence on the Cavendish teaching of practical physics, but after a certain stage he held it back. For him, his was the only way of doing things.

Ratcliffe took over courses from Searle twice, the first time before the Second World War. Searle was very pertinacious, even when he was aged. It took Ratcliffe a whole term to get him out of the classroom. There were two doors to it. Ratcliffe got in first, but Searle came in at the other door, and started teaching. During the war Searle had officially resumed teaching, and when Ratcliffe returned after the war, he had to get Searle out of the class-room again; he was then eighty years old. From 1902 until 1946 Searle brought out in the same handwriting the same sets of questions.

The Cavendish archives are not good. There was nowhere to put anything, so severe was the competition for the meagre space. Just after the war, Aston saw Bragg, and suggested that he should keep his room and original apparatus as a museum, but of course, in the circumstances this was quite impractical.

The Cavendish, like the whole of British science, was sublimely disinterested in its historical aspects. Why British science should have been so non-historical is a significant question. The scientist, engaged in investigation, becomes absorbed in solving the problem immediately in front of him. If he is an experimentalist he is apt to grab anything to hand that will help him in his quest. Humphry Davy had the habit of snatching bits of apparatus from other experiments if he thought they might be useful, even if the rest of the set-up from which they were snatched was smashed. Yet outside the laboratory, and as a writer, Davy had a strong historical sense.

The creation of a museum and the collection of historic apparatus is expensive. To the experimental investigator it seems a negative and barren way of spending resources, especially when these are meagre. How small they were is illustrated by the subject of the conversation between Rutherford and Ratcliffe in the famous 'Talk Softly Please' picture (see figure 24). Rutherford was telling Ratcliffe that they had just received a grant of £200 for a field laboratory for radio research. The whole radio research programme was run on £50 a year. Later on, this programme gave birth to Cavendish radio-astronomy.

In such circumstances physicists did not think of buildings and finance for preserving old apparatus. This is understandable, but what is more peculiar is that very few in the University saw any significance in the history of their researches. Even the time and effort that Clerk Maxwell had devoted to editing and elucidating the papers of Henry Cavendish was grudged. The division between Cavendish physics and the arts and ordinary historical departments appeared so deep that few in the latter thought about the history of science, including Cavendish physics, at all. This was a mark of the persistence of pre-modern scientific attitudes in Cambridge culture as a whole, notwithstanding Isaac Newton, Darwin and the Cavendish Laboratory.

It has been said that in his period, Ratcliffe became the *eminence grise* of the Cavendish. He had great influence with Bragg.

24   *Rutherford and Ratcliffe in conversation*

21

Bragg

The unexpected death of Rutherford precipitated the problem of electing his successor. The great prestige of his school of nuclear research, the construction of a new wing to the Laboratory, and the setting up of large-scale disintegration equipment, created a strong presumption that the successor should be a nuclear physicist. There were difficulties in making such a choice. Several of Rutherford's chief men had very recently left Cambridge for chairs in other universities. There would be a certain embarrassment in returning to Cambridge so soon. The talents of these leading nuclear physicists were, in their different ways, of rather equal weight. It would have been invidious to elect any one of them.

If the successor was not to be a nuclear physicist, the most eminent experimental British physicist of the active generation was W. L. Bragg (see figure 25), but he worked in the different field of X-ray crystallography. He had succeeded Rutherford at Manchester; should he not succeed him at Cambridge? There were those who believed that a new main line would be refreshing for Cavendish physics. Besides this, there were those who wished the Cavendish to concentrate on the physics of solids and fluids for national and industrial reasons. This stream of thought was remarkably like that expressed by Rutherford in his 1919 memorandum on the history and need of the Cavendish, which had so little effect at the time it was written. The desire to improve the research in the government laboratories had prompted the invitation to Bragg in 1937 to leave Manchester, and become Director of the National Physical Laboratory. The aim was to revivify research there, and make it more acceptable to industry. Bragg accepted the directorship in that year.

He quickly made an impression by discussing with a number of the

25 *Bragg*

staff their research problems, not only as an administrator, but as a working research scientist. Besides this, through his personal industrial connections, he was able to improve their sense of working contact with major industries: they began to feel a fresh inspiration. However, Bragg was not attracted by civil service administration; he was fundamentally a personal investigator. When Rutherford died, Bragg had been at the National Physical Laboratory only a few months, but when it became clear that he was the first choice as Rutherford's successor, he left the N.P.L.

After his election to the Cavendish professorship in 1938 he was provided with information on its finances and administrative problems, and began to study these. Before any radical decisions could be taken the Second World War started. Cavendish scientists had already been drawn into preparations for it, and major changes in the normal working of the Laboratory had to wait until the war was over, effectively until 1946.

Besides his eminence and skill in personal research, Bragg brought an outlook that was of equal importance in the direction of the Cavendish. As with his predecessors, his approach to this task was influenced by the social attitudes and conceptions he had acquired through his origins and upbringing.

William Lawrence Bragg came from Australia. Like Rutherford, he arrived in England and Cambridge from a distant part of what was then the British Empire. He was born in Adelaide, and he graduated in mathematics at Adelaide University in 1908. His mind was formed in Australia, and in this respect he was an Australian.

His father had been born in Cumberland, the son of a retired sea-captain, who had commanded small sailing ships trading with India. W. H. Bragg was sent to Market Harborough Grammar School with the help of an uncle who lived nearby, and later to King William School in the Isle of Man. He won a scholarship to Trinity College. At the age of twenty-three he had been recommended as professor at Adelaide by J. J. He married a daughter of Sir Charles Todd, F.R.S., the Post-Master General and Government Astronomer of South Australia. Their son William Lawrence was born in 1890. Thus both sides of the future Cavendish Professor's family contained distinguished scientists.

W. H. Bragg took his teaching duties very earnestly, and became a most accomplished expositor. He played tennis, painted in water colours, and led a broad and cultivated life. The conditions at Adelaide were not well developed for research. However, a wealthy citizen presented a quantity of the newly discovered element radium to the local hospital for medical purposes. W. H. Bragg took the opportunity of studying the properties of the Adelaide specimen, and in 1904 discovered that α-particles had a definite range in air, which presently

271

led to the rule that the stopping power of an atom was proportional to the square root of its atomic mass.

Bragg, who was now over forty years old and had not hitherto published any research, moved swiftly into the forefront of world research on radioactivity. Furthermore, his comparison of the ionising effects produced by radioactive radiations and by X-rays led him to suggest in 1907 that X-rays were particles consisting of neutral doublets travelling at high speed. When these neutral doublets travelled through matter, some of them were broken up, bits appearing as electrons. This would explain why X-rays produced electrons in passing through matter.

Thus, in 1904–7, W. H. Bragg suddenly became not only a foremost authority on radioactivity, but also on X-rays. There are few examples of a physicist who had previously published very little, suddenly, after the age of forty, moving into the front rank of investigators. In 1909 he was offered the chair of physics at Leeds.

W. L. Bragg entered Trinity College in the same year, and graduated with first-class honours in natural science in 1912. His father, with his special interest in X-rays, had been following the discovery, published in 1912, of X-ray diffraction, arising out of the theory and experiments of von Laue, Friedrich and Knipping.

He drew his son's attention to this work, and in the autumn of 1912 W. L. Bragg began to study it. Von Laue had given a complex interpretation of the diffraction pictures. While walking through the courts of St John's College and thinking about the problem, a simpler way of interpreting the pictures suddenly flashed into Bragg's mind. While attending C. T. R. Wilson's lectures in the Cavendish he had learned his method of treating the diffraction of light by a grating, and this gave him the clue. Wilson himself suggested that he should test his idea by specular reflection from cleavage sheets of mica. Bragg's interest in optics, deepened by C. T. R. Wilson, not only led to his invention of X-ray crystal analysis, but to his crucial appreciation of Martin Ryle's application of interference methods in radio-astronomy, and Perutz's development of X-ray crystal analysis in molecular biology. The extent of the Cavendish's achievement in the application of optical principles to research in fundamental physics has not been sufficiently recognised.

The object of von Laue's original work was to discover the nature of X-rays. He and his colleagues demonstrated that X-rays were diffracted by the atoms in a crystal of zinc blende, proving that X-rays possessed wave properties. Bragg saw that the spots in the Laue diffraction pattern could be regarded very simply as due to the partial reflection of the incident beam of X-rays by the principal planes formed by the atoms within the crystals.

The atomic structure of the crystals might easily be deduced from appropriate X-ray pictures. Bragg directed attention to the use of X-rays rather than a consideration of their nature. He discovered that they might provide an extraordinarily powerful method of elucidating the atomic structure of crystalline substances.

He first published this discovery in November 1912. The structures of crystals of potassium chloride and sodium chloride were determined, and the new branch of physics splendidly launched.

Bragg was elected a fellow of Trinity College and University Lecturer in 1914, and the Nobel Prize for physics was jointly awarded to him and his father in 1915. Thus, at the age of twenty-five, Bragg became the youngest of all Nobel Laureates. The fiftieth anniversary of his award was celebrated by his delivery of the First Nobel Guest Lecture in 1965.

The outbreak of the First World War retarded the immediate development of the new science of X-ray crystal analysis. From 1914 until 1919 Bragg was involved in war work, becoming Technical Advisor on Sound-Ranging to the Map Section of G.H.Q. in France.

When Rutherford moved from Manchester to Cambridge in 1919, Bragg was invited to succeed him at Manchester. In 1921 Bragg married Alice Grace Jenny Hopkinson, who was related to the famous engineers John and Edward Hopkinson. They were of Lancashire descent, and had a great deal to do with the development of the Edison–Hopkinson dynamo, and modern electrical engineering.

Bragg's own style of research was of the craftsmanship type: he understood this style with the insight of personal genius. He was not naturally of an organisational and administrative disposition, but his close personal relations with leading industrial scientific families was helpful. They facilitated contact with industry on research problems, and provided a channel through which he could learn how large scientific–industrial matters were handled, organisationally, administratively and financially. Thus, unlike many scientists with his particular gifts, he was not ignorant of these matters.

At Manchester he developed the X-ray study of minerals, especially of the felspars, which constitute a large part of the earth's crust. His pupil W. H. Taylor worked out their structure. Bragg promoted

26 *Cavendish Research men in 1938, with the Mond Laboratory in the background.*

the study of the structure of metals and the explanation of their properties, which are of such great industrial importance, carrying out brilliant work with E. J. Williams. He encouraged research scientists in industrial plants in the Manchester region, giving them facilities in his laboratory, where they could pursue ideas under quieter conditions than in their industrial laboratories.

In 1938, when he came to the Cavendish Laboratory, he was forty-eight, about the same age as Rutherford when he had become Cavendish Professor. Figure 26 shows the research workers at the Cavendish in that year. The outbreak of the Second World War prevented him from promoting extensive immediate developments, but he thought about the problem, and expressed his ideas in an address in 1942 on *Physicists after the War*. He observed that in the Second World War the demand for physicists exceeded the supply. There had been no such general demand for scientists in the First. Physicists had been drawn into the service of the country in a way

274

that had never happened before, and as regards university scientists in particular, this had meant a change of outlook and occupation of a revolutionary kind.

The pure scientist worked in a world of his own, not because he was an impractical dreamer, but because it had usually taken several decades for an important discovery to become of practical use in industry. In the war, however, the pure scientist had had to join the applied scientist, become a technician and devise equipment that would be of practical use within one or two years.

The effect of this was that they saw more of how the country was run. They were somewhat free in their criticism of what they saw, owing to their slowness in realising that a country is a more complex entity to organise than a laboratory. Nevertheless, some were worth listening to, because they came from a body of men whose discoveries had been largely responsible for introducing the new technological era, which had completely altered our way of life.

Everyone was hoping that we would not drift back after the war, but would use our assets of character and brains to make the country a better one than it had been during the previous twenty years. We must take stock of our physicists and make better use of them.

Bragg then made some estimates of the number of physicists that Britain might be expected to produce. From his experience he deduced that Britain produced one good physicist per year per million inhabitants. At Manchester he used to get three or four a year, in a department drawing on a regional population of about four million.

At the beginning of the war about 1200 physicists were recorded in the Central Register of scientists. Assuming that a physicist had a working life of thirty years, this figure roughly confirmed his estimate of one in a million.

This estimate by Bragg inspired D. J. de Solla Price to start his statistical investigations of the numbers of scientists, and of scientific activities generally, which have since become a feature of contemporary scientific development.

Bragg found that only one in a hundred heads of physics laboratories came from the English public schools; most had been educated at secondary or grammar schools.

He then discussed the characteristics of British physicists. Britain, like the United States, Germany, Holland and Scandinavia was a 'physicist-producing' country. It was hard to understand how the British physicist held his own, especially in competition with the

275

wealth of knowledge, mathematical ability, power of philosophical generalisation and industry possessed particularly by the German student.

He thought that the compensating advantage of the British physicist was a curious kind of horse-sense, which made him distrust the too rigorous application of logic to his experimental observations, and gave him a good sense of proportion in what to accept and what to reject. It was like Tolstoy's repeated theme of the superiority of peasant directness over clever brilliance.

He had heard German colleagues speak of British physicists with mingled admiration and irritation, that people who knew so little should be right. One of his famous German friends had told him that he always read some of Faraday's lectures before delivering his annual course to try to capture something of his spirit. Bragg admired the teutonic thoroughness with which his friend was determined to be unteutonic.

Bragg considered it important not to forget the merits and the achievements of the existing system of science. He thought that too much central planning, and putting men on pure research divorced from teaching, had dangers. On the other hand, there were professors who battened on students, using them for selfish ends; this was one of the major scientific crimes. He thought that the purest pleasure in scientific life was to see the germ of an idea planted in a younger man's mind develop to an extent that could not have been foreseen, and to see him get recognition for his work.

Teaching was a great safeguard against getting stale, and good to fall back on when research was going badly.

Central research institutions suffered from the danger of fossilisation; this did not appear in their first decade. The successful central research institution had a nucleus of permanent staff, with a tradition of techniques peculiarly its own. Its main service was as a place open to all for short periods of intense work, and its main population was continually changing. He mentioned international biological laboratories, no doubt having in mind that at Naples, but this description, in 1942, expressed very well the main characteristics of such laboratories as CERN, the Central European Laboratory for Nuclear Research, created at Geneva two decades later.

He thought that Shaw's criticism of the traditional academic system in the words of Undershaft in *Major Barbara* was bitterly correct. If a student 'shows the least ability, he is fastened on by schoolmasters;

trained to win scholarships like a racehorse; crammed with second-hand ideas; dulled and disciplined in docility and what they call good taste; and lamed for life so that he is fit for nothing but teaching.'

In former days the leading position of Britain in industry had been largely created by inventive men, who had gone as boys straight into industry, and he was in favour of honours physics students working for six months in an industrial concern before coming to the university.

Significantly, Bragg compared the placing of pure scientists in well-equipped laboratories, divorced from teaching and application, with the assembly and isolation of poets in a fine house in beautiful country surroundings, where they were then told to get on with it. In such conditions, pure scientists and poets would, as he put it, not be visited by the muse. Bragg had an artistic conception of research and discovery; he approached them in the spirit of a poet, and his genius was for personal insight and creation in an artistic manner. It was on this aspect of science that he was at his strongest.

If the origin of a valuable research development was traced, it was nearly always found that it had arisen from a chance remark made in a scientific discussion, or from reading another man's paper. If we were preoccupied with the details of administration, the muse of inspiration would be prevented from visiting us. The most important factor in research was that researchers should have *brain-waves*.

He thought the ideal research unit should consist of six to twelve scientists with a few assistants, one or more first-class mechanics, and a good workshop. There should also be a good store of material, and especially of *junk*. The latter was a priceless source of oddments, which could be used constructively in rapidly devising a first rough apparatus for trying out a new idea.

He had met a famous scientist who was unhappy in his beautiful clean new building, because he was feeling the lack of junk. It was desirable that every researcher should have at least a small sum to spend on material exactly as he liked, without having to get prior approval.

Bragg thought that after the war there would be an annual demand for two to three hundred physicists in Britain. Their selection for appointments might be carried out by a central organisation rather than individual university boards.

After 1946 Bragg became fully occupied in organising the post-war development of the Cavendish Laboratory. It was a formidable administrative task, owing to the increase in numbers of researchers and scale of equipment. Bragg struggled with this problem while strongly retaining his artistic conception of the nature and method of research.

Meanwhile, the world wanted to know how the Cavendish was coping with the post-war situation. In 1948 he delivered several addresses on the organisation and work of the Laboratory. Many of the problems facing the Cavendish were common to all science laboratories, especially those arising from an increase in numbers. Government and industry were offering more jobs. Before the war the annual production of honours physicists was about 160; this had now increased to about 600.

Some of the Cavendish's problems were peculiar, because of its unique position as the most famous physics laboratory in the British Empire. Many research students came from Britain, but they had also to try to find places for men with scholarships from countries in the Empire, who wanted to come to England for two or three years' research. Besides these, there were many from other countries.

There were 160 in the Laboratory in 1948, compared with 40 to 45 before the war, and they were able to accept only one out of every three applicants. To build up the depleted resources of the Laboratory after the war, and provide proper facilities for such an array of research men was a big task.

Because so many men came to work, a wide range of research subjects and facilities was necessary; it was undesirable to force all these men into one channel. The large number of researchers and the need for catholicity in subjects gave a rather special character to the organisational problems of the Laboratory. In the old days a professor had half-a-dozen men under him, and he could talk personally to them almost every day. One had to have two or three hours' talk with a research man at least once a fortnight, if one was really to understand what he was doing, and be able to make helpful suggestions. Formerly there were few problems of administration, and the professor had plenty of time to look after such students as he had.

In the situation that had arisen after the war it had become impossible to know and help all the men who worked in the department. Bragg had found in his experience that one leader in research could not look after more than six to ten men, in what he had found to be the necessary and proper way. The Laboratory's researchers had therefore to be clearly split up into groups, and some of the larger groups had to be divided. The groups they had defined were in nuclear, radio, low-temperature, crystallography, metal and mathematical physics.

Each group was made as self-sufficient as possible. It had its own workshop, assistants and secretariat. There was a main workshop for the Laboratory as a whole, where jobs requiring special machine tools were done. It was more economical to do as much as possible in a small workshop belonging to a group than in a central workshop, where personal contact was less close, and tiresome problems arose such as whose job should be done first.

This system led to a certain duplication of tools, but it was far cheaper to keep machines than to waste the time of the researchers. They had a 'students' and 'boys' workshop, where research students could learn the use of tools, and young assistants could be trained. There was also a 'special techniques' workshop, and one for dealing with electronic gear.

A considerable secretariat was necessary, with so much 'chasing' and ordering of stores and apparatus, and the carrying out of building operations. Bragg recalled his early days at Manchester, where with great diffidence he asked the University for a secretary–typist. He was told he could have one half-time, if he met a good deal of the expense himself. At the Cavendish they now had a clerical staff of nine, and a Laboratory Secretary who was really a works manager, looking after finance, engagement of assistants, preparation of matter for meetings, payments to students, ordering of apparatus, care of buildings and other administrative tasks.

The teaching of experimental physics was now carried out in five practical laboratories, each with its own staff. Teaching on this scale becomes impersonal, but in Cambridge the situation was saved by the college system. As well as duties in the Laboratory, most of the teaching staff had a college appointment. Each of these, in his own college, had about a dozen students in his charge, who came to him singly or in pairs weekly and could talk over their difficulties.

The elucidation of a point that was blocking a student's progress was very important; it might affect his whole career. Subtleties scarcely explicable in a lecture or a crowd could be conveyed in a personal conversation in a tutor's room.

Elaborating Bragg's point, it might be said that the student's sense of quality as well as information was strengthened, which in turn contributed to the intellectual atmosphere of the Laboratory.

Bragg observed that college teaching was done not only by the regular University staff, but also by some of the senior research men who were interested in it. Direct personal teaching by eminent investigators was invaluable, and it was fortunate that there were enough to contribute to this significant work.

In 1948 about 160 researches were being pursued. The radio group directed by J. A. Ratcliffe were studying the upper atmosphere by radio waves, and the radio waves reaching the earth from outside. One of these researches was on the 'Luxembourg' effect, where signals from one transmitter affected the transmission of those from another on a very different wavelength. They were studying the effect on 400-metre waves from Edinburgh, which were received at Cambridge through reflection in the ionosphere, of 1500-metre waves from an intermediate station, being emitted at the same time. When these were on, the signal from Edinburgh diminished in strength. It was a result of increased absorption of the Edinburgh waves in the ionosphere, owing to the speeding-up of its electrons by the waves from the 1500-metre station. The B.B.C. co-operated with the Cavendish by sending out the signals.

Very fascinating work on one-metre radio waves from the sun was in progress. The Cavendish investigators had concentrated on determining what part of the sun the waves came from. The waves were picked up by two receivers spaced some distance apart. As the earth turned round, the waves reaching each end of the base were alternately in and out of phase, producing interference fringes in appropriate recorders. By increasing the distance between the receivers, and observing the effect on the fringes, an outer limit for the size of the place they came from could be calculated. It had been found that the place was smaller than the sun itself, and was almost certainly a sunspot. This radio research was a new kind of observational astrophysics, free from interruptions by the weather.

The low-temperature researches in the Mond Laboratory had been directed by J. F. Allen, who had just left for a chair at St Andrews. D. Shoenberg was now in charge. They were investigating the properties of liquid helium, superconductivity and related subjects. Liquid helium was available 'on tap' three times a week. One of their subjects was ferro-electrics, which have electric dipoles, and properties analogous to those of magnetic bodies. They were industrially important for condensers.

The crystallography group was directed by W. H. Taylor, and was the one with which Bragg himself was most closely involved. Here it will be convenient to interpose in Bragg's observations of 1948, an account of the Crystallographic Laboratory.

When Bragg came to the Cavendish he had originally brought A.J. Bradley with him. Bradley had collaborated in the notable work at Manchester on the X-ray analysis of metal structures, which had become Bragg's central interest. Unfortunately, Bradley's health began to fail, and the task of running the Crystallographic Laboratory devolved largely on Lipson.

Bragg sought for other help, and in 1943 had asked his former Manchester colleague, W. H. Taylor, whether he would be willing to come. Taylor had been one of Bragg's most brilliant Manchester colleagues, having in 1933–4 determined the atomic structural scheme common to all the felspars, one of the major classes of minerals, which form 60 per cent of the earth's crust. From 1936–45 Taylor was at the Manchester College of Technology, whence Bragg finally persuaded him to come to the Cavendish. Lipson then went to the College of Technology to succeed him. Taylor was sorry to leave Manchester, both professionally and socially.

Bragg's main interest at this time was concentrated on metals. Taylor contributed to the broadening of the lines of crystallographic research. The crystallographers were organised in the Crystallographic Laboratory, which was part of the Cavendish, but had an independent statutory existence; in this respect, its status was different from that of the Mond Laboratory and other groups. These arrangements had arisen in the 1930s, when there were discussions on whether the Crystallographic Laboratory should be in the Cavendish or in the Mineralogy Department.

One result of this was that there was close co-operation between the Cavendish and the Department of Mineralogy in this field. W. A. Wooster, who had done distinguished research with Rutherford and Ellis on α-rays, and subsequently became University Lecturer in Mineralogy, was one of their valuable collaborators, supervising research and engaging in teaching. He was particularly good with students from abroad, especially from India.

Taylor secured Helen Megaw from Bernal's laboratory to work on silicates. No woman had previously been given a staff job in the Cavendish. In order to provide a suitable appointment, a fellowship was arranged for her at Girton. She became Taylor's right-hand colleague.

A letter was received from C. A. Beevers at Edinburgh, recommending W. Cochran, then a talented student. He was taken on, and led a group on X-ray diffraction and lattice dynamics.

Lipson's use of the X-ray microbeam had shown that it was a promising tool for investigating imperfections in the structure of materials. P. B. Hirsch, who joined the Laboratory in 1946, pursued this line. Another line led to collaboration between Jane Brown and Bruce Forsyth, at Harwell. This extended to still wider collaboration, with workers at Grenoble.

A course of instruction in X-ray analysis was carefully worked out. The different aspects were balanced, so that students would be well trained for entering any branch of solid-state physics. This was an evolution in the direction of teaching for a mixed Tripos examination. The course was based on the crystalline state, and was dependent on teaching in mineralogy, but involved increasingly closer contacts with chemistry and metallurgy. Such a training might lead to a degree that would be useful for teachers in schools and polytechnics, as well as for research workers.

Taylor had worked in Bernal's laboratory in the Cavendish in 1934-5. As a visitor from Manchester, he was struck by the wide variety of science in the Cavendish. He described Bernal as 'spraying out ideas', but he left it to others to pick them up and work them out. It may be added to Taylor's description that Rutherford loathed this impressionistic mode of research. He demanded that the author of the bright idea should prove it by elegant experiment and irrefutable logic.

Bernal could not bear Rutherford's heavy insistence on the obvious. For him, he was 'limited'. Rutherford could not abide the untidy prophet of the permissive society, and he transferred his distaste to the sciences that happened to be associated with him. Bernal's laboratory was starved of resources, no doubt partly owing to this situation. Fortunately, Bernal succeeded Blackett in London in 1938.

Returning to Bragg's reports, the Manchester work on the X-ray structure of minerals was continued. Taylor extended his studies of the felspars, investigating such properties as the variations due to the thermal history of the earth's crust. Technique had so advanced that they were now able to get useful results from X-ray pictures made from a microgram of material; when Bragg and his father had started X-ray crystal analysis, they had begged for crystals several centimetres across!

Another line of investigation was on that artificial and industrially very important mineral, cement. These researches were on the X-ray structure of materials consisting of comparatively small molecules.

At the other end of the scale, M. F. Perutz was investigating the structure of a very large molecule, that of haemoglobin, with a molecular weight of 70 000. Bragg commented: 'The goal is very attractive, because if only we could find out the arrangement of the units in a molecule of living matter it is hardly an exaggeration to say that a new branch of science would be opened up.'

The metal physics group was directed by E. Orowan. They were investigating the structural phenomena bearing on the strength of metals: creep, brittle fracture, rolling, plastic flow and other properties. Besides their scientific interest, these researches were of fundamental industrial importance.

Bragg himself was personally interested in making models of crystalline structures, as an aid to understanding and elucidating their properties. One of these consisted of a large mass of very tiny bubbles blown in soapy water, among the advantages of which was that they could be produced cheaply in the enormous numbers necessary (see figure 27). Such a structure was, in fact, rather like that of a metal, and showed many of its properties, such as slip and dislocation. This is an example of Bragg's special gift for pictorial thinking on physical phenomena.

The nuclear physics group had been under J. D. Cockcroft, who had recently left for Harwell, and had been succeeded by O. R. Frisch. It had about fifty members, and was the largest group in the Laboratory. Bragg said that it was the one on which he felt least competent to report.

The vastness of the equipment required for nuclear research was a major problem. They could not match the enormous machines installed in other places, especially in the United States. The project for a new experiment in this field now often cost £100 000, and several years might pass before its results were realised.

They had one- and two-million-volt machines, and 'a cyclotron as temperamental as most are'; and they were putting up a six-million volt Van de Graaff machine. Plans were being made for the construction of a linear accelerator for a considerably higher voltage.

Frisch intended to study the properties of the lighter atoms, for which these very high energies were not so necessary. Bragg thought that, as has often happened in the past, some relatively simple apparatus would in the future be discovered, which would provide a key to phenomena that were at present a mystery.

283

In 1950 Bragg reported to the University that the total number of students in the Laboratory had increased to 513, from 459 in 1948–9. However, the number of first-year students had declined from 234 to 204; the post-war enlistment was beginning to contract.

27 *Bragg's bubble-model of the arrangement of atoms in a crystalline solid*

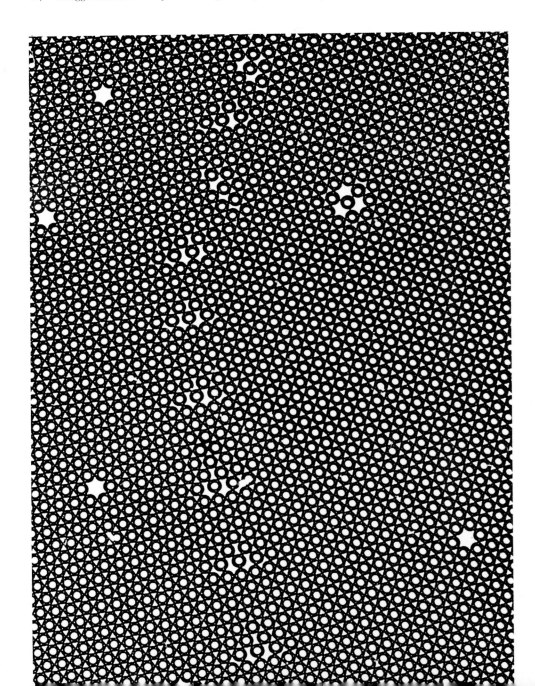

Features in the teaching were a new course in theoretical physics by Hartree and Temperley, attended by fourteen students, and changes in elementary courses which eliminated repetition for students who had arrived already well prepared. This enabled them to proceed more quickly to advanced work without wasting time.

Courses of entirely voluntary lectures on arts were organised. The subjects included the Ancient World, covering Egypt, Sumer, the Indus Valley, and Greece and Rome; and the Growth of English Literature, including Chaucer, Marlowe to Milton, the relations between literature and thought from the seventeenth to the nineteenth century and modern poetry. These lectures were given by well-known authorities, and at times when many physicists were about the Laboratory. The average attendance was 220.

Special science courses were given by C. F. Powell on his cosmic-ray researches, for which he had recently received a Nobel Prize, and V. E. Cosslett conducted a summer school on electron microscopy, which had an attendance of thirty-eight.

Besides the meetings of the Cavendish Physical Society, all research sections held weekly or fortnightly colloquia.

Among the appointments was A. B. Pippard as University Lecturer in succession to S. Devons, who had been appointed to a chair at Imperial College, and T. Gold to be a University Demonstrator.

The staff included three professors, four readers, eight lecturers in experimental and two in theoretical physics.

The 123 registered research students were broadly divided among nuclear physics (37); radio waves (20); crystallography (17); low temperature (10); metal physics (9); and fluid dynamics (3). Including the staff, 161 researchers were active in the Laboratory, and during the year 114 papers of various kinds were published.

In nuclear physics there had been problems with the cyclotron, which had been running on rather low power, owing to difficulties of shielding from the radiation.

In radio, progress had been made in measuring the velocity of the winds in the ionosphere, and in the radioastronomical researches; the number of radio stars located had risen to fifty.

In crystallography, research was continuing on minerals such as felspar and afwillite, and the structures of alloys, especially of aluminium with the transition metals, were being investigated.

A distinct advance had been achieved in the technique of the analysis of the structures of organic compounds.

In work on the structure of proteins, financed by the Medical Research Council, an attempt was being made to elucidate the structure of the very large molecules that constituted living matter. The difficulty of the work was great, but so was its importance. Any features of the polypeptide chain in the proteins that could be established would be of the greatest interest in giving some clue as to why these substances are such universal constituents of living matter.

The collaboration of the Mathematical Laboratory, with its computer equipment, had been of the very greatest service in summing the Fourier series used to express the results of X-ray analysis.

In low-temperature physics the Kapitza helium liquefier installed in 1934 had been replaced in 1949, after long and excellent service, by a new machine with three times the output. Their researches in superconductivity and liquid helium included measurement of the depth of penetration of the magnetic field, and they had obtained interesting new results in the kinetics of transition from the normal to the superconductive state. New methods of measuring the mean free path of electrons in metals had been worked out. Investigations below 1 K had been carried out to decide between two theories of the 'second sound' waves phenomenon.

Temperatures down to 0.1 K had been attained, and two new superconductors discovered, osmium and ruthenium.

In electron microscopy research into its fundamentals and the application of the electron microscope as an instrument had been pursued. Routine examination of specimens for other, especially the biological, departments of the University, had been carried out. The number of photographs taken had risen from 1900 in the previous year to 2560. Research on instrument development had grown to a volume that had not been anticipated. A new electron microscope had been devised for metallurgy, making use of examination by direct reflection of electrons. The great majority of metal specimens were too thick to be penetrated by electrons.

In metal physics, research was being conducted on silver chloride and sodium chloride as models for metal crystals. The phenomena of deformation were being investigated. Other subjects were aluminium single crystals; creep in zinc crystals; creep in crystals of ice; yield and ageing in iron of low carbon content; a revision of Orowan's theory of creep; the high-speed propagation of cracks; plasticity in glaciers; and dislocation theory of plastic deformation.

To find rooms for all this activity, larger laboratories were being divided by partition walls. A factory site in Madingley Road was being considered as a laboratory for the proposed high-energy linear accelerator. Plans for the modification of the Austin Wing, which was only about ten years old, were approved. These involved better arrangements for the unloading of stores, ventilation, and the use of the main hall.

In the summer the Laboratory received about 100 casual visitors each week. A catalogue of the museum of Cavendish materials had been made to assist these visitors.

Lastly, the meteorology section directed by T. W. Wormell had been transferred from the Observatory to the Cavendish Laboratory in 1950, and the Napier Shaw Library had been housed in the old library in the Cavendish.

Such were the activities in the Cavendish when Bragg had come thoroughly into control. Several new features began to stand out, especially radioastronomy and molecular biology. Bragg dwelt on Ratcliffe's researches on the horizontal movements in the ionosphere at heights of 100 km and more, and the extension of the number of radio stars detected to more than 100, all of which were more than a light-year away.

He stressed the general aim of the X-ray work as the improvement of the technique, so that ever more complex substances could be analysed. W. Cochran had determined the position of hydrogen atoms in organic substances; this was a difficult problem, because these atoms produced such slight diffraction effects.

Many X-ray analysis experiments were simple to make, but the interpretation was difficult. Computing the results was like trying to break a very hard cypher. They used punched cards and digital computers, and optical methods of simulating results, as accessories to pure and simple computation. Methods of trial and error, which had been effective in the early analyses of simple molecules, were impossible when there were 10000 atoms in a molecule.

Cosslett had developed stereoscopic microradiography. He had shown the organs in the wing of a *drosophila* fly in three dimensions.

Bragg was attracted by any kind of interesting experiment. He commented on R. M. Goody's meteorological research, in which he had made an infrared spectrometer that could be fitted in the tail of a Mosquito aircraft. With it the proportions of the components of the atmosphere at heights of 12000 metres were investigated. Bragg drew attention to Perutz's experiments on glaciers as well as on haemoglobin, and to the researches of the nuclear physicist D. H. Wilkinson on the migration of homing birds by methods analogous in principle to those used in the analysis of heat flow. Wilkinson attached a radium clock to a quill of a bird's wing to record the bird's flying time. His experiments showed that random search by the bird in finding its way home was far more important than had been realised. While the experiments did not disprove that the birds had some special mechanism for homing, they were not inconsistent with the bird's dependence on purely random search.

Bragg also appreciated the need for the Cavendish to look after its archives. He arranged that D. J. de Solla Price should start putting them in order.

The problems of organisation and finance were very much in his mind. He repeatedly remarked that there had been a tendency in physics to go too far in the direction of central control. A general strategic direction of research was necessary, but detailed tactical control was death to discovery. The research worker should be told what scale he can work on, and then be left to plan and spend his resources as he thought best.

He commented that the cost of providing a research place in the Cavendish was on average £600 a year. A crystallographer's place cost about £350, and a nuclear physicist's £1000 a year.

Rutherford and Bragg both concentrated very much on their own lines of work. For Rutherford the Cavendish was largely a personal institution, in which nuclear research under his general direction was carried out by the hands of others. Bragg also had a strong personal approach to his own line of research, but after the war a variety of other lines quite suddenly became prominent. While approving of them strategically, he left the groups of scientists tactically to get on by themselves.

During the war the Cavendish researches had largely been suspended, but Halban and Kowarski, who arrived from France, pursued work that contributed to the beginning of the atomic bomb.

After the war molecular biology extended naturally through the pursuit of crystallography. It appeared a strange subject to members of earlier generations of Cavendish men. When Crick and Watson made their first model of the double helix, various younger and older members of the Laboratory first heard of it one day at tea. Searle, then eighty-nine, asked if he could see it. When taken along to observe 'the secret of life', he stared at it and commented that if that was the basis of their heredity, 'no wonder we human beings are such a queer lot!'

Nevertheless, when the molecular biologists left the Cavendish, many thought it was a pity; if they had remained, they would have added still more to the Cavendish's fame.

Rutherford and Bragg were the last of the professors who were primarily personal workers. The increasingly complex problems of management required a different kind of direction.

By 1953 discoveries of world importance had emerged in Bragg's own field of X-ray analysis and in radioastronomy. He was sixty-three years old, and was naturally attracted by a post with less onerous administration, for which he had no particular taste, and which was not in consonance with his kind of genius. It happened that a crisis arose in the Royal Institution, where his father had been Director of the Laboratories for nearly twenty years. The conditions there were perfectly suited to Bragg's qualities, and he was pressed hard to accept the directorship. He decided to do so, and for fourteen years continued his fruitful career. In his laboratory at the Royal Institution, Phillips solved the structure of an enzyme, announcing his discovery on Bragg's seventy-fifth birthday, and Bragg himself inaugurated a system of lectures for schools, in which he introduced tens of thousands of school-children to science, with unsurpassed skill and fascination. It was one of the most valuable efforts of his creative life.

Before leaving Cambridge Bragg and his wife were entertained to a farewell dinner in the Cavendish. In his speech Bragg expressed some of his most heartfelt thoughts and principles.

Bragg looked for and attracted good men, and supported them when he saw they had a good idea. This was not confined to advice and the provision of resources. He personally made drawings for Perutz and Kendrew's papers, because he liked doing the job; he supported Ryle when he sought for radio waves from space, supposed by the radio experts to be below the limits of observation. The achievements of his molecular biologists and radioastronomers delighted him.

The aspect of the Cavendish described by Watson in *The Double Helix* was invisible to some of the other eminent observers on the spot. Frisch remarked after reading the book that he realised he had had no idea of what was going on in some parts of the Cavendish.

Bragg appeared to his technical assistants as a man who was very scientifically minded, in that his emotions were deeply involved in science. He was reserved and conventional until the ice was broken, and then he could be most charming. He was interested in the families of his assistants. One day he found a small boy, the son of one of them, in a corridor. Bragg asked him what his name was, and the boy replied 'my name is Stephen, Sir'. Bragg was pleased with him, and whenever he met his father enquired: 'How's that nice little boy of yours?' He enquired about the families, and knew how many there were in each. The Braggs gave a Christmas party, in which he entertained the children with delightful skill.

22

The Technicians' Contribution

The relations between the technicians and the research scientists was a crucially important factor in the achievements of the Cavendish Laboratory.

The contribution of the technicians may be illustrated by that of G. R. Crowe, who was Rutherford's personal assistant during the whole of his directorship of the Laboratory. He joined the staff as a boy in 1907. After he had been working there several years and had learned the technician's job he applied, as a rather young man, for a technical post in New Zealand. Rutherford was the referee, and Crowe went to Manchester to see him. He did not get the job, but Rutherford appreciated his qualities and told him that he would keep him in mind. During the First World War Crowe went to work in the Cambridge Scientific Instrument Company.

In 1919, when Bragg was appointed to succeed Rutherford at Manchester, he told Crowe that Rutherford wished to know whether he would be willing to become his assistant. After seeing Rutherford in London, Crowe was appointed. He went to Manchester to learn the technique of handling radioactive materials.

At the Cavendish Crowe prepared the radon and other radioactive sources required, working under Chadwick's supervision. Rutherford depended on Crowe's great manual skill in the preparation of materials and the setting up of apparatus. Crowe appreciated Rutherford's excitement in experiment, but was apprehensive when he handled apparatus himself, because of his clumsiness. He found the personal experience of working with Rutherford wonderful.

But in 1926 his fingers began to show signs of radioactive burns; it appeared that he had not strictly followed the rules for handling tubes containing radioactive materials, sometimes using his bare hands instead of the stipulated gloves. The skin on his fingers deteriorated, in spite of treatment and repeated skin grafts; one finger was amputated. He was taken off radioactive preparation but was active in other work. However, he could no longer play the piano or shoot as well as he used to.

Crowe was retired in 1959 from his position as principal assistant, although he felt that he was still doing a useful job. He walked out of the Laboratory on a Friday, just as if he was going home in the usual way. According to Crowe, nobody said goodbye to him, and the Professor, then W. L. Bragg, would not have known of his departure if he had not gone to him to say goodbye. Subsequently, Bragg proposed that the University should confer on him an honorary degree, and Crowe was given a Master of Arts degree in consideration of his long and devoted work at the Cavendish.

Crowe was a skilled craftsman who could do most things that were required in metal, glass or photography. He was an artist in handling wax and mica. Probably his greatest single contribution to Rutherford's work was the preparation of the thin glass tubes in which the radon was sealed.

He was a major colleague in the preparation and operation of the experimental illustrations to Rutherford's lectures at the Royal Institution during the 1930s. These were outstandingly well done. Rutherford probably took more trouble over them than over any other lectures. He and Crowe regarded them rather as a holiday, and while the lectures were being prepared, their work in the Cavendish was largely suspended. Oliphant and Wynn-Williams assisted in translating Rutherford's ideas for the lectures into instrumental arrangements, and Crowe carried them out.

In 1934 they set up a portable Cockcroft and Walton accelerator, and counting equipment, for an Evening Discourse in the Royal Institution, giving the first public demonstration of the artificial disintegration of atoms.

Crowe had many stories about collaborating with Rutherford, which were summed up in his account of Rutherford fumbling and failing in a manipulation, and saying furiously to him: 'Don't shake the bloody bench, Crowe'.

He was partially deaf, physically strong, and of a phlegmatic temperament. Bursts of temper passed over him, and as he could not hear well, Rutherford was never quite sure whether this was rude obstinacy or merely because he had not heard him, but he decided it was better to forget it, and get on with the experiment. Crowe was an enthusiastic player in the Cavendish cricket match, in which his calm and unruffled temperament also showed to advantage.

When Bragg succeeded Rutherford he also called on Crowe for assistance with his Royal Institution discourses. These were quite exceptional for clarity, thoroughness, ingenuity and delicacy of technique.

During the Second World War Crowe assisted Bragg, and after it he assembled and operated the first electron microscope in the Cavendish, a line of work which was to contribute to the famous biomolecular discoveries that came a few years later. Crowe was an expert in preparing the specimen replicas needed in this field of research. Bragg's insight in utilising Crowe's gifts was one of his valuable contributions to the continuation of the Cavendish achievement. Crowe was known to a few as 'Crowe', but to most as 'George'.

J. J. had found his personal assistant, Everett, in 1886. He was then known as a boy called 'Ebenezer', who was working in the Chemical Laboratory under Liveing. J. J. started by employing him out of working hours to make urgently wanted apparatus; then, with Liveing's agreement, he engaged him. As the Fourth Lord Rayleigh put it, he was promoted from 'Ebenezer' to 'Everett'.

Everett worked in the same room as J. J., keeping his blowpipe for glass-blowing there. According to Rayleigh, he sat on a high stool as he blew, which professional glass-blowers do not do. He was self-taught, but could produce just what J. J. required. He was mainly occupied with setting up the apparatus, evacuating the tubes, and getting things ready. He made rough tables and stands that served their purpose. When Rayleigh commented on their scratch character, Everett rather tartly told him that the professor was too impatient to wait for neat finishing; he wanted to get on quickly.

Besides assisting the Professor, Everett made apparatus for other research workers in the Laboratory, and taught them glass-blowing. He did not know much science, and did not always understand the aim of an experiment; he even disliked having experiments explained to him. He considered it his duty to do what the Professor wanted, not to understand why. He could not be left with the responsibility of making observations.

However, Crowe learned from Everett, and his own observation, that J. J. was never to be allowed to touch any apparatus; if he did, things inevitably went wrong. For instance, Everett had made some very thin metal discs for an experiment. As J. J. talked about the experiment, he unconsciously picked up the discs, rolled them into balls, and put them down, quite unaware that Everett would then have to make a new set of discs.

Rayleigh has recounted an incident in which J. J. was trying to deflect positive rays with a magnetic field. J. J. was observing the movement of a gold leaf through a scale, to give a reading. He instructed Everett to switch on the magnet. Everett closed a large switch, which made a noticeable click. J. J. repeated his instruction to put the magnet on. Everett replied that it was on. J. J., still peering down the microscope, said it wasn't, and told him again to put it on, and Everett again said that it was on. Then J. J. called for a compass needle. Everett returned with a large compass containing a needle nearly a foot long, used for teaching in elementary lectures. J. J. took it and approached the electromagnet. When he was about a foot away, the needle was so strongly attracted that it flew off, and crashed into the glass globe producing the positive rays.

Those who watched observed Everett glowing with triumph, while J. J. dejectedly stared at the wreckage, remarking that after all it was on.

J. J. did not mind interruption, and Everett tried to establish the principle that his room was private, but when Rutherford and McLennan worked there as research students, the principle faded away. J. J. and Rutherford chatted on and off, discussing the latest papers that had appeared, besides their own work. When the discovery of X-rays was announced, it was Everett who made the X-ray tubes for the experiments that had a revolutionary impact on the minds of the physicists, especially Rutherford, and provided an essential means for the discovery of the electron.

Everett's tubes were also used by Hayles, the lecture assistant and photographic expert. Hayles made many X-ray photographs for surgeons practising in Addenbrooke's Hospital.

One of Everett's tubes was also of prime importance in C. T. R. Wilson's demonstration that the nuclei always being produced in small numbers in the air within the cloud chamber were of the same kind as those produced in large numbers by the X-rays.

In 1926 the annual Cavendish dinner was changed into an occasion of celebration of J. J.'s seventieth birthday. Rutherford took the chair, and others at the top table besides J. J. and Lady Thomson included Langevin, Schuster, Newall, Glazebrook, Threlfall, Lodge and

Larmor. Also present were Everett, Lincoln and Hayles; and Rolph, who had been associated with the Laboratory even before J. J.

Lincoln was appointed superintendent of the workshops in 1899, in succession to W. G. Pye, who had left to found his now-famous business. Lincoln had joined the Laboratory in 1892, as a boy so small that he had to stand on a box to work at the bench. He learned in those days, when expenditure in the Laboratory was extremely low, to be very economical, and he imposed this rule on others. He was an erect man of medium build, with thick dark hair and complexion, and a large moustache with fine waxed points. He had a rather forbidding decisive manner (see figure 10).

If a research worker wanted a piece of wire or rubber tubing, it was measured to the inch. If rare things, such as tungsten wire or sheet nickel were required, he offered scrap from an old tobacco tin, unless the applicant was sufficiently determined to accept only new material. He kept these stores in an assortment of locked cupboards. When he was not about some men unscrewed the backs from the cupboards, and took what they wanted. He was particularly keen on good wood, and as a rule issued only the coarser sorts. J. A. Ratcliffe once wanted a piece of good wood for part of an apparatus. He went to the workshop to ask Lincoln for it, but he was not there, so he helped himself to a nice piece of mahogany. When Lincoln heard of it, he looked gravely at Ratcliffe, and said to him sorrowfully: 'and 'ad it to be a bit of *mee'ogany?*'

Oliphant's work was once held up for some weeks because Lincoln had given him what turned out to be a piece of German silver rod, when he had been assured that it was pure nickel. Nevertheless, Lincoln was very popular, and when Oliphant moved to a chair at Birmingham he took as his personal assistant one of Lincoln's boys, M. P. Edwards, who stayed with him until he retired.

Some of the most interesting and penetrating recollections of the Cavendish have come from the technical assistants. They saw the scientists from a detached and a different social point of view. W. H. Andrews, the senior teaching assistant in 1973, joined the Cavendish in 1931, when he was sixteen years old. He recounted how, just after he joined the Laboratory, he was told that the professor wanted to see him in his study. He was terrified, for sometimes Rutherford seemed like a bully. When he went in Rutherford was holding his tobacco pouch, and said to him: 'You know where I live. You'll find the tobacco drying on the window-sill. Fill the pouch and bring it back.'

Rutherford's friendly confidence in sending him into his own home restored his own confidence, and gained his regard.

1931 was a period of economic slump when jobs were hard to find. Andrews was at the County Secondary School and due to leave. One day, when his cousin went to the bank to collect staff wages he happened to meet Lincoln there. He asked whether there were any jobs at the Cavendish, and Lincoln arranged that the cousin should see Rutherford. Andrews got a job.

At that time the assistants included Everett, Cole, Jack Lindsay, Neill and Edwards, and half-a-dozen junior assistants under the age of twenty. They had had little or no technical training before they came to the Laboratory. Some went to evening classes to improve their qualifications; they sat for City and Guilds certificates in engineering drawing and workshop practice, and the Laboratory paid their fees.

Andrews started at a wage of ten shillings a week, which was increased by half-a-crown at Christmas. After the cost of insurance stamps was deducted, he had 11s 4d. Increases were annual, and at twenty-one he was receiving £2 a week. Assistants were included in the University pension system. In 1939 his income was about £2 17s 6d a week.

Lincoln was then the heart and soul of the workshop side of the Laboratory. He was a very strict disciplinarian, and highly skilled. His primary job was to administer the Laboratory workshops, passing out work, and maintaining equipment and apparatus. Repairs were done in the workshop; coils and transformers were wound; instruments were made. There was just one motor for driving all the workshop machinery. One of Andrews's first jobs was to make wooden clips for holding glass tubes, and to learn hand-lacquering.

Crowe, Rutherford's assistant, was a very able man. Oliphant has recorded that in a moment of exasperation over a research man, Rutherford exclaimed that if only there were thirty Crowes's instead of thirty research workers, research would proceed much faster.

Skill in repairing apparatus was particularly needed in technical assistants. Among them, J. Fuller had a full technical qualification, and Edwards had received some training at college.

Andrews thought that Blackett, whose assistant was Aves, was good at mechanics; Wynn-Williams was very inventive; Oliphant was good in the workshop; and Carmichael's designing of apparatus for high-altitude sounding was impressive.

296

In 1939 there were about twenty assistants. Those who were married, the seniors, were referred to as 'men', and the juniors as 'boys'.

Under Chadwick's persuasion Rutherford presently acquired an exceptionally skilled glass-blower, Niedergesass. To Oliphant he appeared a cadaverous and untidy person, his fingers yellowed with nicotine, but with uncanny skill in the manipulation of glass. He made his own blowpipes. He was skilful and obliging, and free from the temperamental peculiarities of most glass-blowers. He would try anything no matter how difficult, and if a piece of glass apparatus that had cost him many hours of work was broken by the experimenter through carelessness in handling, he did not lose his temper. To Andrews, a skilled man himself, he appeared highly skilled.

During the Second World War several of the assistants went with the scientists to assist them in work in government departments. Andrews, however, stayed in the Laboratory. Teaching still went on, though the work in the new Austin Wing was secret. There were a lot more students than before, owing to the evacuation of various student bodies, in particular from Queen Mary College, Bedford College and St Bartholomew's Hospital Medical School.

One of the professors from Bedford College was W. Wilson. When Andrews first saw him from a distance, he thought he was Chamberlain. He talked about relativity at such length that Searle fell asleep listening to him; when he woke up, Wilson was still talking about relativity.

Andrews used the room formerly occupied by T. G. Bedford. One of their war jobs was training people in radio for the Government.

Bragg installed the first electron microscope in 1940. Crowe was assisted by Chapman in setting it up, and getting it working. Bragg was back, and in full harness, in 1946.

In Rutherford's day, from the technician's point of view, there were two main sections at work: the atomic physics boys, and workers on radio and ionospherics. Bragg set up a section on crystallography, and reorganised the staff. People returning from the war had learned electronics, and the skill they had gained in the war was applied in the workshops. These were increased in size, and students used the older parts. The workshop staff increased to nearly fifty.

Under Rutherford research had concentrated on atomic physics. This was not due to a decision of policy: Rutherford encouraged the scientists to propose and develop their own ideas and lines of research.

But his personal enthusiasm was so contagious that nearly everyone chose to work in atomic physics. Under Bragg the pursuit of several different lines of research became more distinct, and the weight of influence among the various lines was more balanced. Each section wanted to have its own workshop.

In 1972 there was a main workshop, and subsidiary workshops for electron microscopy, crystallography, radio and surface physics, which was very large, and developing very fast.

Rutherford had the welfare of his assistants at heart; he kept his own eye on them. Latterly, with the growth in size, this had to be delegated to others.

Andrews remarked that there were plenty of quips and cracks about Lincoln. Everybody enjoyed themselves, but the work was done. Lincoln kept a strict eye on them.

Rutherford said to Crowe: 'You, Crowe, have your job; I've got mine'. Crowe used to say: 'Nothing is impossible; the impossible takes a little longer'.

28 *Rutherford and J. J. watching the game at the annual Cavendish cricket match*

29   *W. B. Lewis and F. Lincoln at the annual match*

Chadwick, as the immediate superintendent of the Laboratory, was very fair to the assistants; if you wanted a half-day off, he would grant it. But it was no use asking Lincoln.

By the 1970s there was a change of atmosphere in the Laboratory. From the technician's point of view, everything was in a rush. Research men wanted photographs in a day or two. The reason for all this rush was not evident. Perhaps a keenness to get Ph.D.s had to do with it. The technicians wondered whether some of the research students were keener on getting their Ph.D.s than in making discoveries.

The formalisation of the technical assistants led to a different form of representation. It seemed also that in future assistants would be graduates.

The requirements in buildings changed rapidly. It had been expected that the Austin Wing would meet requirements for at least thirty years, but after the Second World War, that is, within less than ten years, it was quite out of date. The situation was mitigated by adding an additional floor for theoretical physics.

In Rutherford's day the Cavendish Physical Society met once a fortnight. It was social as well as scientific. By the 1970s there were many more colloquia.

One must not forget that the Cavendish is, after all, a part of the University: it has to serve the University's purposes, as well as pursue its own interests. There have been times when the University appeared to be an appendage of the Cavendish.

The Cavendish tradition is different from that of other departments. It is connected with its very close-knit system of group research. In it, the technicians and laboratory assistants have had an important place; they were assimilated into the group. Consequently, the Cavendish assistants have shown a special loyalty to the Laboratory, and pride in its achievements. Some of them were men of exceptionally strong personality, stronger than that of some of the quite well-known scientists they were assisting; as men, they were more striking. In other departments, assistants were often treated as employees; in the Cavendish they were part of the team. The tradition has survived among the assistants, owing to the development of the group organisation, which has again provided a closely-knit unit in which they are incorporated.

When the Department of Physics engaged Dibden as administrative secretary, one of his duties was to look after the assistants. This was so successful that he was charged with looking after the assistants in all the University science departments.

Evidently, there was something special about the conditions in the Cavendish, which drew attention to, and understanding of, laboratory assistants. One factor in this was no doubt the craft character of a great deal of research in experimental physics. This is clear from Blackett's description of what was actually done by a Cavendish experimenter.

The common craft factor in the activities of the experimental scientist and the laboratory assistant stretched across social class barriers. Nevertheless, the assistants wished to preserve their own class identity; they did not welcome social familiarity from the scientists.

One of the features of Cavendish life was the annual cricket match between the laboratory assistants and the scientists. This event was proposed by the assistants, and adopted. Rutherford used to declare a general holiday, the Laboratory was closed for the day, and a whole day's match was played. Rutherford and J. J. were among the spectators (see figure 28) and Lewis and Lincoln were among the players (see figure 29). On the occasions when Cockcroft and Massey opened the innings, the scientists usually won.

23

Physicists and Molecular Biology

In 1948 Bragg described J. D. Bernal's development of X-ray methods for analysing the structure of complicated molecules, by drawing conclusive results from a large number of observations each in themselves inconclusive, as the most important since the original foundation of the subject. In 1923, after graduating at Cambridge, Bernal engaged in research in X-ray crystallography under W. H. Bragg at the Royal Institution. He returned to Cambridge as lecturer in structural crystallography in the Cavendish in 1927, and became Assistant Director of Research in this subject in 1934. In 1938 he succeeded Blackett as professor at Birkbeck College, London.

Bernal's insight and development of X-ray analytical methods enabled him to advance in the Cavendish the analysis of the structure of complex chemical and biological substances. He investigated the structure of amines, and made a crucial discovery about the structure of sterols. In 1934 he obtained the first X-ray picture of a protein. Bragg regarded this as the pioneer discovery in protein molecular structure. Bernal and his colleagues succeeded with a crystal of pepsin, owing to their realisation that the crystal must be kept in its mother liquor if it were to retain its structure.

Five years later Bernal reviewed the situation, describing the structure of proteins as 'the major unsolved problem on the boundary of chemistry and biology today'. Direct X-ray analysis of crystalline proteins provided abundant data on their structure, but it was 'extremely difficult to interpret'. An ambiguity was introduced through the impossibility of knowing the phases of the reflections corresponding to the different spots. This ambiguity could 'only be removed by some physical artifice, such as the introduction of a heavy atom'.

Bernal commented that 'the problem of protein structure is now a

301

definite and not unattainable goal, but for success it requires a degree of collaboration between workers which has not yet been reached'.

When Bernal moved to Birkbeck in 1938 he left behind only one of his group, M. F. Perutz, who had entered his laboratory at the Cavendish in 1936 in order to learn the technique of X-ray crystal analysis.

Perutz was then twenty-three. He came from Vienna, where he had graduated in organic chemistry. It was the work and reputation of F. G. Hopkins that had attracted him to Cambridge. When Perutz first entered Bernal's department he had found it housed in a few dirty, ill-lit rooms on the ground floor of an old building. Bernal had a fascinating optimism about the possibilities of the X-ray method. He and his followers lived in a thrilling little world of their own, shut off from the thunderous intellectual reverberations of the nuclear discoveries being made in other parts of the Laboratory.

Before Bernal had left, he and Fankuchen had shown Perutz how to secure and interpret an X-ray picture of a crystal of haemoglobin. In the following year Perutz and Fankuchen published X-ray pictures of crystals of haemoglobin. Perutz had to conduct the biochemical side of this research in Keilin's laboratory, which was some distance away. So, from 1938 Perutz cycled constantly between the Cavendish and the Molteno laboratories.

After Bragg took charge in 1938 Perutz waited for six weeks for him to visit his laboratory, but as he did not come he concluded he had better go and see him, to show him the X-ray pictures of haemoglobin. Bragg instantly saw the significance of the application of the X-ray method to the analysis of giant molecules like haemoglobin, which contains 70000 atoms, and performs such an important biological function as the transmitter of oxygen in the living organism.

At this time Perutz's personal position had become very difficult. His parents in Vienna had been ruined by the Nazi occupation. The money his father had given him for his studies in Cambridge was nearly exhausted. Within three months Bragg secured a Rockefeller grant of £375 a year, and appointed him his personal assistant. This saved his scientific career, and even enabled him to bring his parents to England.

As an Austrian, Perutz had a difficult and lonely time when the Second World War started, but Bragg looked after him and enabled him to continue his research in the Cavendish. His situation was eased by the existence of a discussion group on X-ray research, which had been formed shortly before the war. Its meetings and activities were called the *Configurations of the Space-Group*. It was one of the many informal discussion groups in the Cavendish organised by the scientists in their particular fields.

The first discussion recorded in the minute-book was held on 11 October 1937, and was opened by P. P. Ewald, on 'Umweganregung'; and the last by M. von Laue on 5 December 1949, on 'Absorption of X-rays in Crystals in the Case of Interference'.

Among the early papers before the war Perutz spoke on 'Protein, Enzymes and Viruses'. On 26 October 1938 A. J. Bradley spoke on 'Lattices with Missing Atoms'. His abstract read:

Face-centre cubic and body-centre cubic alloys have the greatest commercial value. It would however be short sighted to restrict all investigations to these materials. More complex structures are found in small quantities in many very important technical alloys (e.g. the carbides), and the study of these structures has a definite bearing on their commercial application. It seems probable that many of the complex alloy structures are related in some way to the body-centre cubic type by the omission of certain atoms from the lattice and the re-arrangement of the remainder.

On January 19 1939 J. D. Bernal spoke 'On Proteins'. On August 15 1939 Ewald gave a 'Tea-Talk on Sine-deformed Lattices', and on November 1 1939 E. Orowan spoke on 'Surface Energies and Surface Tensions in Liquids and Solids'.

The talks before and during the Second World War illustrate both the welcome given to eminent scientists who had left Nazi Germany, such as Ewald, and to the continuation of research during the Second World War.

Orowan described his new microscope, and a reproduction of a picture of it was pasted in the minute-book. In February 1942 L. Kowarski spoke on 'Experiments on Growth of very Thin Crystals'. In March W. A. Wooster opened a discussion on crystal growth. The minutes for meetings 66–71, from August to December 1943, right in the middle of the war, are reproduced in figure 30.

On 7 June 1943 Perutz spoke on 'The Crystal Structure of Reduced Haemoglobin'. Dorothy Crowfoot and Barbara Low-Rogers addressed them on February 21 1946 on 'The Complete X-ray Analysis of Penicillin'. A historical account of the investigation was given by Miss Crowfoot, showing all stages from the first-order Fourier projections to the final three-dimensional molecular model of potassium penicillin II. The X-ray data give the minimised β-lactam structure as the correct one. Over sixty people attended the meeting, and tributes were paid to the authors for their most remarkable achievement.

66   The Broadening of Debye-Scherrer Lines.
20 August 1943   A. J. C. Wilson

67   Structure of Diamond.
11. 11. 1943.   Kathleen Lonsdale.

68   Satellites in the Kα series
19th November 1943 — Adrienne R. Weill

69 Joint Colloquium with Colloid Science Laboratory.
Dr R.F. Tuckett : "Mechanical Properties of Rubber-
Like   Plastics ".
Dr G. Coumoulos "X-ray Investigation of Plastic
Materials."
                        22nd November 1943.

70   S. G. Spiegler   26th Nov. 1943
         Photographic Measurements
of weak X-ray radiation in
Connection with protection problems

71   Joint Colloquium with Colloid Science Laboratory.
Plastic Flow in Metals , E. Orowan ;

Xray effects of plastic deformation
in metals   H. Lipson .
                        6th Dec. 1943

In this way, research discussion was sustained during the war, and the exchange of new information from various countries stimulated immediately afterwards.

Perutz and Bragg attacked the structure of the haemoglobin molecule for fourteen years without much success. Once Bragg had been convinced of a possibility, he was very determined on realising it, but his insight was physical, and what was needed was a chemical idea.

At last, in July 1953, Perutz obtained reflections which showed that, in principle, the structure of the haemoglobin molecule could be unravelled, thus opening up a vast field of protein analysis. Acting on the idea of Bernal and Dorothy Crowfoot Hodgkin, he had made this possible by forming a compound in which two mercury atoms were attached to the haemoglobin molecule. These acted as points of reference for the interpretation of the rest of the structure.

Perutz achieved this success two months before Bragg resigned the professorship. It was the last of many occasions on which he witnessed Bragg's quick and imaginative response to a discovery. However, several more years of research were required before Perutz had completed the analysis.

After the Second World War other young scientists joined Perutz and Bragg in their pursuit of the X-ray analysis of biological molecules. In 1939 the chemist J. C. Kendrew, the son of the Oxford climatologist W. G. Kendrew and his art historian wife, had graduated at Cambridge at the age of twenty-two. He had taken a course that included some biochemistry. This had brought him in touch with the ideas of Hopkins, St Georgy and Needham, and the fascinating chemistry of biological substances.

Meanwhile, he started research on the kinetics of chemical reactions, but with the outbreak of the war he became more interested in securing research work of direct military value. He worked with W. B. Lewis in the Officers' Training Corps. Lewis had already been recruited into secret research on ultra-short radio waves.

Through Lewis, Kendrew joined the Telecommunications Research Establishment, and started on radar research. In 1940 he joined Watson-Watt's staff, working on operational research. This took him to North Africa, where he happened to meet Bernal and Zuckerman. He was sent to South-east Asia, where he became Scientific Adviser to the Commander-in-Chief.

Starting from his early studies in biochemistry and the influence of Bernal and Pauling during the war, he became increasingly interested in biology. At the end of the war he asked Bernal for advice on what line he should pursue. Bernal advised him to see Bragg, and Bragg introduced him to Perutz.

Kendrew had the impression that Bragg was less interested in biology than in solving the complicated molecular structures associated with it.

Cambridge had just got EDSAC, one of the first working computers in the world. Kendrew started to use it, and soon found that it would be very important for working out molecular biological structures. Bragg and some of his colleagues did not immediately see this. When Kendrew suggested he should start a research student on the production of big computer programmes for this purpose, Bragg was not at first enthusiastic.

In his report in 1965 on 'First Stages in the X-ray Analysis of Proteins', Bragg described several of the mistaken lines that he and Perutz had pursued for years. One idea that proved valuable was popularly known as the 'Liverpool Street Timetable'. This was a theorem concerning any random function, 'such as the times of departure of the Cambridge trains from Liverpool Street Station', which was useful in interpreting features of the haemoglobin molecule, and was later of assistance to Caspar, Franklin and Klug in their interpretation of the tobacco mosaic virus molecule.

When Bragg gained one of his insights, especially in optics and the properties of matter, he often worked out the details with lightning speed, producing neatly drawn sketches and a perfectly written description. This capacity reminded Perutz of Mozart's composing of the overture to the *Marriage of Figaro* in a single night.

Bragg was a sensitive artist in watercolours and crayons, and his approach to science was artistic. It inspired him to introduce the annual courses of arts lectures in the Cavendish. While he recognised the importance for large-scale, highly organised science, he had little taste for it.

The development of 'projects' is important not only for the solution of specific problems, but also for promoting the exploration of their various associated aspects. This ancillary activity is often a source of new facts and ideas. If there are no large projects, this ancillary work is not done, and the unforeseen new knowledge remains undiscovered.

When, by 1947, the promise of molecular biology had become clear, there were no adequate funds for its development. Bragg did not secure them from the University, or from other quarters. D. Keilin, the Quick Professor of Biology at Cambridge, suggested that the Medical Research Council should be approached. Accordingly, in the traditional style, Bragg had lunch with Sir Edward Mellanby, the Secretary of the M.R.C., and explained that Perutz and Kendrew were pursuing highly interesting but very difficult researches, which had at the time only remote chances of success. If, however, they were successful they would provide insight into the molecular working of living material. Even if this were achieved, it might well be a long time before such information became of practical use to medicine. Mellanby decided to take the risk of supporting this research.

Perutz was thrilled when he learned that the M.R.C. was prepared to buy equipment for them. He had had to argue with the laboratory steward over every box of X-ray film that he needed; he had been in the laboratory for several years before he was allowed to buy a pair of scissors to cut it. The necessity for an independent laboratory devoted to molecular biology presently became clear. There was strong opposition in the University to the project, and Bragg disliked the academic politics required to combat it. When Mott became Professor in 1954 he persisted in solving these administrative problems, and with the aid of his day-to-day skill in this field, the M.R.C. Laboratory was built, and opened in 1962.

Bragg's greatest contribution to molecular biology was his support when it seemed hopeless. In some instances he took a decisive part in interpreting the X-ray data, but it was difficult in the early days to persuade him to get finance and space; for these there was a continuous battle.

The determination of protein structure proceeded through many attempts, lasting over nearly a quarter of a century, with an interruption of the research for six war years. At each attempt some fragments were gleaned, but there was widespread belief that the goal was unattainable, at least for a long time.

Quite suddenly, a successful approach was found. In 1963 Bragg read a paper on the long preliminary investigations before final success was achieved, entitled 'How protein structures were not worked out'. He said that

The progress of the work may be likened to the scaling of Mount Everest. A series of camps were established at ever-increasing heights till finally a last camp was set up from which the successful assault on the summit was made. I was closely associated with the establishment of each camp in turn up to the highest, from which those

307

brave mountaineers Perutz and Kendrew made their final dash, and I hope that an account of those ventures, which lasted for twenty-five years before success was achieved, will be interesting as a piece of 'scientific history'.

Bragg turned to adventurous and sporting exploration for an illustration of how the investigation of the structure of protein molecules appeared to him. He described the X-ray studies of horse methaemoglobin by Perutz, Boyes-Watson and Davidson in 1947–9 as the first stage or 'camp'. In one paper Perutz deduced that the molecule had a 'pill-box' structure. In another, the data were processed by an electronic computer for the first time in protein analysis. Perutz commented that, 'If the globin molecule consisted of a complex interlocking system of coiled polypeptide chains . . . the Patterson synthesis would be unlikely to provide a clue to the structure.'

On the other hand, if the polypeptide chains were arranged in layers or parallel bundles, the prospect would be more hopeful. 'All the more plausible hypotheses of globular protein structure put forward in recent years have been on systems of the latter kind. (Astbury 1936, Pauling 1940). Hence it was not unreasonable to hope that the Patterson synthesis might lead to interpretable results which would justify the great effort involved in its preparation.'

The idea that the molecule had a regular structure persisted for a long time, and ultimately turned out to be false. If this mistaken idea had not been followed, the whole research might well have been abandoned as hopeless.

Bragg began to get deeply interested in Perutz's results at the second stage, or 'camp'. He thought Astbury's 'Greek key pattern' structure was extremely improbable, and that a helix was far more likely. This was no new idea; Huggins in 1943 had reviewed various possible helical chains. Perutz, Kendrew and Bragg himself tried various forms of helical chain in a paper published in 1950. Bragg described this paper as 'the most ill-planned and abortive in which I have ever been involved'. They invited their chemical colleagues to offer comments on their proposed structures, but they completely missed the real clue. This came with Pauling's perception that the nature of the chemical bond between certain atoms implied that these atoms must be arranged in a helical manner.

Cochran, Crick and Vand in 1952 analysed the diffraction produced by a helical structure, and in the same year Cochran and Crick showed that the diffraction pictures obtained from synthetic polypeptides were explained by the α-helix structure.

308

The third stage was the approximate determination of the outer form of the molecule, by Bragg and Perutz in 1952. This was the first definite quantitative piece of knowledge to be gained.

The fourth stage, also reached in 1952, depended on exploiting a peculiar feature of the shrinking and swelling of haemoglobin crystals. This led to a small but important increase in the definition of the molecule, for, as Bragg says, 'we snatched at such small successes to keep ourselves in heart to carry on with the investigation'.

The turning point came in stage five, when in 1954 Perutz succeeded in applying Bernal's idea of introducing a heavy atom into the molecule. He solved the problem of finding a group containing a heavy atom that could be attached to the molecule at a definite place, and did not alter in any way the arrangement of the molecules in the unit cell. With an expertness that was probably then unique, he measured sufficiently accurately the changes produced by the heavy atom.

Bragg records: 'I remember coming to Perutz in great excitement one day because I had heard from Professor Roughton that an American worker had succeeded in attaching a mercury complex to haemoglobin in stoichiometric proportions, only to have Perutz tell me very coldly that *he* had given this information to Professor Roughton!'

This statement is additionally interesting because it was apparently not the only incident of its kind. A somewhat similar one occurred with Crick, which required considerably more personal adjustment for digestion.

Perutz had learned from a report by Rigg in 1952 that human haemoglobin combined with two molecules of PCMB (*para*-chloromercuribenzoate).

Bragg compared the fifth stage with 'the last camp before the summit'. It was up to this point that he and Perutz collaborated. But it was not yet certain that heavy-atom substitution would give sufficiently accurate information. 'Now came the problem of how to attack the summit itself. The final success was attained, not by way of the haemoglobin investigation, but by Kendrew's study of myoglobin, which at first had seemed to be less promising. It was quite by chance that Kendrew started his work with myoglobin; a colleague offered him some crystals to investigate. It was at first extremely hard to get crystals of sufficient size until the richness of the source in diving animals such as whales was realised.'

It had been found possible to attach the heavy atom at only one site in the haemoglobin molecule, whereas Kendrew had found that it was possible at five sites in the myoglobin molecule. This molecule was also simpler, because its molecular weight was only 17 000, that is a quarter that of haemoglobin.

Was it now possible to determine the positions in the structure with sufficient accuracy? *'It proved to be possible*! I remember well the thrill of that time.' Bragg was now at the Royal Institution, and the mass of data was shared between the Cavendish and the Royal Institution for interpretation.

Bragg became convinced that success was now certain. Meanwhile, Kendrew proceeded with the final analysis, aided by the electronic computer. Kendrew and his collaborators completed their solution in 1958. Bragg said 'It was a proud day when he brought the model to show it to me.'

When Bernal first saw Kendrew's model he exclaimed 'I always knew it would look like that!' It was not a regular structure, as had been so long and deceptively believed.

After the structure of myoglobin had been solved, it was not long before Perutz had completed the structural analysis of haemoglobin. So came the second major Cavendish physico-biological discovery in a decade.

When Mott became Professor, one of his first tasks was to consider moving the molecular biologists out of the Cavendish, for its financial support came not from the usual physical sources, but from the M.R.C.

The new M.R.C. Laboratory for Molecular Biology operated under the chairmanship of Perutz. By the 1970s it had 130 workers, and had become a laboratory comparable in size with the Cavendish in the 1930s.

Bragg had personal inspirations with regard to the visual imagination. He could revolve in his mind models of atomic structures and see significant aspects of them; he felt this so powerfully that he could communicate his vision to audiences. As he talked with his inspired expression, his hearers also began to see in their minds the atomic structures going round, how they looked from different aspects, and what the different perspectives signified for the nature of the structure.

C. T. R. Wilson's lectures on light had profoundly impressed him as a student, and had given him the clue to the invention of X-ray crystal analysis. Bragg retained a genius for appreciating the application of optical principles in the interpretation of nature. It inspired his encouragement of electron microscopy. He perceived that Ryle's employment of interference methods were in principle the same as those used in X-ray analysis, and his enthusiasm for radioastronomy became intense.

When Bragg encouraged and promoted molecular biology and radioastronomy, he was scarcely making a conscious organisational decision. He did not choose to develop these subjects because, after the Second World War, universities, and Cambridge University in particular, were not suitable institutions for the pursuit of science on a large organisational and financial scale. He acted rather instinctively in attacking problems that appealed to his physical insight, and inspired his imagination.

When he changed the main line of the Cavendish from nuclear physics he did not regard himself as demoting it, but his action helped to prevent the Cavendish from pursuing a line that was no longer suitable for one laboratory or one university, or even one country.

Bragg was very much a child of his social and historical period. His genius and his limitations arose from deep social and historical causes. W. H. and W. L. Bragg both had conventional outlooks, but they carried them rather differently. W. H. had had a very different youth, away from home, and had to make his own way. He was more at ease with all classes of men, and skilful in affairs and committees. W. L. had been born in a cultivated and comfortable atmosphere, with a path in life straight and attractive before him, with good social and educational facilities.

He was gifted, but he was also freed from common obstacles. His mind was formed in the Victorian and Edwardian periods. Research was conceived as an activity in which an intelligent independent gentleman could appropriately engage. Bragg had deep sympathy and understanding of things and situations that came within his range of social and intellectual ideas, but less for those that did not. When Bragg left the Cavendish the modes of nineteenth-century thought that had led to its creation definitely receded into the background.

Bragg had imaginative understanding for those whose interests appealed to him. The nuclear physicist D. H. Wilkinson became ill with radiation sickness caused by neutrons. He relieved him of as much work as possible, and encouraged him to pursue his interest in bird-watching. Bragg shared this interest, and Wilkinson spent as much time as possible in the open air, which advanced his recovery.

At the farewell dinner to which Bragg was entertained at the

Cavendish, he was presented with a large pair of binoculars, particularly suitable for bird-watching.

F. H. C. Crick did not engage in research in the Cavendish until 1949, when he was already thirty-three years old. After graduating in physics at University College London, he joined the Navy's scientific research organisation in 1940; he worked especially on magnetic and acoustic mines.

By the end of the war he had become interested in biology, and he sought A. V. Hill's advice on how he should pursue it. Hill secured a small grant that enabled him to join the Strangeways Research Laboratory at Cambridge, which was devoted to research on tissue culture and associated problems. He began an investigation of the viscosity of protoplasm, which he attempted to measure by the time taken by iron filings to drift through it. He was not very happy with this research.

Crick had other anxious periods in his career. He enquired for a place in Bernal's laboratory, and was informed that the Professor had applications from all over the world, and why should he be admitted? Crick subsequently found Bernal a very interesting man, who threw out ideas and started new lines of thought. Many of his ideas were right, but many also were wrong. For example, he concluded that exact measurement of details in the X-ray analysis of biological molecules was not important; Crick came to the opposite conclusion. He found Bernal personally very considerate, which changed his previous belief that geniuses always behaved badly. Bernal was very far-sighted, but while he was brilliantly stimulating he did not personally pursue his ideas to their conclusions, and he was not tough enough in dealing with people.

After Crick's early failure to enter Bernal's laboratory he went to see Bragg, who put him to work on X-ray analysis under Perutz as his supervisor. Perutz's supervision in the Cavendish tradition was very different from that which Crick had experienced in naval scientific research. In the Admiralty establishments you were always told what to do; in the Cavendish you had to find out for yourself.

Perutz had begun research in the Cavendish under Bernal, who started him by rushing into the room, and thrusting some mineral dust, which the professor of mineralogy had given him, into his hands. He told him to analyse it, and then rushed out again. Perutz went through hell finding out how to do it, but he did ultimately, and acquired the habit of independence in research.

312

Crick, accustomed in the Navy to orders, had to demand supervision from Perutz. His scientific development was quite late; he did not receive his Ph.D. until 1954, when he was thirty-eight.

After Pauling had proposed that proteins had a helical structure, speculation arose on the structure of the material of the genes in chromosomes, on which heredity depended. The American, Huggins, had in 1942–3 expressed some advanced ideas on what it might be. Little experimental progress was made until after the war.

Perutz, Kendrew, Bragg and the Cavendish molecular biologists began X-ray analyses of its structure. Bragg pointed out that much of the interpretation of the structure that had been offered was untrue. He did not find the correct spatial interpretation of the X-ray pictures, but he had the correct approach. He was also misled by lack of chemical knowledge; he did not know about chemical structures involving double bonds. Pauling's superior chemical knowledge enabled him to have the deeper insight into the structure of biological molecules. But Pauling failed to work out the DNA structure because his arguments were based on unsuitable experimental material. This consisted of pictures by Astbury, in which the two kinds, α and β-keratin, appeared on the same picture, whereas it was necessary to have pictures of one keratin alone. If Pauling had seen Fankuchen and Wilkins's pictures, he would have worked it out. Perutz, Kendrew and Bragg, on the other hand, had inadequate chemical information. They put into the DNA structure a constraint that was not needed, and left out one that was; they had a near miss.

Their failure served as a guide to Crick and Watson, who saw the mistakes, both of Pauling and their Cambridge colleagues. In Crick's opinion, Bragg was depressed by his failure personally to interpret the α-helix. After that time, his step on the Laboratory stairs was heavier. Bragg, despite all his achievements and years, retained an attractive boyishness. He had gifts for grasping the essence of a problem, and putting things in a graphic way, especially for simplification. Crick learned from him not to allow himself to be cluttered up by details.

Bernal was more interested in conceiving problems and guessing solutions than in working them out, whereas Bragg loved finding the solution of all those spots on the X-ray pictures. During the war, when confronted with an aptitude test for fitting things together, he solved it in 33 seconds, when no one previously had done so in less than 4 minutes.

During 1946-7, Pauling had the greater influence in the field, through his perception of the importance of the hydrogen bond in the explanation of the properties of biological substances. Bragg's interest came later.

Physicists do not like chemistry, because there are too many details in it that have to be learned. But the entry into the structure and properties of living matter comes more through chemistry than physics. Physical chemistry is of most use to biology. It provides the basis for chromatography, the use of radioactive tracers, ultraviolet spectroscopy, the ultracentrifuge, etc. Such techniques have transformed biochemistry, and made it into a very powerful tool for exploring the properties of living substances.

When Crick started research in molecular biology he was comparatively old as a research student, and had not previously accomplished very much. Bragg was genuinely worried as to whether he was diverting himself too much with outside interests.

Crick, for his part, had a naturally direct critical intellect. If he thought Bragg was wrong, he said so, whether he had or had not been asked for his opinion. Bragg found it hard to accept unsought criticism from a research student whose future did not at that time seem assured. Bragg did not grasp the importance of Crick's work in its preliminary stages, but when he saw the double-helix model he grasped it instantly.

Bragg had simple and old-fashioned views on how a gentleman should behave. One day he was discussing a problem with Perutz and Kendrew, when Crick happened to come in and express critical views on their ideas without being asked. He told Bragg that his ideas were wrong, and he would explain to him why in the morning. This upset Bragg.

While Bragg was uncomfortable with behaviour that did not belong to his own conventions, he did not necessarily disapprove of the science of unconventional men. Thus, he gave Bernal great intellectual support.

The inspiration in the Cavendish was his. While he supported the work of Perutz and Kendrew, some of the older professional X-ray men thought their work was wrong and stupid, and would not be successful. Bragg went on supporting them, even though they produced a whole series of fallacious papers.

While Crick was at the Strangeways Laboratory he read a good deal of biology. He had had a better training than Perutz and Kendrew in mathematics, and when he read some of Perutz's papers he saw that they contained errors. He came to Perutz and Kendrew with the intention of putting them right. At the time, Crick's mathematical papers seemed more destructive than constructive.

James D. Watson came from a social and cultural tradition different from the British. He was thirty-eight years younger than Bragg, and eleven years younger than Crick. He was born in Chicago in 1928, the son of a business man, and grew up in the modern American Middle West. He had become interested in biology when a boy through bird-watching. After graduating at Chicago he went to the University of Indiana, where he was influenced by H. J. Muller, who had made the classical experiments on producing gene mutations by X-rays. Watson became interested in the problem of the nature of genes. He wrote a thesis under S. E. Luria on the effect of hard X-rays on bacteriological multiplication. He was awarded a fellowship that enabled him to study with H. Kalckar at Copenhagen, where he worked on the activities of DNA in bacterial viruses.

In the spring of 1951 he went with Kalckar to the Zoological Station at Naples, where he met M. H. F. Wilkins, who showed him the first X-ray pictures of DNA crystals. These excited Watson's imagination, and he diverted his attention to the structural chemistry of nucleic acids and proteins, which might throw further light on the structure of DNA. In order to learn more about the X-ray technique for determining the structure of biological molecules, Luria arranged in 1951 for Watson to work in the Cavendish Laboratory under J. C. Kendrew.

There he met Crick and found that, like him, he was interested in trying to solve the structure of DNA. They decided to collaborate, and later in 1951 proposed a structure for DNA, which was not, however, the answer. With the aid of more experimental data, and deeper study of the chemical and physical literature on nucleic acids, they succeeded, in March 1953, in formulating the complementary double-helical configuration of the DNA molecule.

Continuing in Cambridge, Watson worked on the structure of the tobacco mosaic virus, and with the aid of the new rotating-anode X-ray tube constructed in the Cavendish, was able to prove the helical structure of the virus.

Watson's influence was important for several reasons. He was one of the first Americans to work in the Cavendish not primarily as a learner. A number of distinguished Americans had previously worked there, but they had come in the main to learn, and did not question the Cavendish tradition. Watson did not have that attitude; he carried out the work he was doing in the Cavendish in an American way. When he went to the Cavendish he had the weight of the modern American tradition behind him.

The account he gave in *The Double Helix* was written according to codes different from those that had governed the development of the Cavendish. His work and his book were marks of the increasing weight of American science in relation to the Cavendish. They were a sign that the centre of scientific influence formerly established in the Cavendish had now moved towards America. In the future the Cavendish might be influenced more by America than America by the Cavendish.

These indications were apart from the technical arguments in his book.

A feature of the original Cavendish tradition was illustrated by Bragg's reaction to Crick and Watson's discovery, and Watson's account of it. When Bragg grasped Crick and Watson's interpretation he instantly appreciated it. He contributed an enthusiastic foreword to Watson's book, in spite of its implied reflections on the Cavendish and himself.

His review of the book in the *Times* on the day of publication was comparable with T. H. Huxley's review in that newspaper of Darwin's *Origin of Species* on its publication in 1859. He said that their discovery of the double helix, which revealed the molecular mechanism by which hereditary characters are transmitted in living organisms, 'has been acclaimed, with justice, to be the most important single scientific discovery of the twentieth century'. Like almost all great discoveries in science it was 'both completely unexpected and so dazzlingly simple' that it carried immediate conviction. Their success had come

from a flash of inspiration, not as the result of a long investigation. I remember the occasion vividly. I was away from the laboratory for about a fortnight with an attack of influenza. When I went down with it, there was no hint of a solution. When I returned, there was the model of DNA complete in its details. They had pieced together from a number of sources—the fine X-ray diffraction pictures of DNA which Wilkins and his colleague Rosalind Franklin had succeeded in obtaining at King's College after a long and patient investigation, and chemical evidence about the constitution of DNA and its behaviour—and they had divined the answer.

Bragg described the book as of the deepest interest for the history of science, because of its detailed account of the steps that led to a great discovery, and of the hopes and fears of the scientist pursuing them. It gave a mind picture of the impression on a young American then only twenty-three years old, who had come to work in Cambridge, and the foibles, eccentricities, and weaknesses of colleagues seen through the eyes of uninhibited youth.

Bragg commented that it was not entirely a record of fact. The author had often based his account of what A or B had said on his estimate of what he considered would have been consistent with their character. The facts were not always directly available to him, and he did not always fully understand the issues, and the relations between the persons, involved.

He had been asked by some who figured in the book to join in asking that its publication should be delayed a decade or two. Though he had had considerable sympathy with them, he had been in favour of immediate publication, and he had consented to write a foreword to it. The blemishes and inaccuracies were unimportant compared with its value.

The resolution of the structure of DNA was indeed a magnificent conclusion to the period of Bragg's Cavendish professorship, and Bragg's reception of the discovery and Watson's book, was even more distinguished than these two things in themselves. It represented a high order of scientists' social behaviour. While Watson, besides making his discovery, had also revealed some of the weaknesses of scientists, Bragg had exemplified the greatness of a tradition.

The incident made doubly clear the need for scientists to be disciplined and responsible, and that unless they were, they would fail in their social responsibilities as scientists and citizens. If scientists were as Watson depicted, they were among the less responsible elements in the population, and it was high time that this should be remedied.

The ambitions, quarrels and weaknesses that Watson described were existent among the nuclear scientists in the 1930s, but they were kept in their place.

When Perutz, Kendrew, Crick and the Cavendish molecular biologists were moved to their Institute of Molecular Biology in 1962 they established a regime that was different from the Cavendish. Perutz was not their Director, but their Chairman. They introduced more democratic manners. Everyone addressed each other by their familiar names. Bragg never did so; in his boyhood he even signed letters to his father 'yours affectionately, W. L. Bragg.'

At the Molecular Biology Laboratory all parking spaces were open, none being reserved for personalities. These influences were probably of American origin in the main. They are also to be seen in German laboratories, which are developing freer methods and styles of research. Some think that, in general, American scientists are more earnest than the British, who leave more scope for a little frivolity. This, perhaps, is a residue of the aristocratic factor in the British scientific tradition.

The third figure in the double-helix discovery was M. H. F. Wilkins. He came from New Zealand, where he was born in 1916, the son of a doctor. He was brought to Birmingham at the age of six, and educated there, proceeding to Cambridge to read physics and graduating in 1938. In his course at the Cavendish he was influenced by Bernal, under whom he received some training in X-ray crystallography.

Wilkins then went to work in the physics department at Birmingham, where Oliphant had recently become professor after leaving the Cavendish. He began research under J. T. Randall, who was at that time studying luminescence and underlying physical phenomena, such as the movement of electrons in crystals. When Wilkins was at the Cavendish the dominant concern was with elementary atomic particles, but he had already become more interested in the organisation of particles in the solid state, and the special properties of solids arising from this organisation. It was perhaps a sign of the direction of his future interest towards the larger biological molecules, whose highly specific properties enabled them to provide the fundamental mechanisms of living processes.

Research on phosphorescence formed the thesis for his doctorate, which he received in 1940. He applied his knowledge of this subject to the improvement of the cathode-ray tube screens used in radar. Then he joined Oliphant in research on mass-spectrographic methods of separating uranium isotopes for atomic bombs, and thereafter went to work on the Manhattan Project in California.

Like Randall and some other physicists, Wilkins became disturbed by the uses to which physics had been put in the war. Randall decided to work in future on the application of physics to biological problems. He had been the chief inventor of the magnetron, the most important scientific device used in the Second World War (the atomic bomb was dropped only after the war had in effect been won). With his high reputation he received substantial support for his pursuit of the application of physics to biology. The M.R.C created a Biophysics

Research Unit under his direction, which Wilkins joined. In 1946 Randall was appointed Wheatstone Professor at King's College, London, the chair Maxwell had occupied before he became Cavendish Professor.

The Biophysics Research Unit was attached to the physics department at King's. Wilkins worked on the development of reflecting microscopes for studying nucleic acids in cells, and investigated aspects of the structure of tobacco mosaic virus.

He then started X-ray studies of deoxyribonucleic acid, DNA. These provided the main data from which Watson and Crick derived the molecular structure of DNA. Further work by Wilkins proved that the Watson–Crick structure was correct. Their great discovery utilised the brilliant X-ray technique of preparing and taking X-ray photographs developed by Rosalind Franklin, a Cambridge graduate who was awarded a research appointment at the Biophysical Laboratory at King's College London, in 1951. Her technique and application of Patterson-function analysis was an essential contribution. Unfortunately she died in 1958, so she was unable to share in the Nobel Prize awarded in 1962.

Thus, various intellectual influences, scientific traditions, social outlooks and talents interacted and flowed together at the Cavendish, in the discovery of the molecular structure on which heredity is based.

24

Final Report

In Bragg's time reports on the Department of Physics were regularly made to the University. These were very useful for posterity, but latterly the University has rather discouraged such reports, because of the heavy cost of printing them.

Bragg made his final Report, for the year 1952–3, rather as a rendering of account for his period of stewardship of Cambridge physics. He observed that the number of students reading physics in that year was 527. One of the first things he mentioned was the course of arts lectures. These had been introduced to widen the outlook of the physicists, and had continued most successfully, with an average audience of 200. One subject had been 'The Novel, from Conrad to Virginia Woolf', and another 'Science and the Modern Novel', the latter being delivered by D. Daiches. Another subject was 'The History of North America'. In previous years there had been courses on 'Economic and Social Development since the Middle Ages', 'Music' and 'Man and his Environment'.

The number of research workers was 85, consisting of 30 in nuclear physics, 17 on radio waves, 10 in low-temperature physics, 16 in crystallography, 4 in meteorological physics and 4 in fluid dynamics.

Twenty-three of the research students came from overseas, forming more than a quarter of the total number, and of these ten came from the Dominions. It cost about £750 to give each of these men a place, and they forged links between Great Britain and the Dominions.

The department of nuclear physics under O. R. Frisch accounted for 'a very substantial part of the whole output of nuclear physics in this country'. After much thought and discussion, they had decided to build a linear accelerator; construction had started in 'the old canteen' on the Madingley Road factory site.

Research on radio waves, under J. A. Ratcliffe, included the discovery of wave reflections from the ionosphere from heights as low as 75 kilometres; and considerable regularities in the 'wind-like' movements of the electrons in the ionosphere. The investigation of radio emission from the sun and galaxies had been extended. The degree of radio 'brightness' of emission across the sun's disc had been determined. The outermost layer of the solar envelope at distances of one-quarter of Mercury's orbit had been investigated. The occultation of radio stars by the solar atmosphere had been observed.

The increase in accuracy of the determination of the positions of radio stars had led the astronomers of Mount Palomar to notice two optical objects that appeared to be identical with the two most intense radio sources. One of these appeared to be a cloud of material in the Galaxy; the other, two galaxies in collision, at a distance of a hundred million light-years. These discoveries suggested that objects beyond the reach of optical telescopes might be observable.

A new interferometric radio telescope was being constructed on the Old Rifle Range. It had an area of 4600 square metres, and would allow data to be collected in quantities to be sufficient to lend themselves to statistical analysis. A new field laboratory was in construction on the Grange Road Rifle Range. Bragg commented:

> This new brand of astronomy is one which, it would seem, would repay development at Cambridge. Large telescopes which make use of the light from heavenly bodies must be sited at high altitudes remote from the glare of towns and in places where the weather is favourable for the greater part of the year. On the other hand, radio-astronomy is independent of atmospheric conditions; since radio waves are not affected by cloud or mist, reception can be at ground level and merely requires a large flat area in which to erect the aerials. It has its own troubles in electrical interference by transport, television and power lines, but it is possible to site the aerials so that this interference is not too serious. The new knowledge of the universe which it is yielding is proving to be of intrinsic interest, and as the Cambridge unit under Mr. Ryle has already established a leading position, the opportunity to develop this new science should be exploited vigorously.

With regard to their low-temperature research, he remarked that 'the existence of superfluidity constitutes an illustration of the necessity for a quantum mechanical approach to problems of macroscopic dimensions as well as to atomic phenomena.'

The department of crystallography, with which Bragg was personally most closely involved, was directed by W. H. Taylor. He summarised its work under six headings. The first concerned advances in the analysis of complex organic structures of great chemical interest, some of which were related to nucleic acids. Advances in the determination of the location of hydrogen atoms in such substances were of particular importance.

The second heading dealt with the analysis of complex alloy structures, and the third with the cold-worked state in metals.

The fourth dealt with the structure of coal. For the first time a soundly based picture of the physical nature of coal had been obtained. This research had been supported by the National Coal Board.

Fifthly, there were the researches in the classical X-ray line on the structures of minerals, natural and synthetic. Sixthly, there were researches that had led to a completely new approach to the problem of ferroelectricity.

The Laboratory had collaborated with Todd and the Department of Chemistry on the analysis of the structure of complex organic molecules, and with the Department of Mineralogy. W. A. Wooster and others from the latter department had carried out research and teaching in the Cavendish. Lectures on the metallic state were given to students of metallurgy.

On the Cavendish researches in the molecular structure of biological systems, under M. F. Perutz, Bragg reported:

This work is carried out by a unit financed by the Medical Research Council. It is investigating the structures of crystalline proteins, muscle, and nucleic acid. The main attack is by X-ray analysis, but other methods are also being used.

For many years Dr. M. F. Perutz has been working on horse haemoglobin as a typical representative of a physiologically active globular protein. This work has continued in collaboration with Sir Lawrence Bragg and E. R. Howells. During the last two years an accurate picture of the shape of the molecule has been obtained, and progress has been made towards discovering the phases of the diffracted X-rays, which must be known before the density distribution of the haemoglobin molecule can be calculated. Quite recently the investigations of Perutz have made it possible to turn the last corner and determine all the phases with certainty. The first pictures of a protein molecule have been made; though these are still under very low resolution, there would seem to be no serious obstacle to refining them in the hope that they will show how the molecule is constructed. The first step was made by attaching two mercury atoms at definite points in the molecule and noticing the changes in diffraction. [A model based on the X-ray photographs of horse haemoglobin is shown in figure 31.]

Dr. J. C. Kendrew and his collaborators have continued their studies of myoglobin, and have found a large number of interesting new crystalline forms from a variety of marine animals and birds. Great difficulty had previously been experienced in crystallising myoglobin from the horse, and this had held up the research, but it has been discovered that myoglobin from the whale crystallises very well indeed, and this has proved to be a rich source of new material. [A model of myoglobin is shown in figure 32.]

31   *The X-ray structure of haemoglobin*

Dr. H. E. Huxley made X-ray studies of living frog-muscle. He discovered that the ultimate protein fibres in muscle form a regular hexagonal array. The structure of this array changes drastically when muscle goes into rigor, but, disappointingly, he could discover no structural change related to the state of extension or contraction in relaxed muscle.

J. D. Watson and F. H. C. Crick had built a model of deoxyribonucleic acid (see figure 33). The model consisted of a pair of intertwining helical chains held together by hydrogen bonds between specific pairs of purine and pyrimidine bases. The model was in striking agreement with X-ray diffraction pictures of nucleic fibres. It predicted a 1 : 1 ratio of the specific pairs of bases, which was borne out by chemical analyses. It also implied a possible mechanism of gene duplication.

Other developments include calculations of the X-ray scattering from helical polypeptide chains by F. H. C. Crick, studies of the form birefringence and polarisa-

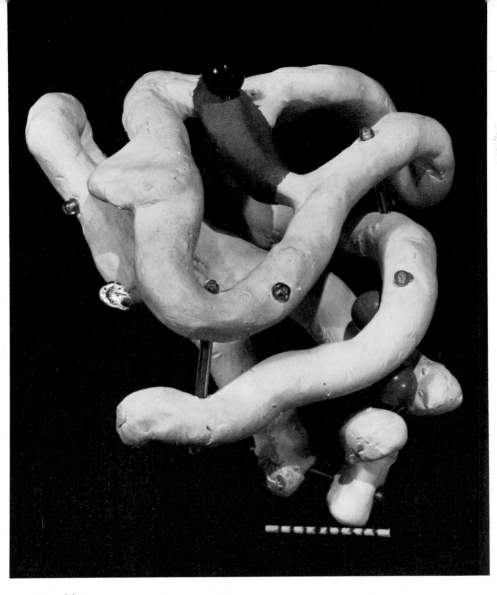

32 *Myoglobin*

tion dichroism of haemoglobin crystals by M. F. Perutz and its theoretical interpretation by Sir Lawrence Bragg and A. B. Pippard, and an analysis of the structure
of tobacco mosaic virus by J. D. Watson, and the construction of a high power rotation anode X-ray tube by D. A. G. Broad.

The department of electron microscopy under V. E. Cosslett,
besides pursuing its own research programme, provided services
for other departments, such as a reflection electron microscope for

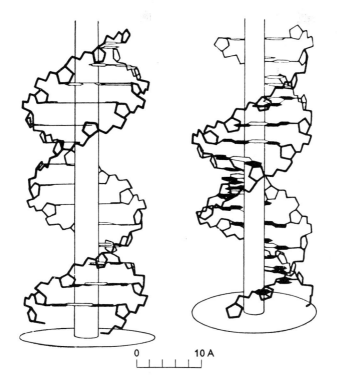

0   10 A

33   *The double-helix of* DNA

examining the surfaces of metals directly. Advances had also been made in preparing bacteria for examination.

Stereoscopic electron microscope images of small insects had been obtained, in which the inner structure was revealed at all depths. The living insects had been photographed after being chloroformed to keep them still during exposure.

The department had taken 3100 photographs during the year, compared with 2500 in 1949–50.

Meteorological physics under T. W. Wormell had investigated the absorption of ozone by infrared rays in the atmosphere, and its dependence on pressure. The mean height of the ozone, and its day to day and seasonal variation had been measured. Radar had been applied to the detailed study of rain clouds, lightning atmospherics and the electric charge on raindrops.

Fluid dynamics under Sir Geoffrey Taylor had investigated heat convection, the kind of convective flow that occurs, for example, above a bunsen burner, and the vastly more complex circulation of air in the lower atmosphere. The role of turbulence in the transfer and redistribution of heat in fluids, and the mechanical energy absorbed and dissipated by turbulence were investigated.

M. F. Perutz had also investigated the plasticity of ice and the flow of glaciers.

D. J. de Solla Price had begun to order the Laboratory's Archives, and a museum of its pieces of historic apparatus.

The total annual income of the Laboratory had been, in round figures, £150000. It was made up of £92500 from the University Education Fund, £7000 from investments and endowments, £5000 from bench fees and sundry sources and £45000 from outside bodies.

The main expenditures were £44752 on salaries, £33419 on supplies and £25694 on wages.

The average cost per research student was £755, of which £257 went on salaries and £148 on wages.

The average cost in all science departments of the University was about the same, but that of the arts research student in 1952 was £516.

As about one-third of the Cavendish's income came from outside bodies, the average cost to the University of a Cavendish research man was about £500.

During 1952–3, 184 papers were published from the Cavendish. These included 24 on the molecular structure of biological systems, by Bragg, Crick, Dorothy Hodgkin, H. E. Huxley, Perutz and Watson. 32 papers on crystallography were published, containing several by W. Cochran. 19 were on nuclear physics, including five from D. H. Wilkinson. There were 13 on radio astronomy, 19 on the ionosphere, 36 on low-temperature physics, 12 on electron microscopy, 9 on meteorology and 10 on theoretical physics.

At the Farewell Dinner in the Cavendish in 1953, Bragg made some general reflections on his experience as Cavendish Professor.

It had not all been easy going when he first came to direct the Cavendish. He had had to succeed Rutherford twice, and had felt smothered by the gigantic folds of his mantle, from which he had had to emerge. Then there was the threat of war, followed by the war itself. The staff Rutherford had collected had done magnificent work in the war. But the Cavendish was spent in the effort; it had flown to pieces. Great work had been done in radar on sea and land, and in anti-submarine research. There had been no small effort in the Cavendish itself, such as Feather's work on atomic energy, and the training of physicists, especially in radar.

After the war they had been hard hit, too; but they had had the benefit of Searle's clear thinking, and Cockcroft, Dee, Feather and Bretscher had gone forward with reconstruction. The period of R. H. Fowler's long illness had been very black. The loss of his dynamic personality had caused anxiety. In 1946 they had secured Hartree, and in 1947, Frisch. Shire was accomplishing fine work in developing high-energy physics.

What a contrast there was when he looked around in 1953! He felt they were living in proud times. Led by Shoenberg and Pippard the Mond was producing a stream of papers, and was at the centre of the superfluidity problem.

He felt pride in referring to the radio research directed by Ratcliffe, and Ryle's development of radioastronomy, which had been foreseen by Ratcliffe.

In crystallography they had developed the foremost methods of attack. It was perhaps not realised outside physics how much they were a world centre of this subject, and a unique international body.

The application to biochemistry had been one of his greatest pleasures, and the cup had been brimming. It had come to fruition under Perutz, and experience would show what great things it would lead to.

His last report on the department, his swansong, he confessed was shamelessly long, but it was a not inconsiderable document. He would have liked to think that he had planned it all himself, but he was too much of a realist not to know how mistaken such a notion would be.

All of this activity was the Cavendish in process of gaining momentum again. It represented something greater than any of the individuals who participated in it.

He could not think of anything that he had really planned himself. One after another, the movements had come from one member of the staff after another. The advance had come through a process of natural selection.

His own contribution had been to try to understand and recognise the importance of their ideas, and help to realise them, with all his energy.

He particularly thanked Ratcliffe for contact on all 'affairs of state', teaching and framing of plans, since his own thoughts did not flow readily in negotiations. Ratcliffe knew how to hold his fire, and deal with general boards of examiners. Their last battle had been for a course in the history and philosophy of science.

Since Dibden had become General Secretary of the Laboratory, and in charge of administration, how different life had been!

All through his period as Director, he had been conscious of the support he had received, of the appreciation of what he could do, and forbearance on what he was bad at doing.

Shyness of personal problems, and inability to remember names, were two great handicaps, and the atmosphere of 'affairs of state' was not one he found easy to breathe.

The Cavendish had grown so. It was now larger and more complex than some university colleges, or even some universities, not long ago. This precluded intimate personal contacts throughout the Laboratory, which he missed keenly, and which came so naturally in a smaller place.

He was handing over to his successor a Laboratory that was one of the happiest, and active in every part. He was not a planner, or he might have wanted to go on planning. On the other hand, his pride and admiration of all that was going on need not, and would not, cease when he relinquished his post.

25

The Mond

In 1953 D. Shoenberg reviewed the history and achievement of the Mond Laboratory since its foundation in 1933. He described how it arose out of Kapitza's first researches in the Cavendish, and his invention of powerful machinery for producing very intense fields for deflecting α-particles. Besides drawing on his review and some of the Mond records, a personal recollection of the opening ceremony is added.

This machinery, and the experimental instrumentation needed, required a considerable amount of space, not only on account of its size, but also because of its effects on neighbouring apparatus. The violent discharge produced a mechanical shock, which disturbed delicate recording instruments, so these had to be placed at a distance where they could record observations before the shock-wave reached them. As the recording took about a hundredth of a second, the instruments had to be at least thirty metres away.

Kapitza began a systematic study of the magnetic properties of matter in very strong fields, and became particularly interested in their effects on magnetostriction and susceptibility. These became much more striking at low temperatures. Instead of buying a hydrogen liquefier he began to construct one himself, and then thought about the construction of a helium liquefier.

It became evident that much more space would be required for the complex assemblage of apparatus, and its efficient utilisation. Kapitza convinced Rutherford of the necessity for a new, specially designed laboratory.

Rutherford persuaded the Royal Society to provide the funds from its Ludwig Mond bequest. Kapitza was appointed the Royal Society's Messel Professor, and Director of the new laboratory.

The Laboratory building was designed by the architect C. H. Hughes according to Kapitza's ideas. It was an original and attractive building, decorated, at Kapitza's request, with two carvings by Eric Gill. One, on the outside, was a large curling crocodile. It has been said that in Russian folklore a great man is sometimes described as a crocodile, and that the carving was a symbol of Rutherford. Inside the entrance hall there was a bas-relief of Rutherford in profile; it looked like an Assyrian carving and aroused a good deal of adverse comment.

The Laboratory was opened in February 1933 by Stanley Baldwin, then Chancellor of the University, and Prime Minister. There were more than a hundred guests, including a number from abroad. Rutherford was chairman of the opening ceremony. He began the proceedings with a speech, and then called on Baldwin, who repeated almost word for word what Rutherford had said! Rutherford had helped him with notes.

Kapitza sat on a chair facing the audience, and looked increasingly uncomfortable as Baldwin went on, unconsciously repeating Rutherford. The audience became slightly uneasy. The situation was saved by J. J. Thomson, then seventy-six years old. When called on to speak, he began a long story about a lady and an egg. He presently seemed to become aware that it was not quite suitable for the occasion, and began to make it more and more obscure. After a while the audience was so curious as to his meaning that they forgot all about the earlier contretemps. Then, J. J. remarked that in his earlier days there were two sorts of scientists in Cambridge, those who made discoveries, and those who got the credit for them. This seemed to be a reference to the feeling among some of J. J.'s pupils that he borrowed their ideas without sufficient acknowledgement. The audience laughed, and the occasion ended happily.

The cost of entertainments was modest in those days. The caterer quoted a price of 1/6 (7½p) per head for tea for the large number of guests. It included tomato, egg and anchovy sandwiches, buttered scones, pancakes, white and brown bread and butter, cherry cake, dessert biscuits, and tea or coffee.

The Laboratory cost £15319 11s 9d. Running costs were kept down. In December 1933 the Laboratory Committee decided that 'the stipend of Dr Cockcroft be £26 per annum'. In the following December, a young lady was engaged as secretary at £1 a week. In December 1936 her wages were raised from 35/- to 37/6 a week. The wages of laboratory assistants were raised from £3 10s to £3 15s for senior men and £2 2s to £2 4s 6d for juniors, respectively. The Laboratory started on its distinguished career most economically.

The Laboratory had been designed specifically to house the equipment so that it could be used conveniently. This was placed in a large central hall, into which smaller research rooms and workshops opened. Research workers had to go the minimum distance to reach the larger equipment, and in the smaller research rooms they could set up experiments free from disturbance and distraction. All of the experimental rooms and workshops were on the ground floor, so that research workers could walk easily from one place to another.

On the first floor were a conference room and some administrative offices and studies. There was a balcony from which the people and various activities on the ground floor in the central hall were visible and audible. It was easy to call from the balcony to those working on the ground floor in the central hall. One of the most valuable properties of the new laboratory turned out to be this possibility for direct communication between the workers, and the arrangement of the small research rooms round the large central open hall made it easier for the scientists to meet casually. This fostered the communal spirit of the laboratory, and the casual discussions, which are often so fruitful when a research man, emerging from his room with a problem very much on his mind, immediately meets just the man who has the particular expertise or knowledge that he requires; if he has to go along a long corridor, or up stairs, to find a man who is shut away in his room, he is less liable to meet him at the right moment.

The Mond men discovered that this design of laboratory was pleasanter and more stimulating than the conventional design in which men are isolated in small rooms opening onto long, narrow corridors. This social feature proved to be valuable, even after the large equipment originally in the central hall was dispersed.

The experience with the Mond Laboratory has influenced the design of the New Cavendish, in which attention has been devoted to arrangements promoting accessibility and social contact.

After the Mond had been opened with its original equipment of machinery and liquefiers, Kapitza proceeded with his helium liquefier. This consisted of an expansion engine, which needed no liquid hydrogen for preliminary cooling, even though he already had a hydrogen liquefier. His machine was the forerunner of the Collins helium liquefier, which was subsequently manufactured commercially, and contributed to the great expansion of low-temperature research in the United States.

When Cockcroft succeeded as director, Rutherford continued to take a keen personal interest in the Mond Laboratory, frequently visiting it, and discussing matters with the research workers. It was decided that in future the Mond should concentrate on low-temperature research.

Up to that time, cryogenic research at liquid-helium temperatures had been dominated by the Kamerlingh Onnes Laboratory at Leiden, with developments at Berlin, Oxford and a few other places.

The Leiden investigators had discovered superconductivity and the λ-transition in liquid helium, but these remarkable phenomena were as yet little understood. It appeared that there was still plenty of scope for comparatively simple experiments to explore points of principle. This policy was adopted, and proved fruitful.

J. F. Allen and his collaborators investigated the properties of liquid helium below the λ-transition point. Their investigation of its

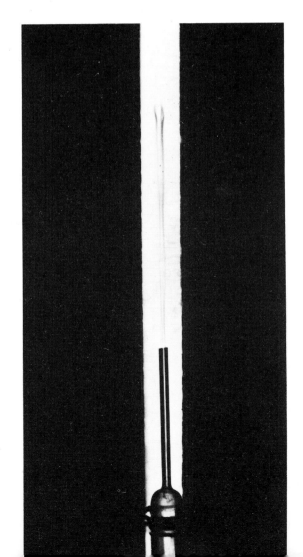

34 *The 'fountain effect'*

anomalously high heat conductivity led to the discovery of the 'fountain effect' (see figure 34). This showed that the flow of heat and the flow of material were intimately connected.

Then it was discovered that the liquid helium could flow through fine channels, showing almost no viscosity. This led to the conception that liquid helium contained a superfluid component. Kapitza, now installed in his new laboratory in Moscow, independently made the same discovery at about the same time.

D. Shoenberg investigated the magnetic properties of superconductors. His studies of the magnetic susceptibility of small superconducting particles in mercury colloids, in parallel with the experiments of E. T. S. Appleyard and others on superconducting thin films, confirmed the superconductivity theory of F. and H. London. The Londons had forecast that a magnetic field would penetrate only a very short distance into a superconductor. This proved to be about one hundred-thousandth of a centimetre, the actual distance being strongly dependent on temperature.

Ashmead designed water-cooled coils that could produce fields of five tesla for use in the recently developed technique of producing temperatures below 1 K by adiabatic demagnetisation.

This period of development was then interrupted by the Second World War, and the Mond was assigned to research for the war ministries. Low-temperature research was revived at the end of 1945 under J. F. Allen's direction. After he was appointed professor at St Andrews in 1947, D. Shoenberg became director.

The most striking early post-war work was A. B. Pippard's application of microwave techniques, which had been developed for radar during the war, to superconductivity problems. Superconductors were exposed to the very high frequencies produced. Under these conditions, the electrical resistance of the superconductors was no longer zero. It dropped only gradually as the temperature was lowered. Pippard also studied the depth of penetration of magnetic fields into superconductors, by developing methods that were more accurate than those used earlier, and were more versatile in dealing with variations in physical conditions, such as crystal orientation and purity.

Pippard's results revealed the limitations of the London theory, which had formerly given such valuable guidance, and indicated the lines along which it required development.

His work on superconductors led him to study the anomalous skin effect in normal metals, which occurs at very high frequencies and low temperatures, when the electron mean free path becomes of the same order as the skin depth. His semi-quantitative explanation was developed in detail by G. E. H. Reuter and E. H. Sondheimer, and led to new information on the electronic theory of metals.

The periodic variation of magnetic susceptibility, the de Haas–van Alphen effect, was investigated by Shoenberg in a number of metals. He developed a new method of investigating the effect by discharging a large condenser through a coil, in the style of the Kapitza tradition.

Research on liquid helium was carried on by K. R. Atkins and others. The measurement of the speed of the 'second sound' in liquid helium was measured at temperatures down to 0.1 K, giving results that exactly confirmed the prediction of Landau's theory.

Further experimental investigations on the helium film, the hydrodynamical properties of liquid helium and its absorption of sound, indicated the need for further theoretical developments.

The advance of the researches called for more complex experiments, larger liquid helium vessels, and more liquid helium for the increased number of research projects. The original Kapitza liquefier, which had had such a useful and long life, was replaced in 1949 by a larger machine built by Ashmead.

This worked on the more conventional cascade principle, but yielded 3.7 litres of liquid helium per hour, which was about twice that of the original machine. The gain in yield was obtained by making use of the laboratory's hydrogen and helium compressors. Particular attention was paid in the design to ease of servicing, so that the running of the new machine was rarely held up for this purpose. This enabled helium to be supplied in as many as half a dozen 'helium runs' on the two days in the week when they were carried out. The supply was later improved by the provision of a large liquid-helium storage tank.

The first twenty years provided much interesting new information, but it also revealed how very much more remained to be discovered, which was a challenging and inspiring situation for the low-temperature physicists.

Pippard extended his researches by detailed measurements, at high frequencies, of the surface resistivity of single metal crystals. He analysed the variations that occur in different directions, in relation to the axes of the crystals. He showed that his measurements on copper crystals enabled a plot to be made of the metal's 'Fermi surface', that is, the contour showing how the wavelength of the quantum waves governing the motion of the most energetic electrons depends on the direction in which they are moving. He constructed a model giving an approximate representation of the 'Fermi surface' in copper (see figure 50).

Pippard's high combination of experimental and theoretical research method had much inspiration and influence in the Mond. W. F. Vinen, who became a research fellow of Clare College in 1955, conducted an elegant series of studies on the hydrodynamics of superfluid helium. These were particularly concerned with the development of turbulence and the frictional effects arising from the interactions of the thermally excited 'normal fluid' with the vorticity of the ground state.

The work of Vinen and H. E. Hall established quantum hydrodynamics on an experimental basis. Vinen directly demonstrated the quantisation of circulation in the superfluid in an elegant experiment, comparable with the discovery of the fountain effect by Allen, twenty years before. Vinen was appointed to a chair at Birmingham in 1960, at the age of thirty-one.

H. E. Hall, who entered the Mond in 1952, followed the Laboratory tradition by combining experimental skill with theoretical analysis. He collaborated with W. F. Vinen in showing that 'second sound' was attenuated when propagated in rotating liquid helium. This directly demonstrated the existence of vortex lines. Hall was appointed to a chair at Manchester in 1961, at the age of thirty-two.

The most outstanding contribution from the Mond in the 1960s was the discovery of a remarkable effect by B. D. Josephson, which has been named after him, and is of great theoretical and practical importance.

In 1962 he published a letter containing extraordinary predictions of what might happen at a junction between two superconductors. One was that a current would flow across such a junction in the absence of a bias voltage. Another was that if a voltage in excess of a small critical value were applied to the junction, it would generate a current with an alternating frequency component equal to $2eV/h$, where e is the electronic charge, and h is Planck's constant.

These two predictions are the most striking examples of quantum effects on a macroscopic scale, and have been confirmed by experiment. The effects were looked for by the Cavendish, but were first discovered and demonstrated with perfect clarity by Rowell at the Bell Telephone Laboratories in the U.S.A.

Josephson then considered the effect of a steady magnetic field on the junction. He found that it would modify the current according to an oscillatory function of the field. Consequently, the variation of the strength of the current in relation to the strength of the field would be similar to the variation of intensity of light across a diffraction pattern. This was also confirmed by experiment.

Josephson is interested in critical phenomena where the sets of laws governing the properties of matter change from one to another. In these regions the laws of physical nature are, as it were, undefined, and a special kind of physical intuition is required to discover the way without the usual signposts. He is one of those physicists who is primarily interested in difficult problems. The solution of such problems is often found to be exceptionally significant and fertile.

Josephson was born in 1940. After graduating at Cambridge, he started research in experimental physics, working in the Mond under Shoenberg. Pippard was his adviser on the research for his doctor's thesis.

At this time, in 1961–2, the American theoretical physicist P. W. Anderson of the Bell Telephone Laboratories was on a sabbatical year in the Cavendish. He acted as nominal head of a group working on solid-state and many-body theory. Josephson attended his lectures on these subjects. He habitually sat with Pippard and Anderson at the table used by Mond workers in the Cavendish tea-room. As Anderson has remarked, it was probably no coincidence that Josephson's discoveries 'occurred in the thoughtful and stimulating atmosphere of the Cavendish'.

The most important immediate practical result of Josephson's discovery was the provision of a very exact method of measuring the ratio between Planck's constant and the charge on the electron. It provided the most accurate standard for measurement of voltage, yielding comparable accuracy in the field of electrical measurement with that of light interferometry in the determination of standards of length. In fact, Josephson interferometry might in the future be used for the definition of the major part of the system of scientific units. It would probably be very important in ultrasensitive electromagnetic measurements, for example in liquid helium, and in gravitational phenomena.

The picovoltmeter developed by John Clarke could measure the extremely minute voltages that arise across a one-micrometre copper film between two superconductors.

Another promising line of application was in the detection of high-frequency radiation, and its demodulation.

Some thought Josephson junctions might be used as a means of generating high-frequency currents. Novel switches based on it were soon devised, and the prospect arose of its being utilised in the construction of computer elements, in which it offered great advantages in size and speed. As Anderson put it, the technology as well as the science of the future would probably depend greatly on the macroscopic quantum interference of matter waves, to the discovery and exploitation of which Josephson contributed so much. By the age of twenty-two, he had conceived the phenomenon, worked it out theoretically, and predicted exactly how it could be proved experimentally.

Before Josephson's discoveries physicists were in the position with regard to superconductivity of possessing the theory before they had the experimental means of proving it. Their situation would have been like that of the creators of the wave theory of light, if they had worked out the theory before they had invented slits, half-silvered mirrors, and the apparatus required for performing the interferometric experiments, by which the wave theory of light was proved experimentally.

The theorists had a theory of superconductivity, but until Josephson's work they had no instruments to verify it. They had postulated that superconductivity consisted of a coherence of the de Broglie waves representing pairs of electrons inside the superconductor. But they believed that the phase of the macroscopic waves was unmeasurable in principle, by the same kind of false reasoning by which the existence of ferroelectricity was thought to be disproved.

It presently became clear that Josephson's equations were equivalent to the theoretical understanding of superconductivity. One led to the prediction of zero resistance, and the other to the Meissner effect and flux quantisation.

The completeness of Josephson's discoveries struck the applied scientists as powerfully as the theorists. When Anderson and John Rowell consulted the able patent lawyer of the Bell Telephone Laboratories, he expressed the opinion that Josephson's paper was so complete that no one else was ever going to be very successful in patenting any substantial aspect of the Josephson effect.

From the Cavendish point of view, Josephson's achievement, in its combination of intuition, theoretical skill, experimental understanding and practical implication, is one of the most Maxwellian of recent times. At the age of thirty-three, Josephson was awarded a share in the Nobel Prize for physics for 1973.

26

'One Good Laboratory Among Many'

By the time Bragg had moved from the Cavendish in 1953, two major advances in science had been accomplished during his directorship, in molecular biology and radioastronomy. Probably neither of them would have been achieved without his special understanding and encouragement.

By the time of his departure, however, this personal style of research direction, which had been established by his predecessors, was increasingly difficult to adapt to the post-war condition of science. The growth of science after the Second World War, and its greatly increased role in civilian and military affairs, involved new problems of management and policy. Scientists such as Cockcroft and Blackett were deeply engaged in these problems. Cockcroft was building and directing the U.K.A.E.A.'s nuclear science laboratory at Harwell, the national institution for solving the scientific problems underlying the utilisation of atomic energy; Blackett was developing the large physics department at Imperial College.

Besides the work in their own institutions their advice was frequently required on national science policy. Freedom from teaching commitments or residence in London facilitated this work. The direction of the extended and rapidly growing Cavendish Laboratory, in addition to the increased advisory work expected of leading scientists, might have demanded more than could have been practicably expected from any individual.

Nuclear and high-energy physics could no longer be studied comprehensively in a university, owing to the size and expense of the equipment now needed to investigate large parts of the subject.

In the Cavendish, academic as well as scientific problems required attention. The University, as well as science, had demands on the Cavendish.

339

N. F. Mott (see figure 35) was elected in succession to Bragg, and took up the appointment in 1954. Mott had had a strikingly successful part in building up a new school of physical research, besides pursuing distinguished personal investigations.

Before the First World War Cambridge and Manchester had dominated British physics. Mott contributed a great deal to the creation of a strong new school of physics at Bristol, which improved the balance of physical research in Britain as a whole. Later on the balance was further improved by the movement of Rutherford's colleagues, Blackett, Chadwick, Oliphant, Dee and Feather to chairs in various universities.

Nevill Francis Mott was born in 1905, the son of C. F. Mott, a noted Director of Education for Liverpool. Both of Mott's parents had been Cavendish students. Mott was educated at Clifton College, Bristol, a public school with a strong scientific tradition. He entered St John's College, Cambridge, and in due course engaged in research. This quickly attracted attention.

After the discovery of wave mechanics he applied it to the explanation of the phenomena of atomic collisions. He was the first to show that Rutherford's law of scattering of particles through collisions remained exactly valid when expressed in terms of the new mechanics. He was also the first to apply Dirac's relativistic quantum mechanics to collision problems. He pointed out the importance of symmetry in such problems, and that under certain conditions electrons could be polarised by double scattering.

Mott was a very original theorist. When he started research in atomic physics attention had been concentrated on atomic structure, largely as a consequence of Bohr's theory. In contrast, Mott attacked the problems of atomic dynamics. His work on the polarisation of electrons was very important. Later on, he concentrated on the theory of the solid state, in which he had always been interested.

His appointment to the chair at Bristol presented the opportunity for change, and for devoting most of his attention to this subject. He was a real theoretical physicist, not an applied mathematician attending to physics. It was his custom to roam round the laboratory, discussing the problems of the men at the bench, at a time when this was not common, so that he was always closely in touch with experiments.

He published *An Outline of Wave Mechanics* in 1933, and initiated a series of works, written with collaborators, on the theory of several branches of experimental physics. With H. S. W. Massey he wrote *The*

340

Theory of Atomic Collisions, with H. Jones *The Theory of Metals and Alloys* and *Electronic Processes in Ionic Crystals* with R. W. Gurney. In 1971 he published with E. A. Davis *Electronic Processes in Non-Crystalline Materials*.

Mott's experience and versatility in collaborating with a variety of physicists were of particular value in dealing with the situation in the Cavendish when Bragg departed. It is the kind of situation that now confronts all large physics departments. In the Rutherford period it was still possible for the director to have an intimate knowledge of most of what was going on. Bragg had to cope with a situation where that was becoming impossible. He did this by the development of the group system. When the group system had itself grown, the succeeding director found himself in a situation different from that in which Rutherford and Bragg had started on their directorships.

When a new director comes into a big department it is very difficult for him to master all that is going on. He is not in the situation of a man around whom a large department has grown up, and in which he may know everything, having initiated most of it. It is unlikely that he will have the same interests as the old, so he may not even be generally familiar with a good deal of the research being pursued. Changes in direction of big laboratories involve serious problems of changes in lines of research and of the personnel associated with them.

The new director must be able to cope with these problems, as well as ensure that his own particular line of research advances. In this situation, a physicist with Mott's qualities had special advantages. Helping a variety of experimenters with their theoretical problems was part of his method of working. He was a participating director.

Mott expressed his approach to physics in a course of lectures to a general audience in the University of Virginia in 1956. He observed that today everyone was concerned with science. It had become so extensive and bore on so many aspects of life, that it was essential that administrators, political leaders and business men should have a general grasp of its nature and implications. It was important that scientists and the other citizens should not form two misunderstanding and mutually suspicious classes.

Many of the great achievements of modern physics, such as the discovery of radio waves through the researches of Clerk Maxwell and Hertz, of the electron by J. J. Thomson, of the atomic nucleus by Rutherford, of the quantum and relativity theories by Planck and Einstein, and of the breakdown of causality, had strengthened the belief that physics deals with the new and unsuspected aspects of nature, of which our ancestors knew nothing.

Yet until recent times many familiar and practically very important properties of matter, which were discovered or known before written history, were even less understood than radioactivity. The liquidity of water and the ductility of metals are examples. Their explanation in terms of atomic behaviour was not advanced until just before and after the Second World War.

One of Mott's aims was to make the investigation of these familiar

342

properties of matter as esteemed and proper a subject of university physical research as the new and unexpected. He chose to work in a field of fundamental research that was closely related to a large and important part of technology.

Mott observed that the discoverers of the new and unexpected, like J. J. Thomson and Rutherford, were literate and famous men, who worked in the tradition of the Greek philosophers. The technologists are the heirs of a very different tradition. The men who found out how to make the steels used in jet engines were intellectually the descendants of the anonymous smiths who began to discover how to strengthen metals 6000 years ago.

It may always have been true that the problems of technology are as intellectually exciting and as difficult as those of pure science; it is certainly true today, when the nuclear engineers are making increasingly stringent demands on the designers of materials, which cannot be satisfied without deep physical investigations.

The strengthening of ordinary pure metals by alloying, the hardening of metals by cold-working and quenching, and their softening by annealing, have been explained in atomic terms only during the last thirty years. The fundamental investigation of these familiar properties of materials presents intellectual problems as challenging as any in physics, and at the same time throws light on practical knowledge of vast social importance. By pursuing it, the Cavendish has drawn closer to the immediate concerns of the people, besides discovering new knowledge and training the new generation.

Mott's appointment as Cavendish Professor was a break with the primarily experimental tradition of the chair. It also reflected a change in the University, represented by his election to a fellowship at Caius College. Hitherto, all the Cavendish professors had been fellows of Trinity. In such a university as Cambridge a change of this kind has social and academic significance, and indicates that the situation in future will not be the same as before.

When the Second World War started Mott found that his quantum mechanics had no immediate war application. With his versatility he turned to lines of research of the most immediate practical importance, and in 1943 became superintendent of theoretical research in armaments at the Government's laboratory at Fort Halstead. Here he made an outstanding contribution to the theory of the fragmentation of shell and bomb cases under the effect of their explosive charges.

His severely practical war research strengthened his realistic approach to scientific matters. Whereas in his early days at Bristol he had appeared to some of his pupils to occupy an elegant but isolated ivory tower, when he returned to academic work after the war his practical spirit seemed equally striking. He extended the researches on the theory of the mechanical properties of metals. He personally worked on the explanation of plastic deformation of metals, in terms of microscopic derangements of the crystalline structure.

When Mott became Cavendish Professor he promoted these lines of research in the Cavendish, personally extending the theory of amorphous materials, such as glass, and the investigation of why some atoms form metallic substances, while others do not.

One of the most exciting moments in Mott's development of metal physics was when P. B. Hirsch rushed into his room in 1957, and asked him to come to see a moving 'dislocation' in the arrangement of the atoms in a metal, which had been revealed for the first time by an electron microscope. This established, on a more adequate basis, the theory of 'dislocations', on which the explanation of many of the properties of metals and materials depends.

Hirsch, who was born in 1925 and became a naturalised British subject, arrived at the Cavendish in 1946. He wanted to work in low-temperature physics in the Mond, which at that time had the highest prestige among the groups. This was probably due to the exciting nature of liquid helium, and the possibilities of adiabatic experiments at very low temperatures. The chief Mond interest presently shifted to Fermi surfaces. The Mond kept its supremacy for a long time, and Hirsch failed to get into it. Then he tried to join Orowan's group in metal physics, but was also not accepted there.

He finally joined W. H. Taylor's crystallographic laboratory. Taylor secured a grant of £180 for him. He was assigned to work on Bragg's bubble-model of metallic structure, which, like all Bragg's ideas, was pictorial, simple and elegant.

Bragg had conceived the microbeam X-ray analytical technique, and its development had been supervised by Taylor. It was thought that it might be used to resolve a controversy between Lipson in England and Wood in Australia about the effect of cold-working on metals. The broadening of the lines in the X-ray pictures due to the cold-working was difficult to interpret. Bragg suggested a beam so small that rays from neighbouring grains would not overlap. A spotty picture would then be observed, and the sub-grain size could be determined. Bragg related the strength of the metals to the size of the grains. J. N. Kellar, a graduate of Reading, who was the leader of the group, and Hirsch built a big rotating-anode X-ray tube to produce the microbeam, which became operative in 1948. They succeeded in showing that in aluminium the grains break up into smaller ones, but that aluminium was the only metal for which the technique worked. They obtained the spotty pictures for aluminium that Bragg had forecast.

At this time Bell Telephone Laboratories produced pictures of very small grains, which suggested that the electron microscope might be a more suitable instrument for their purpose. Bragg then lost interest in the research, and to their disappointment and annoyance, tried to divert their equipment to the research on proteins, to which he had transferred his attention. Taylor, however, kept apparatus for Hirsch, who was presently awarded an I.C.I. fellowship and substantial support from the National Coal Board. With the electron microscope it was possible to get very narrow and intense beams. J. Menter obtained for Hirsch electron-diffraction pictures from beaten gold.

This was the starting point for the research on the electron microscopy of metal dislocations. One reason why they studied metal defects by electron microscopy was because they were X-ray crystallographers, and knew something about metal physics. They worked out a crucial contrast theory for studying the structure. They could do this because they were diffraction experts, not microscopists; their contrast technique was unique. They were able to prove the existence of dislocations to the practical metallurgists.

Hirsch's research student, M. J. Whelan, and R. W. Horne, who was operating the Siemens electron microscope in Cosslett's group, were using double-condenser illumination for the first time in these experiments. On 3 May 1956, while Horne was operating the instrument for Whelan, they switched over to double-condenser illumination, and immediately saw the dislocations moving. It became possible to demonstrate the existence and movement of dislocations in all metals, and: 'Seeing is believing'. Hirsch made a film of moving dislocations which ran for several minutes. When it was shown to a visiting American physicist he commented: 'That is sure the most glamorous film I have ever seen.'

New kinds of defects in metals were discovered, which had not been previously imagined. The calculation of dislocations in metal physics was advanced, and important researches were started on the defects in quenched and neutron-irradiated metals, which could only be studied in this way. The general development of thought in metallurgy was stimulated. The research on neutron damage is perhaps the most important in metal physics, for it is of crucial importance in the operation of nuclear reactors.

One of the features of the Cavendish crystallographers was their solidarity. A number of them, including J. N. Kellar, attended a conference on crystallography in Holland. They were taken for an outing in a boat, which ended in a tragic accident, in which Kellar was electrocuted. Mrs Kellar was engaged by Taylor as his secretary, and the crystallographers subscribed among themselves to contribute to the education of the Kellar children. Hirsch married Mrs Kellar in 1959, and became professor of metallurgy at Oxford in 1966.

Mott's organisational approach agreed with the pattern that had evolved rapidly since the Second World War in most of the countries where advanced science was pursued. The Cavendish grew more like other great physical laboratories that had developed in other places after the war. All of them had similar problems of organisation, which had become a much more important factor in their direction, owing to the increase in scale, numbers and expense. In this situation the Cavendish, as Mott put it, became 'one good laboratory among many'.

Mott found himself confronted by several problems. One was the difficulty presented by the old Laboratory site; there was no room for big machinery. Should the Laboratory move from the old site to a more spacious one away from the centre of the city? Mott had a very important part, in which Pippard joined, in making the decision to move out.

After it had been made, the number of students, which had been a factor in deciding to move to a larger site, began to decrease. But then, after a while, it began to go up again.

The problem of running such a laboratory as the Cavendish is felt especially when appointments are being made. The difficulty is not so much in recognising and backing men of outstanding ability in researches of obvious importance, but in dealing with men and subjects where the evidence of comparative ability and importance is not so clear. In Cambridge it is complicated by the college system.

While the research atmosphere in the Cavendish and, say, in the Bristol physics department might be much the same, the existence of the colleges creates differences in, for example, the teaching. At

346

Bristol the physics department had to look after all the teaching of the students. In Cambridge the student generally does not come entirely under the Professor until his third year. Until then, he may as an undergraduate belong more to his college than to the physics department. Consequently, there is a difference in commitment.

Another effect of the college system is to increase the difficulty of securing new people from outside. Mott tried repeatedly to secure good men from London, but this was difficult because an adequate salary depended jointly on the University and a college. For this reason it was necessary to find a college as well as a university position for a suitable man. In dealing with this problem a Cavendish professor had to be more than just a physicist; he also had to be skilled in Cambridge academic life.

Mott was happy to secure the appointment of P. W. Anderson from America, a welcome infusion from a different tradition.

Another daunting Cavendish problem was the achievement and fame of Rutherford. He was a fantastically great man, and under him research had been enormously exciting. His successors naturally tended to feel a little sour about what could be accomplished in comparison with him. In Rutherford's time the Laboratory was small, and the man at the top could know and decide on most of the things that were being pursued. The growth in size made that form of organisation no longer possible.

When Mott arrived, he found the group system functioning, and he consolidated it. The head of the Physics Department had to be a chairman deciding between conflicting claims rather than telling people what to do. It was very important to avoid conflicts of personality between leaders of groups, for this can be very damaging.

With the development of the group system the unitary feeling of 'the Cavendish' became less strong. Now there is the 'radioastronomy group', the 'solid-state group', and so on. The feeling of the 'Cavendish' comes up especially when they sit down together to discuss teaching. For instance, the radioastronomers teach physics in the Cavendish. If they became an independent department they would have very few students. Teaching is of real value to them, and to the Cavendish; it integrates them with the institution and its traditions.

The announcement of the Mullard endowment in 1955, and the large development of radioastronomy leading to immediately fruitful results, raised the question of the status of radioastronomy. Discussion in the University's governing bodies led the Council of the Senate to report on 18 May 1959:

1. Shortly after the last war Mr. Martin Ryle, who had come to Cambridge as an I.C.I. Fellow, began to take a leading part in work on Radio Astronomy, which was at that time a new subject. The work was carried out in the Cavendish Laboratory and the aerials needed were set up to the west of Grange Road in the area of the Rifle Range. On 1st October 1947 Mr. Ryle was appointed to a University Lecturership in Physics. It soon became apparent that he was engaged on work of great promise and importance. The work could not however be extended or even continued for long in the area of the Rifle Range, and in 1955 Mullard offered to provide under covenant sums which would yield a total of £100 000 towards the cost of establishing a Radio Astronomy Observatory on a new site. A satisfactory site was leased from the Air Ministry at Lord's Bridge, and a development plan, of which the capital and recurrent cost over a period of ten years was estimated at £180 000, was approved by Grace 4 of 20 May 1956. In addition to the generous gift of £100 000 from Mullard Limited a grant of £40 000 was received from the Department of Scientific and Industrial Research, and the General Board agreed to propose the provision of £40 000 from the Development Fund of Groups III and IV of the Faculties.

2. Since the summer of 1956 the development of the site at Lord's Bridge has progressed satisfactorily, the importance of Mr. Ryle's work has increased and been widely acclaimed, and the subject has become a major activity of the Cavendish Laboratory, and one for which it is well known to the world at large. Indeed Mr. Ryle's work and the work of Professor Lovell at the University of Manchester together represent a large part of all the best work that is now being done in this subject. A Readership was established for Mr. Ryle with effect from 1 Oct. 1957, and in their statement of needs for 1959–60 the Faculty Board of Physics and Chemistry gave high priority to the establishment of a Professorship for him. The General Board having considered that statement, have advised the Council to recommend the University to give effect to the Faculty Board's wishes. The Council agreed with the Faculty Board and with the General Board that the status and authority associated with a Professorship would now be more appropriate both to the personal distinction of Mr. Ryle and to the outstanding importance of the work which he is doing in Cambridge, and they therefore propose that a Professorship of Radioastronomy be established for his tenure.

At the same meeting a fourth floor for the Austin Wing of the Laboratory was proposed by the Cavendish Professor, for the accommodation of theoretical physicists and applied mathematicians, at a cost of £45 000, and was approved.

The signatories of the Report were: Adrian, Vice Chancellor, W. Ivor Jennings, Henry Willink, Brian W. Downs, H. Butterfield, N. F. Mott, B. C. Saunders, J. H. Plumb, C. E. Tilley, R. B. Braithwaite, C. O. Brink, F. Wild, Peter R. Ackroyd, G. F. Hickson, J. S. Boys Smith, R. E. Macpherson and C. L. G. Pratt.

This list of names is a reminder that in a university, representatives of many faculties, in the arts as well as the sciences, must agree on important scientific innovations, and that these are not just for the decision of scientists alone.

Whereas radioastronomy remained in the Cavendish, molecular biology moved out. After the discoveries of 1953 Perutz and his colleagues might have pressed at once for the construction of a special laboratory for molecular biology. They did not do so because the work of Watson and Crick was not immediately accepted as significant by many biologists, and Perutz and Kendrew still had much to do before their analyses of haemoglobin and myoglobin were complete. By 1957, three years after Mott had become professor, their position had strengthened; Kendrew's research had progressed most promisingly.

They had been joined later by S. Brenner, who had created a bacteriophage laboratory, in which progress was made in the resolution of the nature of the genetic code, and Ingram, who discovered how the sequence of amino acids in an organism's proteins is affected by genetic mutation. Besides their own group the eminent biochemist Frederick Sanger, who had worked out the chemical structure of insulin, and who had also received support from the M.R.C., said that he would like to join the molecular biologists.

The growth of molecular biology had made steadily increased demands for space in the Cavendish. This had been met by assigning more and more rooms scattered over the Laboratory's accumulation of odd buildings off Free School Lane. It was eased by the provision of further rooms in the adjacent Zoological Laboratories, and a prefabricated hut in the courtyard between the Austin Wing of the Cavendish and the Zoological Museum, which had recently been vacated by the Metallurgy Department. Between the hut and the Austin building was a long rack for bicycles, and around the hut cars were tightly parked. In this hut as late as the first month of 1962, were to be found several of the most famous leaders of the new branch of science. One summer, some time before, a Russian delegation arriving in Cambridge had asked to see the 'Institute of Molecular Biology'. Perutz, duly flattered, took them to the hut. The Russians stared at it in bewilderment, and asked: 'But where do you work in winter?'

Though the accommodation was scattered and shabby, the molecular biologists had superb equipment, mostly bought for them by the M.R.C. The close contiguity of the researchers with those in all the other varied branches of physics pursued in the Cavendish, and with the many other scientific departments clustered adjacently, proved an

349

invaluable intellectual stimulation. However, substantially more accommodation became imperative. The molecular biologists proposed to the Medical Research Council that they should build a laboratory for molecular biology. This was immediately approved by the Council, at a single sitting.

The relations of the laboratory to the Cavendish and the University, and its site, raised many problems. There were those who regarded molecular biology as more biological than physical, and thought that it should be moved out of the Cavendish. There were others who thought that the co-operation between physics and biology should be increased rather than lessened, and a move out of the Cavendish would be a retrograde step to the older, narrower conception of physics; it would be against the historical and philosophical development towards a more unified science, no longer restricted to the traditional compartmentalisations. It would also take from the Cavendish one of its most fruitful and inspiring lines of research, which would not be good for the other lines.

Among the University authorities, however, there was strong opposition to a new independent research laboratory that would have no traditional function in the University teaching. A university was an educational institution. Laboratories entirely devoted to research did not make a contribution to this primary function of a university.

What was the University status of the distinguished members of such a laboratory to be? Were they to be made professors? Could that be proper, if they were not official University teachers? If a string of eminent men in one subject were given professorial rank, would it upset the balance of influence between subjects in the University? On the other hand, it was embarrassing to have eminent men in the district who did not have appropriate status in the University.

After agreement was at last obtained that there should be a laboratory, a site was sought. The molecular biologists hoped it would be near the complex of science laboratories in the centre of Cambridge, but this area was already congested. Finally, a site was found in the area two miles outside the town assigned to the development of a Postgraduate Medical School.

The now independent Laboratory for Molecular Biology was opened in February 1962, with Perutz, Sanger, Kendrew, Crick, H. E. Huxley, A. Klug and J. D. Smith, and working visits from J. D. Watson. The molecular biologists had left the wing of physics and

350

were now neighbours of the medical scientists. Their new situation was materially much more comfortable, but it was also more conventional. Was it as original and interesting as the old one?

Some of the Cavendish physicists had regarded Perutz, Crick, Watson and Kendrew as intruders. They were frowned upon, and some people spoke of 'getting rid' of them; when they went, some threw their hats into the air. In 1952 Kendrew was not allowed to sit in the annual Cavendish photo.

To many Cavendish men, J. D. Watson appeared very quiet; he kept to himself. Crick on the other hand was always colourful, with his resounding laugh and pink shirts, riding about Cambridge on a moped with almost acrobatic skill. He devoted much of his Nobel Prize to the purchase of scientific equipment.

Mott's most unhappy moment as Cavendish Professor was probably when it had become clear that he could not find the money and resources to keep the molecular biologists in the Cavendish. But he had no options; the University had neither the money nor the places. He tried very hard to secure a chair in biophysics for Perutz.

In administration, Mott was unsentimental. He saw himself as a chairman, and was a good one. He was impressive in arguing for a scientific project in a university academic committee. Whereas Bragg disliked interdepartmental and university administration, Mott was keenly interested in it, though he had to leave a good deal to the departmental secretary. He was delighted to win battles in University politics on behalf of the Cavendish.

In 1960 A.B. Pippard was appointed John Humphrey Plummer Professor of theoretical physics. Like Mott he was educated at Clifton College. He was one of the younger physicists who had been in favour of Mott's election in 1953. Mott and Pippard worked together in close understanding. In 1963, when Mott was abroad, Pippard was appointed Deputy Head of the Cavendish.

As the need for a new Laboratory became more pressing, they collaborated on plans for the New Cavendish, with Pippard becoming Chairman of the Building Committee. Many of his ideas were incorporated in the scheme, and he clearly became a strong candidate as Mott's successor. After Mott's retirement Pippard was elected Cavendish Professor in 1971. Like Mott, he departed from the earlier tradition of association with Trinity College, and was a fellow of Clare College. This confirmed the spread of the Cavendish connection with other colleges, and the extension of some of the characteristics that had developed during Mott's professorship.

One of the chief developments in the Cavendish during Mott's tenure was the incorporation in it of F. P. Bowden's laboratory of surface physics. Bowden did not join the Cavendish until 1957, when he was already fifty-four years old. He had started as a physical chemist, and steadily moved towards physics, until he and his accomplished group transferred from the Department of Physical Chemistry, at the time when the physical chemists moved into their new large laboratory at Lensfield Gardens. Bowden's laboratory was adjacent to the Cavendish, and having become closer to the physicists than the chemists in thought and interests, he did not wish to move away from them.

An important factor in Bowden's originality was his Tasmanian boyhood. His father had been a telegraph boy and his mother a postmistress. They had strong cultural interests. Their son was an able student, but no mathematician. He did not pass the matriculation examination in mathematics, and could not go directly to the university. He became a junior laboratory assistant in the laboratories of the Electrolytic Zinc Company, which made use of Tasmania's considerable hydroelectric resources. He immediately took to the experimental work, and started to find himself.

He retained for the rest of his life a sympathy and understanding for industrial research; it had given him his chance. This ultimately enabled him to make his most important contribution to the Cavendish, which has been described as 'his success in forming a research group within a University Physics Department with a positive attitude towards industrial problems, a willingness to help industry, an eye for the interesting fundamental issues within the practical problem, and a feeling for applying fundamental work to practical affairs'. He created a new and significant link between the Cavendish and industrial research.

The outstanding feature of his personal research was its concrete character. With his non-mathematical, almost anti-mathematical, outlook he sought for vivid physical images of what was actually happening in the phenomena he was investigating. In this respect he resembled Rutherford, as he also did as a leader of an enthusiastic team of collaborators. His combination of visual physical imagination and ingenuity enabled him to make original discoveries in surface phenomena by simple and elegant experiments.

His work on friction and lubrication led, on the one hand, to the introduction of dry metal-backed polymer bearings into industry, and on the other, to the improvement of performances in skating and skiing.

Another remarkable line of his researches was on the damage produced by water-drops projected at very high speeds. It was found that their impact produced extremely high pressures due to compressible deformation, which was able to deform the hardest materials. These researches threw light on the erosion of turbine blades, and the effects on aircraft of flying into rainstorms.

Bowden became a Reader in Physics when his group joined the Cavendish in 1957. Before he died in 1968 his range and style of work had been incorporated in the Cavendish. His final identification was completed by his election as Professor of Surface Physics in 1966, two years before his death. By that time his line of work was completely established, and his colleagues Yoffe and Tabor were there to carry it on.

Bowden had first come to Cambridge in 1927, when he worked on electrochemistry under E. K. Rideal. In 1931 he showed that if one of W. B. Hardy's classical experiments on lubrication was performed with scrupulously clean surfaces, his conclusions about the range of the molecular forces controlling friction had to be modified. He became increasingly interested in what exactly happens at solid surfaces. He said that 'Putting solids together is rather like turning Switzerland upside down and standing it on Austria—the area of intimate contact will be small.'

Bowden postulated that high temperatures leading to melting were due to violent rubbing between the small areas in contact. This led him to explain the easy sliding of skates and skis as due to temperature melting, superseding the theory of Osborne Reynolds, that the melting of the ice that facilitated sliding was due to pressure.

Bowden was visiting Australia when the Second World War started. He decided to stay there, and began research on the causes of accidental explosions in munitions factories and magazines. He was joined by J. S. Courtney-Pratt, who invented a simple apparatus for measuring bullet speeds. Later, in Cambridge, Courtney-Pratt invented a high-speed image-converter camera, which took pictures in a nanosecond. With this apparatus the reactions between materials could be followed pictorially.

Bowden's colleague D. Tabor invented the term 'tribophysics' (from *tribos* for rubbing), to define their range of activities. Tribology is the term that has now come into common use.

353

In 1944 Bowden visited England, before returning permanently in 1945. He wrote an official report in which he noted that 'war experience has shown clearly that in pre-war days the academic and practical research were too widely separated and suggests that, in future, contact between the two should be maintained'.

Back in Cambridge, he resumed research on friction, lubrication and the initiation of explosions. His work indicated that explosions in solids, as in liquids, arose from hot spots where energy was produced more quickly than it could be dissipated. His colleague J. D. Blackwood gave the first satisfactory explanation of the decomposition of gunpowder.

In 1953 Bowden was recommended to the Tube Investments Company as a consultant on research. Under his influence a research laboratory was set up at Hinxton Hall, ten miles from Cambridge, and a close relationship was established between the two research groups. Presently, Tube Investments had an urgent metallurgical problem that had arisen in one of their works, and required an instrument that could carry out a point-to-point non-destructive chemical analysis of surfaces. Enquiry showed that P. Duncomb in the Cavendish had an experimental apparatus with which this could be done. Duncomb and D. A. Melford then proceeded to work out in the Hinxton Laboratory the first fully engineered electron-probe X-ray scanning microanalyser, based on the experimental Cavendish instrument.

This was then manufactured and sold by the Cambridge Scientific Instrument Company. It was a striking example of the close interaction of industrial and university science that Bowden had envisaged.

J. W. Menter became interested in the microstructural world opened up by electron optical techniques, and later, in collaboration with P. B. Hirsch and A. Kelly of the Cavendish, studied the microstructure of cold-worked gold foil. This was the precursor of Hirsch's work in the Cavendish on dislocations.

Though Bowden came to the Cavendish late in his career, his contribution to the Laboratory was very notable. For a long time the close relation of the work of Bowden and his colleagues to practical and industrial matters did not recommend it to some, but in the end its positive advantages were generally recognised, to the benefit of the Cavendish.

Bowden was an astute business man, besides being a talented investigator. He was a technical adviser to companies, and had a private office in London. He used to visit America, and return endowed with dollars for his researches. When his laboratory was incorporated in the Cavendish it brought in £50000 from outside bodies.

Mott highly esteemed Bowden's work, and in his retirement worked in a room in the surface physics section of the Cavendish, in which to pursue his own researches, with as much devotion as ever.

In the 1960s the Cavendish widened its tradition by inviting a professor from abroad to join its permanent staff. Since 1961, P. W. Anderson, the theorist of the solid state and superconductivity, has been connected with the Laboratory. He became the first visiting professor with permanent tenure. He preserves his appointment with Bell Telephone Laboratories, spending part of each year in the United States. The General Board has since said that, in future, only in exceptional circumstances will they recommend to the University that a visiting Professor be appointed to the retiring age. This attitude, manifested by most of the University towards visiting Professors, is regarded by many scientists as reactionary.

Anderson published a paper in 1958 on the theory of the structure of disordered substances. It gave working explanations of some of the properties of such materials, and Mott based substantial theoretical developments on it.

Anderson first met Mott at a conference in Japan in 1953. He was then twenty-eight, and the youngest man there. The others were very senior people, and he felt a little isolated. He went on a conference visit, and found himself in the train with Mott, who was very friendly; Anderson greatly appreciated his kindness. Then, a few years later, when Mott was going through the Bell Laboratories, they met again, and exchanged their ideas further. In 1959 a conference on superconductivity was organised in the Cavendish. Pippard was interested in his attendance.

Anderson became more engaged in the theory of superconductivity. Mott, who was even more active in the theory of the solid state after his retirement than before, tried to draw him back to this subject. Mott seemed to him to have a fantastic grasp of the totality of the field; he was the master of so many different threads that were converging. His ideas are clearer than some people think, because they are profound and not easily understood.

In Anderson's view, one of the strengths of the Cavendish is its tradition; it has a long history, and a longer tradition of good physics than any other institution in the world. It is an aspect of the total

English scientific performance, which has been consistently high for three centuries.

Anderson is inclined to Ben-David's opinion that while French science in the eighteenth century, German science in the nineteenth and American science in the twentieth, led the world, British science through the modern period as a whole has given a more sustained performance.

Anderson considered that the Cavendish has been particularly distinguished by intellectual organisation and authority. It has the tendency to select a major problem, and organise a whole group to attack it. Bragg's encouragement of research on molecular biology is an example. The single-minded concentration of the Mond Laboratory on superconductivity, of Shoenberg and Pippard on the Fermi surface of metals and of Mott on the theory of amorphous materials, are others.

Owing to the tradition of intellectual authority in the Cavendish, it is easier than in other places to tell people what to do. Intellectually, science is not a democratic activity. Anderson believes that the citation test of merit gives significant results: the number of people who are continually quoted is rather small.

He thought that in the 1970s physics was a little bit ill almost everywhere. People expected a great deal of it, but permanent expansion is impossible. The easy problems have been tackled very fast, and now physicists are faced with an accumulation of difficult ones that have been left over. The decline in numbers and the unemployment of graduates affects the minds of many people, though Anderson did not feel it himself. The contraction in science would not contract the good scientists.

In the 1950s and 60s many good scientists had been attracted to the United States, but in the 1970s Germany was attracting scientists from all over the world. Both the old and young generations in the Cavendish appeared to Anderson to be excellent.

He worked for some time in Japan, and found it a fascinating experience. The Japanese had accomplished a great deal in technology. Their professors had high prestige and honour, but the Japanese think that that is enough; they do not want to burden them with salaries and resources. The intellectual is much distrusted, especially by business and the military. He is expected to train their sons, but not to influence them. The Japanese are very conservative, and so is Cambridge, but the two kinds of conservatism are very different.

Japanese social ideas are particularly apt for industrial society; they are not so apt for their scientists, who are very good, but do not get as much support as in other countries.

Anderson believes that the Cavendish has a good chance of a relatively strong future. Even if government financial support contracted, the Cavendish would still have a teaching function. At the worst of times, the Cavendish still gets the better students.

27

Current Researches

In 1970, near the end of Mott's professorship, a broad summary of researches was compiled. It presented a cross-section of the activity in the Laboratory. It is drawn upon here as a guide to the balance of research interests at that time.

There were fourteen research groups, comprising the crystallographic laboratory, electron microscopy, fluid dynamics, high-energy physics, liquid metals, meteorological physics, metal physics, radio-astronomy (and ionosphere), slow-neutron physics, solid-state theory, surface physics, electronics service section and crystal-growing unit.

In the crystallographic laboratory the distribution of electron and spin density in the unit cell of crystal structures was being investigated. The electron density was derived from measurements of X-ray scattering, and results compared with the predictions based on various theoretical atomic models of the solid state.

Spin distribution in magnetic crystals was derived from measurements of neutron magnetic scattering cross-sections. In ionic crystals these measurements were used to determine the degree of covalency in the ionic bond, and the variation of the spin direction in the unit cell. In metallic materials the values of the individual atomic moments were derived, and in some cases, the complete spin distribution in the unit cell.

This information, combined with knowledge of the electron density, provided a powerful means for investigating the electronic structure of solids.

The neutron-diffraction measurements were made at Harwell. Two years later, a good deal of this work was being transferred elsewhere, especially to Grenoble, where Jane Brown, one of the Laboratory's leading crystallographers, went to work.

The electron-momentum distribution in solids was investigated by studies of incoherently scattered X-radiation, giving the Compton profile. Measurements of the profiles in several light elements revealed deviations from theoretical predictions.

Research was pursued on the structures of ferroelectric materials. A very complex series of structural transitions was revealed, which could be correlated, in part, with the known dielectric properties of the materials.

Investigation of felspar minerals was continued. The order–disorder relationships and anomalous effects due to stacking faults in a variety of felspars was examined, with the aim of developing principles likely to be of wide application, especially in nonmetals.

The electron-microscopy section was concerned with five main subjects. The first of these, electron optics, was engaged in researches on the improvement of the performance of electron-beam devices, such as reduction of distortion by projective lenses. Experiments were being made with quadrupole lenses, and with ultra-high-frequency electric fields.

In electron physics an extensive experimental programme was being carried out, in order to obtain reliable data on electron penetration and the efficiency of X-ray production, to enable more accurate predictions to be made on the X-ray microanalyser that was to be constructed.

By 1973 the possibility of a high-resolution microscope, carrying resolution down to atomic dimensions, had been envisaged. Such an instrument might cost a quarter of a million pounds, and be established as a national facility. In the future the Cavendish will probably be less engaged in the design and development of instruments.

Research on X-ray microanalysis was concerned with chemical microanalysis by means of the spectrography of X-rays excited within the specimen. The development of quantitative analysis of thin specimens was also pursued.

These researches, especially by V. E. Cosslett and T. A. Hall, were very important for biology. The section was collaborating in the application of the technique with biologists in Cambridge and elsewhere.

Electron microscopy and diffraction were applied to the study of the domain structure of ferromagnetic films. An extensive programme was pursued on the structure and electrical properties of amorphous alloy films.

359

The application of the electron microscope as a small-angle electron-diffraction camera of very high angular resolution was applied to a variety of problems in physics and biology.

Theories of the structure of the surfaces in liquids have suggested that they possess a quasi-crystallinity. Experimental research was being prepared to obtain further confirmation of these theories.

The major development in high-voltage electron microscopy was the construction of an instrument that operated between 100 and 750 kV. Four programmes of research with it were being carried out: on metal foils thicker than those investigated with less powerful instruments, and closer in properties to bulk metal; on the mechanism of electron transmission in crystalline layers much thicker than the limit at which kinematic theory must be abandoned for dynamic theory; on the quantitative evaluation of the response of photographic emulsions and of fluorescent viewing screens as a function of incident electron energy; and the exploration of the limits of observation of thick sections of biological material at high voltages. This was for the study of micro-organisms, and possibly of tissue cultures in the wet state.

The work on fluid dynamics, which included researches by Townsend and G. I. Taylor, was mostly directed to the development of a coherent account of the many kinds of turbulent flow, in which the more important properties could be predicted from general principles. Their view was that these principles should be based on a sound understanding of the interaction of the eddies of the turbulence with the mean motion, and with each other.

Experimental researches on turbulent flow were being conducted by hot-wire anemometry. One subject was the mechanics of the entrainment process in free turbulence, that is, when turbulent flow is surrounded by fluid at rest or in uniform motion. In those conditions, turbulent eddying fluid is separated from ambient fluid in irrotational motion by a remarkably sharply defined bounding surface, and the spread of turbulence into the ambient fluid occurs by a process of folding-in or engulfment.

Optical methods for the study of turbulent flow were being further developed, and applied to the study of the bounding surface in a boundary layer.

In co-operation with the meteorological physics section, the velocity and temperature fluctuations in the atmosphere at a height of two kilometres were measured in order to detect and measure the internal gravity waves that might be radiated by the turbulent boundary layer on the earth's surface. Preliminary observations had revealed the presence of temperature fluctuations, of a period of 24 seconds, which might be caused by such waves. Energy and momentum might be carried by the waves without the movement of the air, which might account for some of the observed occurrences of turbulence in clear air.

This line of research presently ended when the meteorological section was closed down.

The high-energy physics section was concerned principally with bubble chambers, and theoretical research. This included Frisch's work on the 'Sweepnik', his device for the automated measurement of the hundreds of thousands of particle tracks photographed in bubble-chambers at CERN, and elsewhere (see figure 38).

The counter section worked with counter and spark-chamber techniques at the Rutherford Laboratory, Harwell. The research was especially in the field of weak interactions, and the decay of K-mesons. As much of this work was done at Harwell, it could not easily be combined with teaching, and has since been closed down.

The theoretical research, in which R. J. Eden was prominent, dealt with high-energy collisions of strongly interacting particles, developments in Regge theory, collision amplitudes and S-matrix theory, the relation of resonances and elementary particle interactions, symmetry and group theory of elementary particles and electromagnetic and weak interactions.

The liquid-metals section had been engaged on experimental studies on the transport properties of alloy systems of the sodium–potassium type. Much work was done on the conductivity of liquid metals at high frequencies, as deduced from their optical properties in the visible and ultraviolet regions, and this was extended to the infrared.

The use of ultrasonic techniques to measure compressibility as a function of concentration in liquid alloy systems was also being explored.

The section on meteorological physics, which later was to be closed on T. W. Wormell's retirement, had been interested mainly in atmospheric radiation, the physics of clouds and other particulate matter in the atmosphere.

Among the topics investigated was the possibility of studying the large-scale vertical motion of the atmosphere by using ozone as a tracer. Combustion products in the atmosphere, and the photochemi-

361

cal action of sunlight on certain trace gases, form particles that carry most of the radioactivity in the atmosphere, whether natural or man-made. They disappear ultimately by a slow process of coagulation. The phenomena are of importance in pollution problems, and laboratory experiments were made to elucidate them.

Studies of thunderstorm electricity had continued, in the long Cavendish tradition made famous especially by C. T. R. Wilson. The current experiments showed that the recovery of the electric field after a lightning flash varied markedly with the distance from the storm, which implied that the theory of one of the accepted basic parameters of a thunderstorm required modification.

The metal physics section extended research on the more detailed explanation of metal properties in terms of imperfections, such as dislocations, stacking faults and point defects. A better knowledge of the precise configurations underlying these imperfections was required. The development of the application of electron microscopy and X-ray investigation required a great deal of experimental and theoretical effort. On the theoretical side, the dynamical diffraction theory of the propagation and scattering of fast electrons in imperfect crystals was developed, so that the maximum amount of information could be extracted from the experimental observations.

Extensive experimental researches were made by electron microscopical and X-ray methods on dislocations and single crystals. Among other effects observed were striking periodic arrangements of parallel dislocations after push–pull fatigue deformation.

In studies of the detailed structure of lattice defects, the highly strained region of crystal near the dislocation core where Hooke's law fails and the structure verges on that of a two-dimensional liquid, was an object of investigation.

Surface processes were investigated with the aid of ultra-high-vacuum electron microscopy. As a result, the role of interfacial dislocation in the deposition of thin films was considerably clarified. The process is of wide significance in physics and biology, and in the industrial manufacture of semiconductors.

Studies of inelastic scattering and propagation of high-energy electrons in crystals provided information on the frequency and localisation of plasma oscillations in imperfect crystals and inhomogeneous alloys. It also assisted the interpretation of experiments on the penetration of high-energy electrons through crystals.

Collaboration in phase-contrast electron microscopy was carried

362

out by A. Howie with Professor Valdrè of Bologna on the construction of a phase-contrast electron microscope, in order to improve the visibility of very small crystal defects, or of small atomic structures, such as clusters of atoms or molecules.

The Mond and magnetic laboratories had been active in detailed studies of electron behaviour in metals, superconductivity and dielectrics.

The work on the determination of Fermi surfaces, initiated by Pippard twenty years before, had continued. The use of superconducting solenoids had allowed precise determination of the almost spherical Fermi surfaces of the alkali metals, and more accurate determination of other metals. With this information it had become possible to make detailed investigations of phenomena involving the dynamics of electrons in a magnetic field.

In superconductivity studies the implications of Josephson's effect were being explored. The 'slug', in which the effect was applied by J. Clarke for detecting extremely weak magnetic fields and potential differences, was being used for investigating electrical conductivity in superconductors, in conditions where conventional instruments were inapplicable.

J. R. Waldram had developed a new experimental technique of measuring the surface impedance of superconductors in a microwave field. It was being used to obtain information about the response of both the supercurrent and the normal electrons to electromagnetic fields.

A dielectric research was initiated to discover whether suitable insulating materials were available for high-power superconducting cables. Quantum-tunnelling in polyethylene was noted, and similar effects looked for in other substances. Later on, this work was closed.

Radioastronomy was summarised under the work of the Mullard Radio Astronomy Observatory. The study of discrete radio sources, such as the pulsars they had discovered in 1967, and the quasars, was pursued in order to reveal the overall structure of the universe, and elucidate the physical processes occurring within sources that provided an extensive range of conditions quite unattainable in terrestrial laboratories. Their principal instrument had been their one-mile radio telescope.

The pulsars, the discovery of which had been described as the most important in astronomy in the previous five years, appeared to be rotating neutron stars with a density of 10^9 grams per cubic centimetre.

A second large-aperture synthesis telescope had been built for interpreting radio emissions from nearby galaxies other than our own. It was being used to study the distribution of magnetic fields and of neutral hydrogen in spiral galaxies M31 and M33.

Radio emissions in the Solar System included thermal radio emission from the Moon and Venus. The measurement of the polarisation of the radiation provided information on the electrical properties of the surface materials.

The diffraction effects of the radiation from quasars were being studied in order to obtain information on the interplanetary medium, and the dimensions of weak sources of small angular size.

Other experimental work included studies of galactic 'halos', occultations of sources by the Moon, measurements of atmospheric effects on the propagation of microwaves, and design of a complex digital radio-line spectrometer.

Theoretical work included the explanation of the nature of radio sources. It seemed that the radio emission from both supernova remnants and extra-galactic sources was due to synchrotron radiation by relativistic electrons.

Other topics included the propagation of waves through turbulent media, and of radio waves in the ionosphere and magnetosphere.

Future development included the construction of the very large 5-kilometre radiotelescope.

The slow-neutron physics section was studying lattice vibrations in metals by the method of inelastic scattering of slow neutrons. These yielded information about atomic forces, and especially the role played by conduction electrons in these forces. The experiments were carried out on nuclear reactors at Harwell.

Measurements on transition and divalent metals were made over a range of temperatures from liquid helium to 1800°C. Members of the section spent about one third of the time at Harwell, preparing apparatus and making measurements.

The solid-state theory section dealt with the electronic properties of materials, in close co-operation with several experimental sections. There was also research on many-body systems, involving the strong interaction of large numbers of particles, which leads to magnetism and superconduction, to critical fluctuations near second-order phase transitions and to the superfluidity of liquid helium.

One of the main themes was to find a complete and consistent interpretation of all the properties of metals that depend directly on electron energy levels and electron scattering. For simple metals, semiconductors and alloys, these can be expressed in terms of a model of the forces between atoms. It had already become clear why mercury, gallium and indium have complex crystal structures; they are mostly distortions of the simpler structures of their neighbours in the periodic table.

They had started on the calculation of the structures of alloys. The programs for the calculations for thirty elements, to be made on the Titan computer, were being worked out.

Increased attention was being given to surface physics. With joint experimental work it was felt that the first non-trivial structure analysis for a surface should not lie too far in the future.

The concepts used in this work were being extended to molecular problems. Quantum chemistry tended to emphasise either gigantic *a priori* calculations on tiny molecules, or the roughest empirical approach to all other cases. The Cavendish investigators were attempt-int to occupy the middle ground, creating methods that were theoretically rigorous yet simple, and aimed at *understanding* the properties of the chemical bonds, rather than calculating the last decimal place.

In modern many-body theory the section had recently achieved an exact mathematical solution of one of the simplest cases, and they were actively applying their solution to the problems of impurities in semiconductors, catalysis and magnetic atoms in large organic molecules, such as haemoglobin.

They had been active in the studies on the surface properties of liquid helium, impurities in liquid helium and the relationship between magnetism and superconductivity.

Finally, they were continuing research in a field in which, historically, the Cavendish had made the most important contributions: the electronic properties of amorphous and liquid materials. The problem on which they were concentrated at the moment was the localisability of electrons.

The surface-physics section worked on surface properties, strength of solids, solid-state physics and decomposition in the solid state. On the first of these topics it had been applying and developing physical techniques for studying surface topology irregularities and defects.

Experiments on the contact between optically smooth rubber and glass had provided a direct means for measuring the forces between electrically charged double layers. These are relevant to adhesion, lubrication and the stability of colloids.

Other investigations concerned: the structure and chemical nature of ultra-thin surface films; the formation of the very thin coatings of polymeric fluids involved in lubrication at high temperatures; and the configuration and orientation of polymer molecules at interfaces.

A broad attack had been started on the surface properties, adhesion and friction of polymers. It had proved possible to form surface films providing effective lubrication at 1000°C.

These experiments were also extended to very low temperatures, giving results with a direct bearing on the action and behaviour of rubbing surfaces in spacecraft.

Research on the strength of solids included experiments on single metal crystals. These showed that the material may exhibit the full theoretical strength, and that a single crystal of gold may be as hard as tool steel.

Experiments on the strength properties of solids included the investigation of the strength of refractory materials over a temperature range of 1000–2500°C. Carbides, though very hard at room temperatures, suffer a very rapid fall in strength at high temperatures. Much of their research had been on this effect in titanium carbide. They had also studied the strength of ice and polymeric solids.

Experiments on the deformation of solids at high rates of strain were conducted by subjecting metals to impact velocities of 7 kilometres per second. Theoretical and experimental researches were made on high-speed liquid impact. High-speed cameras working at framing rates of up to 10 million a second, with exposure times of one thousand millionth of a second, were used for recording events.

The effects of particles one ten thousandth of a centimetre in diameter shot at solid surfaces at a velocity of several kilometres a second were investigated.

The damage produced in solids by laser-beam impingement was studied.

The experiments on high-speed liquid impact were relevant to the erosion of turbine blades in steam, and to the erosion of the wings of aircraft when flying through rain. Their experiments also bore on the impact of space vehicles with micro-meteorites.

The solid-state physics section had been active in studies of semiconductors, transition metals, glasses and other amorphous systems. An initial comparison of the electrical, structural and optical properties of amorphous and crystalline arsenic triselenide had been completed.

Researches on decomposition in the solid state included a study of the role of the surface; it had been shown that slow decomposition in azides is predominantly a surface phenomenon. When thermally unstable materials are broken up, thermal decomposition ensues. High-speed photography had revealed the role of defects in the material, when deflagration changes into detonation, or burning into explosion.

Finally, there were two small service sections, one for electronics and the other for crystal growing. The first gave advice on the choice of instruments, the design of instruments that the research worker may build for himself and the design, development and manufacture of new instruments. The second of these sections produced single crystals of metals and alloys, and the production of pure specimens of metals, such as barium and its alloys, for the experimenters in other sections.

Such was the broad cross-section of research in the Cavendish in 1970. When compared with the activities in the Rutherford and Bragg periods, the changes in subject and emphasis become plainer.

Firstly, main lines of research have considerable continuity and tend to be associated with senior workers in the subject; when they retire, the line becomes less prominent, or may be closed down altogether. Secondly, many individual researches have been completed, or discontinued as unpromising.

A research laboratory is a dynamic institution with an inherent stability, even if many individual activities in it may be out of key with the majority, like molecules in a gas with very much higher or lower velocities than the mean.

28

New Instruments

The factors upon which the progress of physics depends include the invention and development of instruments. The motive for this generally comes from within existing subjects of laboratory research. The experimental physicist desires to obtain a clearer view or a more precise measurement of certain phenomena.

Among those in the Cavendish who have made notable contributions in this direction are V. E. Cosslett in electron microscopy and O. R. Frisch in nuclear instrumentation.

The development of electron and X-ray microscopy was encouraged by Bragg, because it was needed for molecular biology, biophysics and metallurgy, in addition to the general physics of the structure of materials. The Cavendish had been one of the centres in Britain to receive a R.C.A. electron microscope in 1942, under the American lend-lease plan. It was first in the charge of G. R. Crowe, Rutherford's former personal assistant.

In 1946 Cosslett, after having been engaged during most of the Second World War lecturing in the Electrical Laboratory at Oxford, applied for an I.C.I. Research Fellowship, with a proposal to work on electron microscopy in the Cavendish Laboratory. He received no answer to his application, so after about a month he wrote to enquire what was happening. It appeared that his papers had been lost, and he was invited to come to the Cavendish for an interview. He was told that the fellowships had already been awarded. However, Bragg liked the idea, and said that perhaps an extra fellowship could be arranged. This was done, and Cosslett came to work in the Cavendish later in the year.

The provision of electron microscopical aid to other departments in the Cavendish, and elsewhere in the University, was one of their first aims. In addition to this they conducted research into the development of the technique. In 1947 news came from Philips in Holland of the operation of the first successful high-voltage microscope in Europe. This functioned up to 400 kV.

The Cavendish researchers were stimulated to consider whether an electron microscope could be hitched to the 2 MV Cockcroft and Walton generator, which had been bought from Philips, and installed in the nuclear physics department just before the war. Irradiation experiments on living organisms appeared to show that there was little chance of their surviving the exposure that would be required to focus and photograph them, even at the high speed permitted by using a voltage of 2 million.

Cosslett pointed out in 1950 that though the effect of an electron beam on an object, such as a cell, decreased with increasing voltage, reaching a minimum between one and two million volts, living organisms would probably be killed before they could be photographed. In the light of this conclusion, the idea of designing a high-voltage instrument in the Cavendish was dropped.

Independently of the Cavendish research, the Metropolitan-Vickers Electrical Company had become interested in high-voltage microscopes. One was constructed in 1952, operating at voltages up to 500 kV. But at this time, the ultra-microtome was introduced. This enabled biological materials to be cut into sections so thin that they could be examined effectively by microscopes operating at voltages as low as 80–100 kV. This seemed to have removed the main biological demand for high-voltage microscopes.

Some years later the 2 MV generator was sold very cheaply, since it was no longer considered useful for nuclear research at the Cavendish. A few years after that, towards the end of the 1950s, interest in high-voltage microscopes was revived by news from Japan of success in examining metal foils at voltages up to 350 kV. In France the Director of the French National Scientific Research Organisation (CNRS), G. Dupouy, had drawn a different conclusion on the implications of Cosslett's analysis of 1950. He thought it showed some hopeful possibilities, and he inspired a project for a 2 MV instrument at the University of Toulouse, where he had originally graduated.

By 1960 such progress had been made with the instrument that diptheria and other bacteria had been micrographed. They also presently obtained detailed pictures of dislocations in metals. This especially excited the interest of the Cavendish metal physicists, and their desire for a high-voltage instrument.

The Cambridge electron microscopists began to regret the cheap sale of the 2 MV generator, now that they found it desirable to return to the development of high-voltage microscopes.

After the news of the Japanese and French advances in electron-microscopy arrived in the late 1950s, the metal physicists in the Cavendish and the metallurgists at the National Physical Laboratory pressed for high-voltage instruments for studying the structure of metals. A committee was set up with Cosslett as chairman, which sent two emissaries to Japan. They returned with clear evidence of the increase in penetration at 300 kV. The committee considered all aspects of high-voltage microscopy, and recommended in 1961 that two projects be started, one for a 1 MV instrument to be made by Associated Electrical Industries and the National Physical Laboratory, and the other for a 600 kV instrument by the Cavendish Laboratory.

The British Government was not prepared to support the AEI–NPL project, so the Cavendish applied to the Royal Society, and received a grant ultimately amounting to £64 600 from the Paul Instrument Fund. This, together with University funds, enabled them to proceed.

The Paul Fund had been set up under the will of R. W. Paul, one of the founders of the Cambridge Scientific Instrument Company, for the development of new and unconventional instruments. The grant, which was the largest yet made from the fund, was very much in line with Paul's wishes, for he had specifically mentioned the electron microscope as an example of the kind of instrument he had in mind. The first had been built in 1931 at the Technical University in Berlin, to operate at 70 kV.

After discussion, it was decided that the operating voltage of the Cavendish instrument should be 50 kV. The design was carried out by K. C. A. Smith, who also co-ordinated the assembly. About half of the manufacture was done in the Engineering Department of the University, and half in the laboratories of Tube Investments at Hinxton and the Newport Instrument Company. The main feature of the design was flexibility, which probably made it the most versatile instrument of the kind, when it was completed in 1966.

Cambridge University subscribed £30 000 to the capital cost, and the Science Research Council granted £17 600 for the first investigations to be carried out with it. These sums indicate the scale of the instrument development in which the Cavendish was now engaging.

Besides the foreseen extensions in the microscopy of metals, new

370

regions of research were opened up in non-metal microscopy, such as that of ceramics and cement, in which it is almost impossible to make films thin enough for effective examination at 100 kV. Within a short time the demands on the microscope were so great that it was kept running for 12 to 16 hours a day, frequently including Sundays.

The success of the Cavendish instrument stimulated the production of a 1 MV commercial model by the G.E.C.–A.E.I. Company. A design project was started under the auspices of the Ministry of Technology, in which the firm set out to produce a model based broadly on the Cavendish instrument. This involved close co-operation between the Cavendish and A.E.I. teams, under the overall supervision of A. W. Agar.

By 1968 three instruments were being constructed: for the Atomic Energy Research Establishment at Harwell, the National Physical Laboratory and the Metallurgy Department of Imperial College.

During the first twenty years of its development, the electron-microscopy section of the Cavendish acquired several different makes of microscope, besides designing its own. Their activities stimulated knowledge of, and interest in, these instruments, so that presently there were about forty in the neighbourhood of Cambridge, creating one of the largest concentrations of its kind anywhere in the world.

The section made fundamental investigations of new types of electron lenses, and various novel electron-beam instruments. The projection X-ray microscope, constructed by Cosslett and W. C. Nixon in 1953, was the first of these. Then a scanning electron-probe micro-analyser was built by P. Duncumb and Cosslett in 1957. This became the prototype of the 'Microscan' commercial instrument produced by the Cambridge Scientific Instrument Company. Without this experience gained in designing electron optical instruments, the designing of a high-voltage instrument would scarcely have been possible.

The first research programme dealt with tests of the theory of electron penetration; the distribution of atomic dislocations in metals; the properties of magnetic thin films; metal and other films at very low temperatures; radiation damage in metal foils by electron bombardment; quantitative studies of electron scattering; and the examination of biological material, especially micro-organisms in the wet condition.

Specimens from other laboratories were examined, such as metal from A.E.R.E. (Harwell), the National Physical Laboratory and the

Tube Investment Laboratory. Polymer films from Bristol University, and cellular membranes used in the artificial kidney machine at Nuffield Hospital, Oxford, were also early subjects for examination.

Other investigations included the particles involved in causing asbestosis; it is now possible to distinguish the individual particles: which are asbestos, which carbon, etc. Another application was in the investigation of corrosion in nuclear reactors. This was of major practical and economic importance in the generation of nuclear power.

The innovation involved a three-sided collaboration between university research, industry and government, such as had not previously been pursued in a developed manner in the Cavendish. In the period

36   *High-voltage electron microscope*

from Maxwell to Bragg, the Cavendish approach was strongly individual, even in Rutherford's time. Rutherford's authority was not authoritarian; it arose naturally from the distinction of his work and personality.

From the later 1950s the collective aspect became more marked in Cavendish research. Peak discoveries did not cease, but tended to be displaced by a plateau of steady advances.

The Cavendish development of electron microscopy is a good example of innovation in a university. As such, it was studied by the Science Policy Research Unit of Sussex University, and the Manchester University School of Business. A Cavendish electron microscope is shown in figure 36. Figure 37 illustrates the high-precision photography of which it is capable at magnifications far greater than the range of optical microscopes.

One of the interesting features of the Cavendish electronic instrument development was that the patents they were able to join in, and take out, have earned a considerable sum. The Royal Society as a provider of grants has had a share in some of these, and has found itself in the unaccustomed position of receiving substantial sums instead of the usual one of handing them out.

Frisch's instrumental invention arose out of the situation of nuclear physics in the Cavendish, when he succeeded Cockcroft as Jacksonian Professor in 1947. With Cockcroft's departure, and the building-up of large-scale nuclear physics at Harwell, a new style of nuclear research was needed in the Cavendish. Frisch pursued experimental research in nuclear and high-energy particle physics in the spirit of an individual artist rather than a directive organiser. He concentrated on important experimental problems that could be solved elegantly with small apparatus.

His high place in the history of physics had been assured long before. In 1939 he had given the name 'fission' to the uranium phenomenon discovered by Hahn and Strassmann, and with R. E. Peierls he had composed in 1940 the memorandum 'On the Constitution of a "Super-bomb", Based on a Nuclear Chain Reaction in Uranium', which made them the chief authors, in so far as it could be ascribed to any individuals, of the conception of the atomic bomb. In three foolscap pages they showed that such a bomb was theoretically possible, and that only about one kilogram of uranium 235 would be needed. After explanatory investigations at Liverpool and Oxford, Frisch joined the American atomic bomb laboratory at Los Alamos in 1943.

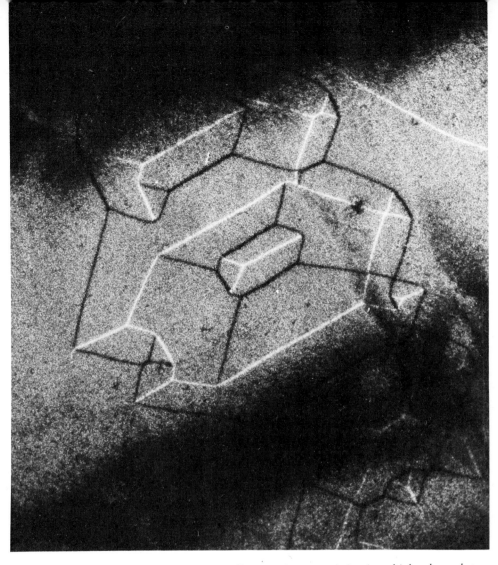

37 *Magnetic domain cells in a thin iron film, seen in transmission in a high-voltage electron microscope*

Three years later he went to Harwell, as officer in charge of nuclear physics.

Frisch's gentle and artistic personality was in remarkable contrast with the powerful influence of his work on history. He brought to the Cavendish an unusual combination of qualities. He was born in Vienna in 1904. His mother was the sister of the eminent physicist Lise Meitner. After graduating in Vienna he engaged in research on photometry and interferometry in Berlin.

374

Then he collaborated with Otto Stern in Hamburg in classical experiments on molecular beams, measuring the magnetic moment of the proton, and investigating the diffraction of helium atom beams by crystals. Since he was of Jewish descent he was presently forced to leave Germany, and joined Blackett's laboratory at Birkbeck College.

At Hamburg Frisch worked quicker and more easily than anywhere else. Stern was more of a theoretician than an experimenter. His experiments were very carefully ordered and arranged; there was no botching. He worked out exactly what result was to be expected, and what the deflection should be. He kept a tight rein on his experiments, but once they had been prepared they were often quick and easy to carry out. His style was not typically German. In an adjacent department experimenters were wasting two years over making a cloud chamber, by devoting too much attention to irrelevant finish.

The contrast between Birkbeck and Hamburg was sharp; all the lecturing at the former was done in the evening. Blackett's laboratory was known as 'the League of Nations', because he welcomed students of all nationalities. The material means were poor; there was only one mechanic; there was no supply of pieces of rubber tubing exceeding fifteen inches in length. Then there was a deafening bang every hour or so, when Blackett's cloud chamber was triggered off by a cosmic ray. Frisch himself made notable cosmic-ray experiments at the bottom of the old British Museum Tube station.

In 1934 he went to Copenhagen. While he was still there Hahn wrote to his life-long collaborator Lise Meitner, who had taken refuge from the Nazis in Scandinavia, of his discovery. She told her nephew of it, and together they explained the phenomenon in terms of Niels Bohr's liquid-drop theory of the nucleus. Frisch suggested that the splitting of the drop might be appropriately described as 'fission', by analogy with the division of cells in the process of growth. They calculated the energies of the two parts, and Frisch very quickly confirmed their results by experiment.

He had been in Copenhagen for five years, and intended to become a naturalised Danish citizen. Niels Bohr had nearly got his naturalisation papers through when, in the summer of 1939, Frisch visited Birmingham to see Oliphant. While he was there the Second World War started, and he decided to stay in England. He wrote an article on nuclear fission for the Annual Report of the Chemical Society, in which he concluded that the construction of a fission bomb was probably impossible. This was published, and may have misled the

Germans, for shortly afterwards he and Peierls went through the arguments again, and came to the opposite conclusion. This was kept secret.

When Frisch came to the Cavendish in 1947 he had only to look after nuclear physics. One of his first impressions was how Lincoln, the famous chief technician, resembled the mechanic he had met at Birkbeck College fourteen years earlier. Frisch never felt cramped in the Cavendish for lack of materials; unlike the Americans, he did not have expensive tastes. When he wanted a special computer costing £15,000, Mott secured it for him quickly.

Among the post-war nuclear researches carried out in the Cavendish Frisch particularly esteemed the work by S. Devons on the angular distribution of the products of nuclear disintegrations.

Frisch became keenly interested in instrumental developments, which, like those in electron microscopy, manifested a new feature in Cavendish research: the invention and development of instruments of scientific and industrial importance.

After the war a considerable time was taken to decide what Cavendish policy on high-energy physics should be. The Laboratory's site in the centre of Cambridge was not suitable for big machines. The cyclotron had always been cramped by the proximity of other buildings, and there were difficulties in shielding against its radiation. It was presently handed to Birmingham, where it did useful work. Cockcroft had been of the opinion that the Cavendish ought to have some kind of big machine, and there had been plans for, and some progress with a linear accelerator; but Mott decided to discontinue it. Though it was a correct decision, it severely shocked the nuclear physicists.

As the Cavendish withdrew from using large expensive high-energy machines, research on smaller and cheaper instruments was intensified. Frisch was fired by the example of the American physicist Robert R. Wilson, who had built an electron accelerator on a shoe-string. He met him by mere chance in Paris. Wilson was also a brilliant sculptor, and this combination of science and art particularly appealed to him. As Frisch modestly put it, he began to fancy himself as a little Wilson. He went to Ann Arbor, Michigan to join in 'a brain-storming session' on the possibilities of strong focusing. When he returned to the Cavendish, he decided not to build a machine.

Then he became acquainted with Glaser, the inventor of the bubble-chamber. Glaser also particularly appealed to Frisch, for he played the viola, and Frisch could accompany him on the piano. Glaser pointed out in 1955 that bubble-chambers one foot in size were already operating. They could take pictures every one or two seconds, and about one in ten registered events of interest. With the increase in size of bubble-chambers to fifteen feet, the number of pictures and events that they could take became enormous. Manual examination became impracticable, and mechanised examination of pictures became absolutely necessary.

The invention of a device for this was exactly in the line of Frisch's interests. He began to think about it, and then built a hand-operated machine, with which the events recorded in the pictures could be measured automatically.

The development was really in the classical Cavendish line. C. T. R. Wilson had introduced the cloud chamber in 1911. His invention probably did more than any other single discovery to make the atomic world appear real, and not merely a chemist's fiction, but it was not easy to use. It remained sensitive after expansion for only a tenth of a second, and needed a period of several minutes for nursing and recovery before it was ready to take another picture. Altogether, it was sensitive for less than one minute in a whole day, and the chances of catching a cosmic ray with it were very small, though Skobeltzyn got one in 1921. Blackett greatly increased the power of the cloud chamber by making the expansions much more rapid, and arranging that any cosmic ray that passed through the chamber also passed through a battery of Geiger counters, the discharge from which triggered off an expansion.

After this the photography of cosmic rays was no longer a matter of chance. Besides greatly increasing the rate at which pictures could be taken, Blackett's device revealed showers of particles, which were found to contain positive electrons. It led to the discovery of the muon, and at least two other unstable particles.

The big accelerators that came into use in the 1950s produced bursts of particles at accurately predictable moments. Interest in the original Wilson type of chamber revived. It was built with heavier walls containing gas under high pressure, in order to stop the faster particles, so that their energies could be measured. The chambers got bigger and heavier, until they became reminiscent of prehistoric monsters ripe for extinction.

This led Glaser to investigate whether the high-pressure gas could be replaced by a liquid. Aware that superheated liquids could be made to boil explosively by exposure to a radioactive source, he suspected that the boiling was promoted by bubbles formed along the path of a fast electron. Making use of these ideas, he succeeded in 1952 in photographing the track of a cosmic ray, marked by a row of coarse bubbles in a glass bulb the size of a thimble.

After this fundamental experiment, bubble-chambers rapidly increased in size to the order of fifteen feet. Many of them worked with liquid hydrogen at $-250°$C. In this medium, penetrating particles collide practically exclusively with protons, for the electrons are just pushed out of the way. If liquid deuterium is used then collisions with neutrons can be investigated. A big accelerator and bubble-chamber may produce 50 000 tracks in a day.

Frisch first happened to meet Glaser at Ann Arbor in 1955, when the first practicable bubble-chamber had been built. Glaser was already worrying about the coming problem of analysing vast numbers of pictures. Frisch's technical interest in his problem was reinforced by their musical tastes and pleasure in playing duets. On returning to the Cavendish, Frisch constructed his first semi-automatic Track Analysing and Recording apparatus, with financial support from the U.K.A.E.A.

The pictures of events recorded on photographic film were projected onto a ground-glass screen bearing a reference mark, and moved by hand through a pantograph, to bring points of interest onto the mark. The pantograph carried a plate bearing a square lattice of dots. Four photocells were arranged to mark off the co-ordinates of points in numbers. The two stereoscopic images of a point could be merged, so that the co-ordinates of its motion in space could also be recorded.

After the setting of each point, the numbers containing its measurements were punched onto tape. This could then be fed into a computer, which rapidly performed the required calculations. This first modest apparatus, accommodated on a table about 1.3 metres square, was found to perform usefully. Frisch published an account of it in 1960.

In 1965 he decided to design a machine that could be considerably faster than one that employed a servo-mechanism. He used a beam of light to move along the picture of a track, at the same time performing a small circular scan about fifty times a second. This enabled the co-ordinates of fifty track points to be stored in every second. The circular sweep, flitting about the film like a little sputnik, suggested that the machine should be called a 'Sweepnik' (see figure 38).

The beam of light was provided by a small helium–neon laser. A start was made in 1968 with useful track measurements, and then an engineered model was designed.

A company was founded to manufacture the machine. The principals in it were Frisch, J. G. Rushbrooke and G. S. B. Street. They thought of calling themselves 'F.R.S. Ltd', but felt that that might have been misunderstood, so the firm was named 'Laser-Scan Ltd', with Street as chairman. When Frisch retired from the Jacksonian chair in 1971, he was happy to find his former pupil, Street, now his 'boss', and himself with a very busy and interesting post-retirement occupation.

The first complete manufactured Sweepnik was installed in Helsinki University, and a second was scheduled for delivery in 1972. They cost about £30000 each.

Thus a line of development started by C. T. R. Wilson in the Cavendish in 1911 with a total expenditure of about £5 was still

38   *The 'Sweepnik'*

bearing fruit in the Cavendish more than fifty years later, with the new feature of providing an industrial, besides scientific, product. This engagement in innovation was another example of one of the newer features of the Cavendish in the 1970s.

What, for Frisch, with his wide experience, was the outstanding characteristic of Cambridge and the Cavendish? 'The self-propagating characteristic of excellence.'

29

A New Line

When Frisch retired in 1971, A. H. Cook was elected to the Jacksonian chair. He took up quarters in the Cavendish, and began to organise research in geophysics, a new line in the Cavendish.

In recent times geophysics has generally been regarded as allied more with geology than laboratory physics. One of the reasons for this is its value in applied geology and mineral prospecting. The science has, however, a fundamental physical history, especially in connection with Cambridge. Isaac Newton contributed major mathematical discoveries in the subject, such as the explanation of the flattening of the earth at its poles, and the implication of this for the constitution of the earth.

Even more apposite to the Cavendish tradition, Henry Cavendish himself made one of the greatest of physical experiments in his determination of the gravitational constant. He published his result in 1798, and the measured values available today are still scarcely more accurate.

The constant is believed to express a fundamental property of the physical world, or of the method by which it is investigated. It is known to be independent to a very high degree of the nature of the materials, the size of whose attraction it defines, but it is not known whether it is dependent on time, or direction in space. Dirac and other cosmologists have considered whether it might slowly change with time. If it does, it might provide new basic knowledge about the universe.

Cook is interested in the question of whether the gravitational constant varies with the rate of rotation of matter. More experimental information on the constant is of fundamental importance. It is difficult to obtain because gravitational forces between bodies are so small.

The electrical attraction between a proton and an electron is 10^{39} times greater than their gravitational attraction, which indicates the extreme delicacy of the required measurements.

Cook and A. Marussi of the Institute Di Geodisia e Geofisica, of the University of Trieste, have been preparing a new determination of the constant by a method proposed by Braun, depending on the measurement of the period of an oscillating system, which is more accurate than the measurement of deflection or amplitude of oscillation. They propose to set up their apparatus in the Grotta Gigante at Trieste.

The grotto is so large that they would be able to use a torsion balance with a suspension fibre 80 metres long. The rotational periods of the balance about its vertical axis would be measured in thousandths of a second. The gravitational forces to be measured are between moving masses of 10 kilograms and stationary masses of 500 kilograms. The attraction would be relatively much greater than in any previous experiment.

Their arrangement does not move in accordance with a simple harmonic equation of motion. They propose therefore to work out solutions to an approximate equation of motion by means of a computer, and then compare them with the observed motions, the measurement of which will be by a gas laser interferometer. The interferometer signals are recorded on tape, providing the computer with the data to be analysed. The measurements of the gravitational constant should provide a new order of accuracy. They promise information for deciding between cosmological theories, and possibly will provide the foundation for a major advance in cosmology.

Cook was born in 1922, and graduated at Cambridge. Like many other young Cambridge scientists he was engaged during the Second World War in radar research, at the Admiralty Signal Establishment. After the war he became a research student in the Department of Geodesy and Geophysics, and then an assistant. He joined the National Physical Laboratory in 1952, and later became Superintendent of the Division of Quantum Metrology. In this he was particularly concerned with the development of high-precision measurement, based on modern experimental and theoretical quantum physics. Among the topics he investigated was the utilisation of the Josephson effect, which had recently been discovered in the Cavendish, for very high-precision measurement.

Physics cannot advance without accurate measurement. The teaching of high-precision techniques in the Cavendish will be strengthened, and the knowledge and skill in this field, particularly cultivated at the National Physical Laboratory, will be passed on.

The Cavendish has since its opening studied matter as found on the surface of the earth. One of the perspectives now is to relate this knowledge more closely with that of matter in more extreme states, such as exist in the centre of the earth, planets and stars.

Studies of the time of travel of waves from earthquakes have shown that the earth is not rigid to the centre, and contains a core, starting a little less than half way to the centre, which is presumably liquid. There is a steady increase of density from the surface down to the beginning of the core. This is followed by a large sharp increase in density succeeded by a slow steady one inside the core.

With increasing depth changes occur in the properties of materials, owing to the changes in crystalline structure that are induced by increasing pressure. Similar phase changes are well known and understood from laboratory experiments at high pressure. In 1936, while he was at the Cavendish, Bernal suggested that one of these changes was in the crystal structure of the mineral olivine, which is in a hexagonal form in the upper part of the Earth's mantle, while in the lower part it is in the more compact spinel form. Thirty years later the experimental support for this idea was very much strengthened.

The state of matter in the major planets, such as Jupiter, is very different from that in the Earth, the Moon and the terrestrial planets. It seems that Jupiter must consist almost entirely of hydrogen, and that the temperature is so low that the hydrogen is for the most part solid.

As hydrogen is the simplest element, its properties can be worked out in some detail from quantum mechanical principles. At a sufficiently high pressure it is transformed into a metallic state, in which the electrons are detached from the protons. These are held together in a lattice, through which the loose electrons move. The properties of Jupiter cannot be calculated exactly from those of pure hydrogen, but if some helium, and perhaps small proportions of other gases are added, a very close model of Jupiter can be constructed.

The theory of the solid state, and knowledge of complex minerals, are not yet sufficiently advanced to make similar calculations for terrestrial planets, but the idea presents an inspiration for experimental and theoretical research.

The radioastronomers are acquiring information of still more extreme states of matter, which appear to exist in quasars and pulsars.

383

Cook wishes to develop a Laboratory astrophysics group, in which these various lines of investigation would be brought together, so that the resources of the Cavendish could be focused on such problems. For example, the expert experimental and theoretical knowledge of the Mond low-temperature physicists would be of great help in analysing the conditions in the major planets Jupiter and Saturn. Are their components of hydrogen and helium changed into metallic forms by the conditions there, and if so, do they form solid or liquid hydrogen metal, and is it superconducting and superfluid?

So the Cavendish acquires a new perspective in experimental and theoretical research.

30

The Depths of Space

Cavendish radioastronomy — in the 1970s an outstanding part of the Laboratory's science — is a product of continuous development in electromagnetic researches, from Clerk Maxwell through Rayleigh, J. J. Thomson, Appleton and Ratcliffe, to Ryle (see figure 39).

Ryle's father, J. A. Ryle, was Professor of Medicine in Cambridge from 1935–43, and was consulted in Rutherford's last illness. His uncle, Gilbert Ryle, was Professor of Philosophy at Oxford. Ryle's intellectual background is notable.

He was born in 1918, and graduated in physics at Oxford in 1939. He was interested in Appleton's researches on the ionosphere, and was accepted as a research student under Ratcliffe to work for his doctorate; but before he could start the Second World War began. Ryle found himself directed to the Telecommunications Research Establishment. He was at once involved in the development of ground radar. Later on, he was engaged in the design of aerials for airborne radar, and the Yagi aerial system for long-range radar for Coastal Command. There was a remarkable concentration of talent in TRE, the work was important, intensely exciting and fertile. Ryle spent six laborious years in this famous establishment.

At TRE Ryle exhibited natural engineering and inventive ability, which was developed by the nature of the work. At the Cavendish he had scope for the exercise of natural scientific insight. His combination of intellectual and scientific qualities with engineering inventiveness and practicality was very unusual. It was a decisive factor in the Cavendish development of radioastronomy. He also had the gift for group leadership, and was able to make his group coherent. He regarded laboratory organisation as primarily a matter of relations between persons, which cannot be well-run on what are called business lines.

39 *Ryle*

When the war was over Ryle was extremely tired and had no definite idea of what he wanted to do, except an almost violent desire to escape from the Civil Service, and any kind of military science.

During the war Ratcliffe had left the Cavendish to work in TRE, and for about a year he was Ryle's chief. Ratcliffe returned to the Cavendish as quickly as he could, to join in the revival of the Laboratory, in teaching as well as research. He invited Ryle to join him.

When Ryle arrived at the Cavendish in October 1945 he found himself confronted by an array of problems. At the TRE he had been engaged in engineering rather than physics, in the development of instruments and apparatus. He had not learned any recent physics, and felt that, perhaps, now he never could. In this intellectual crisis he received crucial moral support and encouragement from Bragg.

Ryle believed, wrongly, that the basic work on the ionosphere had already been done. He wondered whether there was anything he could usefully do in physics, and even thought of retiring altogether from the subject. He did not yet appreciate the profound nature of the lessons he had learned during the war in scientific practicality, in quickly making apparatus work, with imperfect materials and under inadequate conditions. This experience had a determining influence on his future achievement. It helped him to get impressive results soon, and thus ease the provision of greater resources.

As he was not inclined to start on classical radio research Ratcliffe suggested that he should look into some of the peculiar disturbances of radar signals that had been noticed during the war. Ratcliffe was the trigger that released Ryle's development into a radioastronomer.

One of the disturbances was the occasional general fade-out of 4-metre radar anti-aircraft gunsights over a large area of England. This had led to the suspicion that the Germans had invented a new jamming device. Investigation showed, however, that the fading was probably caused by radio waves from the sun.

The precedence of military problems prevented this from being extensively investigated during the war. By the end of the war the increase in the sensitivity and the signal-to-noise ratio of radio receivers made observation of these waves from the sun much easier.

When the general subject of radio waves from space was looked into, the discoveries of the American radio engineer K. G. Jansky were appreciated. He had proved conclusively in 1932 that radio waves were arriving from outer space. His research had been extended by the American radio amateur Grote Reber, who had published striking advances in 1940–2. This brilliant work had not been followed up in America, because radio was regarded as engineering rather than physics, and a side-line of the electrical-communications industry. One of the most important contributions of the British investigators was to develop these remarkable by-products of American radio engineering into a new branch of science based on physics and astronomy: radioastronomy.

This change in point of view promoted the development of radio-astronomy as an independent science, and not as a detail that might improve radio communications. It encouraged the search for information on the nature, origin and significance of the radio waves from space, their bearing on the history of the universe, and the properties and laws of matter in its remoter regions.

Ryle was given one of the new ICI research fellowships, which had such an important part in reviving academic science after the war. He was expected to teach, as well as help in building up the new research.

The Cavendish's equipment was run down and out of date; there was little money, so Ryle and his colleagues went on foraging expeditions to ministries and research establishments, and returned with much valuable up-to-date equipment, including excellent material captured from the Germans.

Ryle and his colleagues began to examine the sun on metric wavelengths in December 1945 (see figure 40). In 1947 they were joined by F. G. Smith and in 1948 by A. Hewish.

40   *The first Cavendish Radioastronomy Observatory, 1945–48*

The sunspot cycle was approaching a maximum in 1948, making conditions for observation particularly promising. Centimetric wavelength emissions from the sun could be detected continuously at all times, but metric emissions only at times of great solar activity, when sunspots were visible.

The first question was whether significant metric radiation occurred at other times, or whether it fell to the level normal for a 6000 K source.

To answer this, they built an instrument that would detect very much weaker signals than had previously been observed. It was set up just outside a pre-war ionospheric experimental test site on the edge of Cambridge. They discovered that metric radiation from the sun was detectable at all times, and much more intense than that corresponding to a 6000 K source.

The next question was whether it was coming from small discrete areas of the sun, or from a vast radio-emitting cloud surrounding it. The best instrument for obtaining the high resolution required was some form of interferometric radio telescope, corresponding to the interferometric optical telescope invented by Michelson.

The first radio telescope operating on this principle had been built in Australia. Radio scientists who had been active there during the war applied their technique to the exploration of radio from space, as Ryle was doing at Cambridge and Lovell at Manchester. Each group developed in its own way.

In 1946 Ryle and D. D. Vonberg constructed an interferometric radio telescope of their own design. It consisted of two aerials directed towards the meridian, so that as the earth rotated the angle of reception was swept across the sun. The respective currents from the two aerials produced an interference pattern on a suitable electrical instrument. The distance between the aerials, about 250 metres, could be varied, and from the variations produced in the interference patterns, the diameter of the area on the sun that was producing the radio waves, could be calculated. It involved the same principle as the calculation of the diameter of a star from the interference fringes in Michelson's interferometric optical telescope. Ryle and Vonberg discovered that the radio sources were very compact at times of solar activity, and much smaller than the sun itself.

One of the dominating features of the Cavendish research in radio-astronomy was the development of the interferometric radio telescope. These were made of increasing size, to give increasing angular

389

resolution, so that more and more sources could be identified, and their shapes delineated. Sensitivity was increased to detect weaker and ever more distant sources. Appropriate electrical instrumentation was developed.

Ryle's experience of intensive research on radio and radio-engineering development was an essential factor in enabling him to lead this advance. It involved not only work in the laboratory and the workshop, but in the field, erecting considerable engineering structures on open ground, and in all weathers.

A series of general and special radio telescopes was constructed, culminating in 1972 with the opening of the 5-km telescope, costing more than £2 million. By that time the Cambridge radioastronomical department had equipment and sites that had cost £4 million.

Parallel with the improvement of the instruments, a more powerful method of using them was developed. This was the 'aperture synthesis' technique, the most important Cavendish technical contribution to the development of radioastronomy. Small aerials are moved, successively, to occupy the different positions corresponding to elements of a gigantic radio telescope, and the signals are combined in a computer.

A dish reflector type of radiotelescope of equal detecting power would be vast in size, very expensive and quite beyond the scope of engineering technology. The drawback of aperture synthesis is that repeated observations are needed for the computer to work out the picture. It is therefore not suitable for observing instantaneous or rapidly changing radioastronomical phenomena.

After the measurement of the angular size of the sources of sunspot radiation in 1946, Ryle and Smith measured the angular diameter of the source in Cygnus.

The first application of the aerial synthetic system was carried out by H. M. Stanier in 1949, in the determination of the distribution of emission across the solar disc.

In 1950 the phase-switching interferometer was invented and worked out, and applied by Smith, Ryle and Hewish. By this means, the background radiation is made to cancel itself, and leave the record of the radio source free from the confusing effects of unwanted radiation, not only from the sky, but from electrical machinery in the neighbourhood. With this device, effects due to ionospheric irregularities could also be distinguished. Ryle, Smith and Elsmore made the first survey of 50 cosmic radio sources in the same year.

In 1951 accurate determinations of radio sources in Cassiopeia and Cygnus by Smith assisted the optical astronomers at Mount Palomar to detect the visible stars with which they were associated.

The first year in which they had more than one research student was 1947. There was a steady growth of the group in the following years, and it presently exceeded fifty.

Their discoveries stimulated further efforts, especially in the search for new sources, the analysis of their distribution in space and the study of their various natures. In 1955 a second survey of sources was carried out by Shakeshaft, Ryle, Baldwin, Elsmore and Thomson. It appeared to show that the density of the sources increased with distance. This stimulated Ryle's interest in cosmology.

A third survey, by Edge, Shakeshaft, McAdam, Baldwin and Archer, was completed in 1959. A fourth survey covering the whole of the northern sky and determining nearly 5000 sources, ninety per cent of which had not been observed before, was completed by Gower, Scott and Wills in 1965. The first results of the fifth survey were published in 1966.

As the promise and achievement of the radioastronomers unfolded, more resources were needed to sustain them.

Though Bragg belonged to the optical strand in the Cavendish tradition, he greatly helped Ryle. Besides supporting him in relations with radioastronomers elsewhere on questions of priority, and in morale and intellectual encouragement, he presently became profoundly excited, when he perceived that the problem of deducing the size and position of radio objects in the sky by interferometric methods was an exact analogue of the determination of the arrangement of the atoms in a biological molecule. The latter was being pursued by his X-ray analysts Perutz and Kendrew. It seemed almost incredible that the structure of the cosmic universe and the structure of the molecules at the basis of life should be investigated at the same time in the same laboratory, by the application of the same optical analytical theory. His own career had arisen out of the application of an optical principle, and here was a related optical principle being applied in his own laboratory, to explore nature from one end to the other. As he modestly put it, he 'felt like a dog with two tails'. The dog had indeed two mighty tails.

By the time of his departure in 1953 Bragg had become a profound supporter of radioastronomy. Then, in the interim between his departure and Mott's appointment, J. A. Ratcliffe was appointed Acting

Head of the Department of Physics, from 1 January 1954 until 1 August, when Mott was to 'enter upon his duties as Cavendish Professor'.

The arrangement was a repetition of that made after Rutherford's death, when Appleton, who had been Ratcliffe's teacher, was appointed Acting Head of the Department until Bragg arrived. The incident emphasised the important place of radio physics in Cavendish history. The fact that it happened twice indicated that one of the Cavendish professors might well have been a radio physicist, and the appointment of Ratcliffe reflected the importance of his influence in the conduct of the Laboratory. This had become conspicuous by his own ionospheric discoveries, and by his support of Ryle. When Ryle's genius became apparent, Ratcliffe placed his own research in the background, and made a point of relieving Ryle of administrative work, so that he could devote himself to his highly original researches with as little distraction as possible.

The swift extension of the importance and scale of Ryle's researches had required increases both in financial support and academic authority. An independent endowment was desirable in order to create conditions in which large financial provision from the Government would be seen to be necessary, in the interests of science and the nation. Ratcliffe and Bragg searched for funds, and initiated talks with the Mullard Company. Just before Mott succeeded Bragg representatives of the Mullard Company discussed the matter at a dinner at Caius College. It was suggested that the Company should provide a large endowment for grounds and buildings, and that equipment should be provided by the Government.

The proposal was clinched by Ratcliffe, when he suggested that the new observatory might be called 'the Mullard Observatory'. The details were worked out, and in 1955, during the year following Mott's appointment as Cavendish Professor, the Mullard endowment of £100 000 was announced. Primarily the endowment was used to set up the Lord's Bridge observatory site, but it also made it possible to carry out certain researches already mentioned, and prepare the way for the great developments since.

Ryle had emphasised in 1955 the possible importance of radio sources to cosmology, and in 1958 he demonstrated that most radio sources are powerful extragalactic objects.

Baldwin and Edge observed in 1957 the radio emission from the supernovae first noted by Tycho Brahe and Kepler.

Scott, Ryle, Clarke and Hewish described in 1961 the distribution of radio sources in depth in space, and its significance for cosmology.

The discovery of the quasi-stellar radio sources, or *quasars* was a result of contributions from Australia, Cambridge, Manchester, the Radar Research Establishment at Malvern and the optical astronomers at Mount Palomar in America. The quasars were very distant, and appeared to be enormous energy sources, arising from explosions at the centres of galaxies.

Cavendish observations led to the suggestion that in the course of about a million years the quasar develops into a typical double radio galaxy.

The red-shifts of detectable quasars showed that they were up to 8000 million light-years distant. In the course of these observations the Cavendish investigators detected the most distant objects yet recognised by man. Since the emission from each source probably lasted only about ten million years, most of the sources of the observable radiations must already have ceased to exist.

The Cavendish observers in the early 1970s were inclined to the view that about 10000 million years ago, the Universe was dense and hot. Their one-mile telescope had probably revealed objects as they were at the very distant time when the galaxies were being formed. It appeared that beyond these very intense distant objects space was empty.

The next step in this research was to acquire a telescope that would reveal more about the physical mechanisms by which the radio galaxies are formed. For that reason they built the 5-km instrument.

The detection of sources, and their distinction from each other, are basic requirements of radioastronomy. The waves coming from these sources have various characteristics, which are affected by the media through which they pass on their way from their source to the earth. The investigation of the effects of these intervening media is another basic requirement. Until it is carried out it is uncertain how far the properties of the arriving waves are due to the source, and how far to the modifying effects of the intervening medium. When the latter has been ascertained, the effects of the medium may be used for discovering more about the source.

One of the pioneer observations in radioastronomy, made by J. S. Hey, was that radio sources flicker in a manner similar to the scintillations of visible stars. The scintillations, both of the radio and optical stars, are due to the effects of the intervening medium, and they provide information both on the size of their sources, and the nature of the intervening medium.

In 1964 Hewish started a systematic research on the scintillation phenomenon. The diffraction of the waves by the plasma clouds in interplanetary space could provide information on the size of the radio sources. He designed a special telescope which was set up in 1967 for observing these radio scintillations. It consisted of an aerial containing 2048 dipoles, and covering 1.8 hectares. It operated on 3.7 metric waves. The observations were organised so that the whole sky could be scanned in one week.

This telescope produced 120 metres of recordings per week, which were examined by the research student Jocelyn Bell. In August 1967 she plotted what appeared to be a scintillating source. It was conspicuous because it was in transit near midnight, when scintillation due to the interplanetary medium is low. It might have been caused by casual electrical interference from local electrical machinery, but she acutely perceived that it was repeated at roughly the same sidereal time, although the position appeared to vary somewhat; this was the very kind of observation that had enabled Jansky to found radioastronomy in 1932.

Hewish and Jocelyn Bell pursued the observations and the analysis of the records. Absence of parallax showed that the object was outside the solar system. It was thought at first that it was a radio-flare star. Then, on 28 November, it was discovered that the waves arrived in a succession of regularly spaced pulses. A few days were taken to confirm that the waves were arriving in this form, for it was extremely surprising; this very precise form was just not the kind of signal expected from a very distant celestial source. A certain scepticism of the genuineness of the observations was inevitable.

Pilkington estimated that the duration of the emitted pulse was of the order of 20 milliseconds, and the source was of planetary dimensions.

The extraordinary precision of these signals suggested that they might be emitted by beings in an extra-terrestrial civilisation. The possibility could not be ignored. However, Scott and Collins showed that the signals made no recognisable code. The period between the individual pulses was incredibly precise: the first they examined gave one pulse every 1.337 3011 seconds. The pulses were comparable in accuracy with the finest crystal clocks.

Intensive study indicated that the sources were extremely small, and contained enormous concentrations of energy. The astonishing regularity was discovered when Hewish attempted to detect a planetary motion of the source by the Doppler effect.

By January 1968 it had been shown that the pulse rate was constant to within one part in ten million. Miss Bell continued the examination of the survey records, which were by now five kilometres long. Within three weeks, three more of these pulsating stars, or *pulsars* were found. It was not clear at first whether the frequency drift was due to dispersion in the interstellar medium or in the source itself.

The pulsars were found to be in the galaxy. They appeared to be white dwarfs or neutron stars. The simplest explanation of the constant period appeared to be radial vibration of the entire stars, and that shock waves in the star's atmosphere might initiate radio flashes by some high-energy particle–plasma interaction.

Their results were published eight weeks after the pulsed radiation had been discovered. Some thought the results had been kept back too long, but the Cavendish radioastronomers felt it would be premature to publish before the Doppler measurements had yielded positive results. Within a year ten more pulsars had been discovered, six from Cambridge, and others from the U.S.A. and Australia.

The pulsars were presumably formed of atoms whose protons and electrons had been forced by gravitational pressure to coalesce into neutrons. The spinning neutron star 'lighthouse' model is now thought to give a better explanation of pulsars than the vibration theory.

In this way, radioastronomy reveals new states of matter, which may require for their explanation the discovery of new laws of nature.

The extraordinary nature of the pulsars, whose pulses are so spectacularly different from emissions from other celestial objects, raises the question of why a phenomenon so startling was not discovered before. One reason was that they are extremely weak sources, below the detection limit in the surveys previously made. Another reason was that they were sporadic, and their spectrum of waves fell away at shorter wavelengths.

One of the reasons for their success was that the recording technique had been designed for observing fluctuating radio sources, so it happened to be ideal for observing fluctuations arising in the source itself, as well as those caused by the intervening medium.

Anthony Hewish (see figure 41) was born in 1924, and educated at Taunton. From 1943–6 he was at the Royal Aircraft Establishment at Farnborough. During this period he was seconded to TRE and there met Ryle. Like so many other scientists of the post-war generation, he learned scientific technique in the serious and urgent atmosphere of defence research. After the war he returned to Caius College, Cambridge, graduating in 1948. He joined Ryle in radioastronomical research, and has supported the development of Cavendish radioastronomy with a sustained competence and equanimity, which has contributed much to its efficient functioning and success. His discovery of pulsars was of the kind that, Pasteur said, comes only to the prepared mind.

The 5-kilometre telescope (see figure 42) was inaugurated by the President of the Royal Society, Sir Alan Hodgkin, on 17 October 1972. The instrument was the latest, employing the principle of aperture synthesis, to be constructed at Cambridge since this principle was introduced in 1957. Its chief aim was to gain more understanding of the physical mechanisms occurring in radio sources in and beyond the galaxy. Early telescopes had resolutions strong enough to reveal the principal features of their structures, but left important details unresolved. As these might contain the keys to the construction of

41 *Hewish*

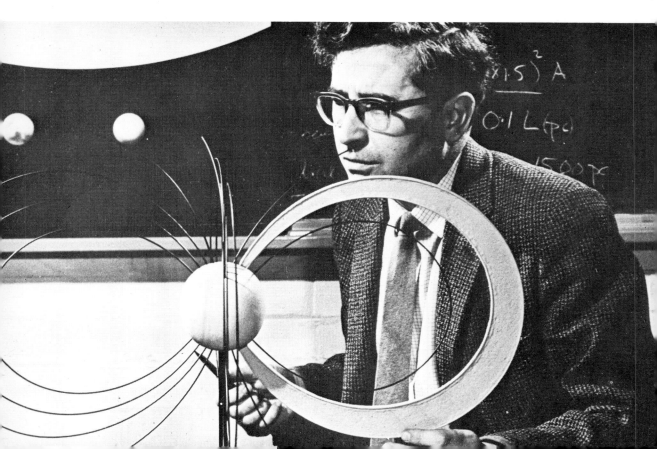

42　*The 5-km radio telescope*

successful theoretical models of the sources, it was essential to possess
an instrument that could extend the resolution to them. It might
enable observations to be made that could decide between the
various explanations of the double structure of the typical radio
galaxy, or quasar. It might provide the observational data for the
development of a better theoretical understanding of the interaction
between an expanding supernova shell and the interstellar medium.

Insight into the problems of star formation might be gained from a study of the compact radio 'knots' observed in dense clouds of ionised hydrogen, and closely associated with infrared and molecular-line sources.

Besides these researches into cosmic physics, the telescope will provide extremely accurate determinations of the positions of compact radio sources associated with faint optical objects. It will in fact provide a primary reference system of celestial positions comparable in accuracy with those already established by traditional optical astronomy, and begin to make radioastronomy as accurate as optical astronomy.

There are several ways of achieving high precision in radioastronomy. One is the use of very accurately made dish-reflectors, operating on very short wavelengths. The limit of accuracy in this kind of instrument is set by the mechanical properties of the materials used in its construction.

A second method depends on making use of occultations of compact sources by the Moon. When this happens it can give very valuable results, but it can be used only for those sources that lie on the Moon's path in the sky, that is, for sources near the ecliptic.

The third method is the aperture-synthesis technique developed by the Cavendish radioastronomers. In the 5-km telescope, fixed and movable aerials, separated at distances of the order of kilometres, can be operated so as to give a resolution equal to that which would be obtainable from a dish-reflector telescope about 5 kilometres in diameter. The engineering difficulties of making an accurately steerable dish reflector of this size are extremely formidable, if not insuperable.

The limits to the resolving power of any telescope on the surface of the earth are set by effects on the radio waves caused by irregularities in the ionosphere and the troposphere. While a good deal was known about effects due to ionospheric irregularities, little was known about those due to turbulence in the troposphere. An extensive preliminary research on the latter was made, showing that it should be possible, for about half the observing time, to obtain angular resolutions better than one second of arc, with an instrument operating with an aperture of 5 to 10 kilometres, on wavelengths of 3 to 6 centimetres.

The original design was based on a total aperture of 5.1 km, but this had to be modified, because a proposal had been agreed to by the Ministry of Housing and Local Government that the M11 motorway should be built along a line that would cross the site. In order to avoid this disruption the total aperture was reduced to 4.6 km. Eight thirteen-metre dish-aerials were aligned on an east–west axis of this length. Four of the aerials were fixed, and four movable. The movable aerials could be shifted forwards and backwards along a straight piece of rail track about 1.2 km long. The distances between the four fixed aerials were equal, also about 1.2 km long. By varying the positions of the four movable aerials along the rail track, 128 different spacings could be made for any particular observation. The angular extent of the sources that can be mapped depends on the number of spacings.

The foundations and structures of the reflectors have to be accurate to within one millimetre. The reflectors depart from the usual paraboloid–hyperboloid curvatures, in order to provide a better distribution of the aperture illumination.

The receiving system used for the initial observations operates on 6-cm waves. Its design and construction were carried out by the Observatory staff, including P. F. Scott, D. W. J. Bly and B. Elsmore. The telescope is controlled by typewriter commands to a computer.

The instrument is built along a section of the old Cambridge–Bedford railway, which runs almost exactly east–west, and lies adjacent to the original Observatory site. British Rail provided the old line, and old railway buildings, at a reasonable price.

The exact determination of the position of earlier telescopes was usually established in relation to compact optical objects. The axis for the 5-km instrument was determined by a precise survey with a first-order precision theodolite, to locate it, with comparable or better accuracy, in relation to the positions of the fundamental stars. It was carried out by the National Physical Laboratory, and involved geodetic levelling, and making a series of astronomical observations from the top of steel towers ten metres high, to determine the precise latitude, longitude and azimuth of the instrument, and the measurement of the separation of the aerials along the baseline.

This complex and large engineering work, costing £2.1 million, was carried out for the Science Research Council by the U.K.A.E.A. It arranged contracts with all the parties concerned, including the Marconi Company, the Mitchell Construction Company and British Insulated Callender Cables. The Engineering Division of the Authority produced detailed designs for the foundations and civil engineering works, and for the control and power-cabling systems.

The work included the designing of the high-voltage electrical power system to supply the eight aerials spaced over 5 kilometres. It operates on 11 kV, and is laid underground to avoid interference. Other nearby area-board overhead lines were replaced by underground cables for the same reason.

Approximately 88 kilometres of low-loss coaxial cable were laid and cut to very precise lengths for the connections between the aerials and the control-room. A network of multi-core cables was required to communicate the received signals to the control-room, and for the computer to send out the necessary star-tracking instructions to the motors driving the aerials.

Thus, as large-scale engineering has moved out of Cavendish nuclear physics, it has returned via radioastronomy. Both of these movements occurred during Mott's tenure of the Cavendish chair.

The development of radioastronomy under Ryle presents an interesting parallel to the investigation of the conduction of electricity in gases by J. J. Thomson. After twenty-five years, each had made and led researches that revealed fundamental new knowledge about physical nature. Both had created a group of research workers sufficiently large to conduct a variety of researches, and contribute different points of view, and yet compact enough for its members to be known to each other individually.

The difference on the organisational and financial side was, however, striking. J. J.'s researches cost very little, while the radioastronomical development involved considerable engineering, management of property, leasing of old railway lines, and resistance to inconvenient local developments, such as new motorways, which are a source of electrical interference disruptive to radioastronomical observations. The background of TRE and war experience was invaluable in dealing with these practical matters.

In 1972 two of the six professors in the Cavendish were radioastronomers. A chair for Ryle had been founded in 1959, and a second chair, for Hewish, in 1971. About a third of the Cavendish's research and technical staff of 91 were in radioastronomy, and about a tenth of the 225 research students.

The transference of the Royal Observatory from the Admiralty to the Science Research Council in 1965, and the appointment of Ryle as Astronomer-Royal in 1972 were significant events in the history of British science. As Christopher Wren had pointed out in 1657, the British development of astronomy was intimately related to the growth of British sea power and overseas trade. The Royal Observatory at Greenwich was founded in 1675 specifically to supply navigational information for the British navy and marine. As part of this, the standard of world time was based on the meridian of Greenwich.

The Admiralty was entrusted with the conduct of the Royal Observatory, because its work was primarily for naval and marine needs, and its director was entitled the Astronomer-Royal because of the national importance of his work.

By the middle of the twentieth century astronomy had long ceased to be the primary British scientific interest; Britain was no longer the premier naval and marine power. Economically the primary British scientific interest had passed from astronomy to physical science a hundred years before, and in this the Cavendish Laboratory had become the pre-eminent British institution.

British research in radioastronomy arose out of physics, and its development promises as much in the realm of physics as in astronomy. The cosmos performs experiments beyond the possibilities of the surface of the earth, and radioastronomy can be developed to report their results. It already provides suggestive evidence about new kinds of solids and liquids in super-dense neutron stars, involving the possibility of new kinds of sub-atomic forces, and intense gravitational fields involving the destruction of matter, and operating according to new laws of nature.

Radioastronomy has become the most important part of British astronomy. In this situation, and since the Cavendish contribution had been particularly distinguished, it was appropriate that the title of Astronomer-Royal should be conferred on Ryle.

The Royal Observatory still contributes a great deal to the purposes for which it was founded in 1675. It carries out about half the calculations required for the Nautical Almanac, and it is the custodian of time. It also provides instrumental resources for optical observations, especially for astronomical research groups in various British universities. The volume of its own research is, however, comparatively less than formerly, in relation to the total amount carried out in university astronomy departments.

The development of radioastronomy in the Cavendish, and the appointment of its leader as Astronomer-Royal is a mark of judgement in the contemporary direction of British science.

These events strengthen confidence in the future of the Cavendish Laboratory, and the development of science in Britain. While the Cavendish may have become different, it will continue to be highly significant.

31

Finance and Administration

The atmosphere and accounting of Cavendish finances, and the way they have changed during the last hundred years, may be illustrated by some details from the 1880s, the 1930s and the post-war years up to 1972–3.

On 7 March 1885 J. J. Thomson started an account at a bank on behalf of the Laboratory. In the first bank-book it is described as for 'Cavendish Laboratory Prof. J. J. Thomson Treas$^{r.}$'.

In the financial year 1888–9, for example, the total expenditure was still only £1091 9s 5d. Payments were generally in items of the order of £1 to £20. Even by 1913 the total expenditure had risen only to £3092 10s 5d.

Forty-four years later, in 1932, Rutherford was still keeping a similar account. It is described as the 'University Imprest Account for Physics'. In January–February 1932, just before the discovery of the neutron by Chadwick, and the disintegration of atoms by accelerated particles in the experiments of Cockcroft and Walton, payments were still being made in the same style. Evidently, physics had been progressing more swiftly than accounting during the previous half-century in the Cavendish.

J. J. had the reputation of being rather remiss in paying the stipends of the departmental staff. These were normally paid terminally, and the staff used to toss up to decide who should remind J. J. that the payments were due. Searle reported that on one occasion when J. J. had forgotten to pay him, they happened to meet in Petty Cury during the following Vacation. Searle told him that he had not been paid for the preceding term. J. J. produced the departmental cheque book from one of the many pockets of his jacket, affectionately known to the staff as 'the poacher's jacket'. He then instructed Searle to bend down, so that he could use his back as a writing desk. Searle commented to a member of the Cavendish ancillary staff: 'I felt such a silly fool, doing that sort of thing in the middle of the road in Petty Cury; it was very busy, being a Saturday morning.'

The methods of accounting in those days were hardly those of a modern school of business.

The kind of questions that raised financial problems, and the rates of expenditure that were common in the 1880s are illustrated by J. J.'s report to the University on the Physical Department in 1889, after he had occupied the chair for five years. It was noted that the most urgent need was for a new classroom for demonstrators. There were 80 students in 11 classes for practical elementary physics. These classes interfered with the preparations for lectures, and more room was needed.

Then there was the payment of the demonstrators, Glazebrook and Shaw. In 1883 they had each been paid £60 a year, for attending 18 hours a week, and this was described as wholly inadequate. 140 students attended their demonstrations. Wilberforce gave lectures on technical electricity, a subject of great scientific and technical importance. He received no payment beyond fees charged to students.

The wages of assistants amounted to £173 11s 0d including £110 0s 0d for R. T. Bartlett; £38 5s 0d for James Rolph; and boys' wages of £25 6s 0d.

Repairs and ordinary fittings cost £4 3s 2d, consisting of £3 15s 4d for the Gas Company, and 7/10 for the Waterworks. Maintenance amounted to £42 2s 10d, including £4 2s 8d and £8 7s 1d spent at ironmongers, and £5 16s 2d at the chemist's. Apparatus cost £40 18s 8d, including £3 13s 6d for a double-blast circular bellows. The total amounted to £260 15s 8d.

By 1890 wages, repairs and maintenance had risen to £290 7s 11d. Painting of the Cavendish Laboratory had cost £22.

The General Board of Studies of the University had learned with regret that Glazebrook, who had demonstrated for ten years, was to resign. They proposed that he should be appointed Assistant Director, at £50 a year.

This proposal shocked some personalities. When it was discussed in the relevant University committee, 'Mr. Mayo said that even if the University were rolling in wealth it ought to look closely at an expenditure of £50 a year. Supposing the retiring Demonstrator were not the man he was he did not think that this would have been proposed. It was only a personal compliment, and this was not a time for personal compliments at the cost of the University.' Such was the financial atmosphere in the 1890s.

In spite of Rutherford's continuation of J. J.'s care over the petty cash, the total expenditure increased considerably during the next half-century. The annual Income and Expenditure for the Department of Physics, published on 5 November 1937, just after Rutherford's death, and just before the Second World War, was £17 321 8s 4d.

By 1945–6, in the first year after the war, when W. L. Bragg was able to resume his full professorial duties, income and expenditure exceeded £50 000. According to the University Accounts, for the year ending 31 July 1946 the precise income was £56 470 12s 10d. This included £25 650 from the University Education Fund. £4500 was transferred from the Cavendish Endowment Fund. The War Emergency Fund provided a non-recurring grant of £13 357. The British Electrical and Allied Industries Research Association provided £1799 7s 1d. The Iron and Steel Federation supplied £375 for X-ray research on the structure of alloys, and £1200 for research on rolling mills. The Rockefeller Foundation provided £501 10s 9d, and Nuffield £1500.

The precise expenditure, which was £53 709 3s 5d, included £3147 7s 4d for administration. Of this, £200 was assigned to Bragg, and £562 10s for I. T. James, the Departmental Secretary. Of the Staff, J. A. Ratcliffe received £626 1s 3d; D. Shoenberg £690; J. F. Allen £437 10s; J. V. Dunworth £107 16s 2d; and S. Devons £128 2s 6d. J. D. Cockcroft, W. B. Lewis, C. W. Gilbert and H. Carmichael were paid from elsewhere.

Wages paid to H. E. Pearson were £489 5s 0d; G. R. Crowe £379 7s 0d; W. Birtwhistle £379 15s 0d; C. E. Chapman £380 7s 0d; and F. Niedergesass £224 12s 6d.

Maintenance of supplies, such as gas, etc., cost £7486 15s 0d, and of the premises £1431 1s 1d.

Expenditure on X-ray research cost £239 3s 4d, and a grant for the X-ray index amounted to £98 11s 5d.

The surplus for the year was £2761 9s 5d.

The Royal Society Mond Laboratory's income was £7525 1s 4d, and the surplus over expenditure was £71 4s 8d. J. F. Allen received a stipend of £100 for administration. The wages of assistants amounted to £1516 10s 11d, and equipment cost £480 5s 4d.

By 31 July 1947 the annual Cavendish expenditure had risen to

£58 026 16s 7d, an increase of only about £5000 in this period of immediate post-war repair and development. Among sums expended were £510 7s 2d on X-ray research; £999 1s 10d on sheet steel, and £3501 10s 4d on rolling-mill research; £3850 by the Cavendish Radio Research Account; a further non-recurrent grant for radio research of £500; and apparatus for practical classes £550.

The total income for 1952–3, given in Bragg's final report, was £150 000.

The rapidly growing expenditure on science, and the large fraction of the finance coming ultimately from the State, prompted in the 1960s the development of more analytical and comprehensive methods of financial accounting.

In June 1966 Mott wrote a report entitled *Notes on Assessing the Cost of a Scientific Department*. He suggested that the value of a scientific department to the community might be assessed under five headings: the education of undergraduates, education for higher degrees, research, service to industry and advisory services.

Any attempt to assess separately the effort spent by academic staff on undergraduate and postgraduate education would be very unrealistic, but some approximate separation in the cost of equipment and technical staff might be possible.

Any attempt to assess the cost of research only in terms of the training it provides was somewhat unrealistic. 'Members of the staff of this department who undertake research believe that the results obtained may be of value; they believe that the discovery of the electron and of the neutron in the Cavendish Laboratory has not made its impact on the world solely through the students trained in the process.'

One of various ways of assessing the 'value' of the published research from the Cavendish might be to work out the cost per paper, and find out how it compared with the cost per paper from other physical laboratories.

With regard to service to industry, at least a third of the current research was very closely related to industry. The Department organised 'Summer Schools' in subjects such as electron microscopy and the physics of metals, which were popular with industrialists, and represented a major teaching effort. Most of the senior members of the Cavendish gave a good deal of their time to general advisory services, serving on research committees of nationalised industries, Defence committees and educational advisory committees.

Research was being conducted under, broadly, eight headings.

These were: experimental and theoretical high-energy nuclear physics; solid-state physics; radioastronomy; physics and chemistry of surfaces; electron microscopy and metal physics; crystallography; meteorology; and fluid mechanics.

Academic salaries at that time amounted to £165 027. £11 100 of this was paid by colleges to University staff for teaching, and about £10 000 to research students for supervision.

Colleges also paid for various tasks, such as conducting scholarship examinations, perhaps altogether about £6000 p.a. But there were fringe benefits; for example, most college fellows got a free meal a day, and sometimes a free room in college. Unmarried fellows, of whom there were very few, got free board and lodging.

The cost of administration was £18 845. The salaries of 111 technical staff amounted to £81 650. Materials and apparatus, travel, etc., amounted to £117 380. The Library assistants ($1\frac{1}{2}$) cost £1121 p.a., and books, etc., £2787. Heat, rates, cleaning and maintenance cost £39 129.

Government supplied funds for 30 graduate staff, amounting to £37 499, and bought apparatus, etc., amounting to £91 648. Altogether, £151 050 was provided. No funds came from charitable bodies and foundations. Industrial concerns provided £76 444.

Estimates of the University administrative costs of the Department might be based on the assumption that they would be 5 per cent of the total University administrative cost. This gave a figure of £12 500.

The cost of refectories and other student facilities was estimated at £50 000.

The Department used about 30 per cent of the available time of the computer services. The share of the cost was £26 000. The total capital cost of the Titan computer was £450 000.

The Cavendish students made small use of the University Library, and the cost of providing them with this service can be neglected.

The total University administrative costs of the Department amounted to £88 500. Major new buildings since 1945 cost £341 000. The conversion of the high-tension laboratory to classrooms and a lecture theatre cost £174 000. The conversion of the examination hall for classes cost £57 000, and alterations to the physics and chemistry of surfaces building, £30 000. The re-equipping of the latter cost £12 500.

Research councils and other government organisations provided £50 050 for student grants. Other sources provided £29 400.

There were 418 undergraduates and 121 postgraduate students. Postdoctoral research workers, SRC, ICI and Exhibition of 1851 fellows, and visiting research workers made a further group of 37.

The total annual expenditure was £1 257 399.

One of the most interesting points in the *Notes* was an attempt to estimate the cost of the facilities provided by the Rutherford Laboratories at Harwell for Cavendish research workers. It was fixed at £220 000.

This figure was deduced from the estimated number of bubble photographs that would be taken in 1966, and the strength of the counter teams. It was thought that two million photographs would be taken, 65 000 of them by the Cavendish men. The cost of their share of using the giant accelerator and bubble chamber would be about £100 000.

113 men were making counter observations, of whom 5 were Cavendish men. They therefore incurred about 5/113th of the total cost of £2.75 million, which came to £120 000.

The cost of the facilities at CERN for Cavendish men working there was estimated at £210 000.

Estimates and accounts vary a good deal according to the way in which they are made. The Consolidated Accounts for 1966–7, prepared for the University Grants Committee, included in the expenditure on the physical sciences £558 588 on the salaries of teaching and research staffs; £224 827 on wages; and £138 781 on consumable materials.

Maintenance of equipment, etc., cost £111 178, and other expenses were £25 821. The total of these expenditures was £1 059 195. In addition, £481 188 was expended from research grants and £79 998 from other sources.

The cost of major buildings completed in 1945–67 was given as £288 367.

According to the Approved Estimates of Expenditure in 1972–3, the Total Chargeable on the Universities Education Fund was £562 934, an increase of £90 014 on the previous year.

Research grants totalled £643 598. Stipends amounted to £33 977 and personal emoluments to £520 642.

There were 6 professors, 10 readers, 15 university lecturers, 7 assistant directors of research, 4 senior assistants in research, 9 senior technical officers, 6 assistant technical officers, and a computing officer, making 74 in all.

407

The total costs of teaching and research were £294036. The administrative cost was £5439, including £4064 for the departmental secretary. The total stipends were £299475. The wages of 86 technical assistants of various grades were £155757. The estimate for a clerical staff of 22 was £29942.

There was £2201 for the library; £18267 for maintenance and cleaning, and £15000 for cleaners. Total wages amounted to £221167, and salaries of personnel £520642. Departmental and Laboratory expenditure amounted to £77634. Telephone, postage, printing, etc., was £14300. Books, etc., for the library cost £6700.

The estimate for the Museum was £100. The size of this estimate has an interesting implication on the attitude of physicists towards the history of science, and the preservation of scientific records. It may be compared with the estimate that the service of the University Library to the Physics Department is of negligible proportions.

The estimate for maintenance of the premises was £10200. The total of these and other charges was £108934.

The Research Grants estimate of £643598 did not include £2 million for the construction of the 5-km radiotelescope.

The estimates included £27273 for bubble-chamber film analysis; £29914 for research on the fracture of solids; £20879 for ultra high-vacuum electron-microscope studies; and an Annual Consolidated Grant from the SCR for radioastronomy amounting to £38050. The estimate for basic tribological studies and metal forming was £32590.

By 1972–3 the direction of the Cavendish Laboratory involved a budget of the order of £2 million, and a scientific staff of 74, an assistant staff of 108 and 265 research students, comprising some 450 persons.

This presented problems in management with which J. J. and Rutherford had not had to contend, though there is no reason to suppose that they would not have found out how to deal with them.

The figures illustrate the peculiarities of accounting in Cambridge, and during the life of the Cavendish. They show the difference in the problem of direction and management in the Cavendish before, and after the 1930s.

The administration of the Cavendish did not become professional until after the Second World War. Bragg instituted departmental secretaries. One of the earliest of these was E. H. K. Dibden who was appointed in 1948. A physicist by qualification, he had worked in naval radar in the Second World War, and became acquainted with the scientists who developed and operated it. He had previously studied physics under C. D. Ellis at King's College London. He returned to King's in 1947 and read for a degree.

One day his wife saw an advertisement in *The Times*, for the post of secretary of the Cavendish Laboratory. He and his wife decided that he should apply for it. There were many applicants, and a short list of three was drawn up. These were invited to lunch by W. L. Bragg and E. S. Shire in the rooms used by the English Speaking Union in Matthews' Café in Cambridge. Whether or not he did not eat peas with a knife, he was selected. He served as secretary from 1948–60, having been preceded in the post by I. T. James, who had been at the College of Education at Hull.

The secretary was really the administrative manager. The Laboratory had virtually disintegrated during the war, and had to be built up again. Costs had formerly been reasonable, but after the war they were on a new scale. Vast sums of money were needed to operate the large, expanded and expensive organisation.

The Laboratory had changed from a family set-up into a large institution, more like an industrial organisation.

One of Dibden's first jobs was to find space for the young Perutz and Kendrew. They were put in a little bit of a hole on the top floor of the Austin Wing. When the metallurgical department was moved out of the Austin Wing he arranged that Perutz and Kendrew should have the vacated space.

Dibden regarded Bragg as a good administrator. He knew what needed doing, and believed in delegation. Dibden felt, after he had worked under him for six months, that Bragg had decided that he was dependable. After this, he trusted him completely, delegating to him all relevant work, and even discussing personal problems.

He thought that Bragg left the Cavendish for several reasons. One was that he considered he had a filial duty to go to the Royal Institution, where his father had been director. He felt it was necessary to rescue the Institution when it had fallen into a period of difficulty. If he had not been prepared to go there, it might never have recovered.

Bragg was a great teacher and lecturer. He taught teachers how to teach, and set up new illustrative experiments.

He was the last of the gentleman-scientists, and he was also a gentle man. He saw that nuclear physics was no longer properly a university subject; the resources and organisation it needed were too large to be properly accommodated in a university.

Despite opposition Bragg diversified the lines of research and thereby saved the Laboratory. He helped J. A. Ratcliffe, W. H. Taylor, V. E. Cosslett, G. I. Taylor and D. Shoenberg in their various directions of research.

Dibden had had dealings with many strong personalities. He put Bragg among the five greatest men he had met, and very probably the greatest.

Mott belonged to a different tradition. He was a scientist in the modern traditional manner. In the earlier days of the Cavendish the assistants were a servant race. Mott treated them as social equals, and was not always understood. Men who had been laboratory assistants to Rutherford and Bragg were taken aback. They had difficulty in understanding such an attitude in the Cavendish Professor.

Mott appealed more to the younger scientists. He fed and fostered them intellectually. He was personally kind and generous, helping men in financial difficulties. He would give men time off to look after a sick wife or family. Mott was generous in granting rises in salary.

Since Dibden's time, with the growth of the Laboratory, the increase in the number of research workers and technical staff, and the size of the budget, the secretarial management is now more like that of an industrial laboratory, with computerised accounting and systematic personnel management.

A. D. I. Nicol was appointed Secretary of the Department of Physics in 1963, subsequently becoming Secretary General of the University Faculties.

In 1962 Nicol made a very effective intervention in the Senate House discussion on the plan to redevelop the Cavendish on the New Museums Site. He observed that the provision of alternative off-site accommodation was required during redevelopment, so that teaching could be carried out in an area well-removed from the building operations. According to this plan

Instead of providing a preliminary release of pressure, we now propose to increase the existing congestion by adding to the 2,000 normal inhabitants an additional array of builders supported by columns of mechanical aids, which even under the most stringent control must occupy areas, at every stage of the plan, considerably less than 30 feet away from buildings in which teaching and research is to be continued. The difficulties and distractions which will ensue need not, I think, be stressed.

If the plan went forward those working on this site had 'little option but to look forward to many years of unrelieved disturbance'. He mentioned that when the High-Tension Laboratory was converted into lecture theatre and class-rooms, 'much of the disturbing noise complained of was due to operations *within* the High-Tension Laboratory'.

Nicol's remarks contributed to the creation of opinion that led to the decision against the New Museums Site, and in favour of the West Cambridge Site, finally chosen.

J. Deakin was appointed Secretary of the Department of Physics in 1966. He graduated at Bristol, and after working in aerodynamics at the Royal Aircraft Establishment, engaged in industrial management. One of his tasks was to carry out the computerisation of the University's accounts for a five-year period.

The handling of the Laboratory's accounts is now a matter of considerable dimensions. These circumstances have been a powerful factor in moulding the Cavendish into a form more similar to that of other great physical laboratories.

The University expects the Laboratory not to overspend, in order to keep within the University Grants Committee requirements, but it does not call for an account of day-to-day expenditure.

There are important differences between management of a large industrial organisation and a large organisation within a university. One of the prime points in industrial management is getting an idea accepted. After that, action is comparatively simple. In a university the problem is more complicated.

In 1968 a Departmental Policy Committee was set up for advice on staff questions and problems in the organisation of research. It consists of a group of members of staff, including six elected representatives and meets twice a term. Thus the voice of each grade of staff has now more independent expression.

32

The New Cavendish

The swift post-war growth of science, and the large increase in numbers of students and research workers, severely congested the University's science departments. After various enquiries and discussions the Council of the Senate published a Report on 19 November 1962 on the redevelopment of the New Museums Site, including plans for the redevelopment of the Cavendish and other scientific departments largely on their old sites. At a meeting of the Senate on 4 December 1962, this Report was discussed.

Pippard made an impassioned speech. He said he stood there not to attack the Report, but to point out what would happen if the University acted upon it, and neglected at the same time to solve the problems it left untouched. There had been no radical attempt at long-term planning.

They had reached a moment of crisis in their development, so he asked indulgence to explain at some length 'the peculiar importance of the Cavendish to Cambridge and to the country in general', so that the urgency of his appeal could be understood. The Cavendish had been developed from the original leadership of Clerk Maxwell, that 'man of genius, up to the time it became, under Rutherford, the leading Physical Laboratory in the world. We who work there are proud of the tradition we have inherited, and jealous to preserve our legendary reputation.'

They were confronted by two dangers, one that they should live on their reputation, and the other that after the death of a great leader they could never be the same again.

Since Rutherford's day the Cavendish had wholly lost its eminence in nuclear physics. Referring to the recent discoveries in molecular biology, he reflected 'on the irony that sheer lack of space should

412

remove from the Cavendish that new branch of science which was initiated there and which alone in post-war Cambridge rivals the great discoveries of Rutherford.'

The Cavendish could not easily maintain a leading position in 'the face of the worldwide competition that has resulted from the post-war industrial application of modern physics'. Nevertheless, their duty as well as their choice was to continue near the forefront of the battle.

Pippard gave figures illustrating how powerful the attraction of the Cavendish was to able students. If the Cavendish were deliberately destroyed by driving out the best teachers and denying it adequate accommodation, it would continue to attract the best students for another ten or twenty years. But in the end, a blight would be laid 'on British science from which our industrial economy might never recover'.

Could they trust only to 'the attractions of the social life, the Cavendish reputation, and the prospect of teaching good students...' and 'remain grotesquely cramped in old and converted buildings...?' Perhaps they had survived so long only because other universities had been equally blind to their needs.

Cambridge might be the last English university to come to terms with the problem, and he had real fears that it might 'mean the end of the leadership of Cambridge in English Physics'.

The truth was that British society was facing a crisis because not enough good scientists were being trained to meet the expanding needs of industry, government and universities. Too many were going to America, especially because of the attractions of better equipment.

The post-war eminence of Cavendish physicists was still very considerable. Out of 78 university physicists in Australia and Canada 27 had worked at Cambridge, 15 at London and 14 at Oxford. One of the pressing reasons for adequate accommodation was that this large-scale high-grade training in physics should continue.

Pippard gave details of the degree of overcrowding in the old laboratories. He pointed out that while in the glorious days of 1934 the number of undergraduates and the floor area were three times smaller, there were $4\frac{1}{2}$ times less staff and research workers and $5\frac{1}{2}$ times less assistants. Thus in those days, in spite of the stories of their squalid conditions, they had more space per individual physicist.

He remarked that 'when the Department as a whole woke up to the realization of what was implied by the Council's first Report, we were prepared to accept that our slowness of reaction had seriously com-

413

promised us'. Many had felt that a move to the west side of Cambridge would be 'too harmful to the interplay of the sciences', and had been inclined to make the best of the old site. Now they found themselves 'dangerously ill-suited to meet competition from outside, with precious few promises of help from within, and a very strong conviction of our duty to survive as a great power in the world of Physics'.

He felt that an intense effort should be made to obtain part of the Addenbrooke's Site, and secure a special grant of £3 million, or however much was needed, for 'a new and spacious Department of Physics'. He wanted it, if possible, to be in close proximity with the Departments of Chemistry and Engineering. The future of the country depended on physics drawing together with them, and training the men who could apply modern science to practical needs. 'In Britain we are in desperate need of leadership in the application of modern science, and Cambridge is one of the few universities that can supply that need... Let us see ourselves starting the next hundred years in the same forward-looking spirit as inspired Maxwell at the very beginning.'

Pippard's speech was an inspired contribution to the decision to change from redevelopment of the old, no longer suitable site to the open, radically new site with many possibilities, in west Cambridge. It was a difficult but courageous and correct decision.

Members of the Department of Physics prepared papers on various aspects of their needs. In January 1964 A. D. I. Nicol prepared a *Memorandum on Space*. The needs of the Department of Physics had been stated in February 1963 by Mott as requiring 15500 square metres, with parking space for 100 cars and 400 bicycles.

The total net floor area assigned to research was at that time 5108 square metres. This excluded administrative rooms, libraries, stores, students' workshops and field laboratories.

The sections occupying from 280 to 740 square metres included crystallography, electron microscopy, electronics, metal physics, Mond and Magnetic Laboratories, nuclear physics and radio astronomy. The physics and chemistry of surfaces group had 1200 square metres.

Teaching space for classes and lectures occupied 1975 square metres and administration, including libraries, stores and main workshop, 877.5 square metres. It was usual to allow a further 30 per cent of space for passages, cloakrooms, etc., but in the Cavendish it was about 20 per cent (some apparatus was being kept in passages, etc.).

In a paper dated 27 January 1964 Mott discussed *Proposals for an Investigation of the Long-term Needs of Cambridge*. The central point was whether redevelopment should be on existing sites together with Addenbrooke's, or on a new site in west Cambridge.

The investigation would involve an enormous amount of guess-work, 'but it should be possible for informed people, who know British and American science and wish Cambridge to remain our leading university for science, to see whether the way ahead is clear *either* on central sites *or* using the central sites and the Shorts Factory site' (in west Cambridge).

Mott made suggestions for the personnel of a committee to carry out the investigation, and people it should consult.

In April 1964 Pippard submitted an interim report to the Deer Committee on *Long-Term Planning for Science in Cambridge*. He remarked that a typical arts view was that the optimum size for a university was 5000, while a typical science view was 20000. The latter was borne out by the scientific repute of American universities, which suggested that the largest had an advantage. They offer a diversity of research groups of viable size.

He thought that 'a true scientific culture will be reached only when a student reading sciences does not automatically expect to practice science'; something of this sort 'seems to be needed if we are to find politicians and leaders of industry whose background education renders them sympathetic and understanding in scientific matters'.

This was a reason for welcoming an increasing number of science students, and making the relevant provisions for science expansion.

The relations between the physical sciences should be closer. There should be the closest possible links between physics, applied mathematics and mathematical laboratories. This group should be connected with engineering, the earth sciences, chemistry, bio-chemistry, biology and medicine.

Within twenty years' time biology might be stimulated by techno-logical needs as yet unnoted, and this should be allowed for.

Pippard then analysed in some detail what was desirable, and what would have to be done, if the Department of Physics was redeveloped on a central site. He concluded that he was doubtful about any scheme involving the central sites, and urged that every effort should be made by appeals to the U.G.C., to the public and to foundations, to secure funds for a free development on an unencumbered site.

He thought that the west Cambridge site was probably the only hope, though the contraction of fruitful exchanges with, for example, research workers in biology and chemistry, who would remain on central sites, was a drawback. 'It should be much cheaper to build what is really needed on a new large site, in an open style that need make no concessions to monumentality and which need not come into conflict with the guardians of the amenities of central Cambridge.'

He did not mean that they should aim at something squalid, but that they did 'not desire pretentiousness or permanence'. They 'could achieve decency and utility at a very moderate price'. A closely knit institute embracing pure and applied science, able to welcome contacts with industry and government, was very rightly what is sought after nowadays by grant-aiding bodies. There would be a better chance of getting money for this, rather than for redevelopment along the old lines.

In June 1965 Mott provided a Revised Statement of Needs for a new building aimed at supplying adequate accommodation up to 1980 at least. He estimated the laboratory population in 1980 at 1500. The present teaching staff of 32 was overworked, and by 1980 it should be increased to 68. In 1965 the total staff was 74; by 1980–1 it might be 138. The number of professors would have increased from 4 to 10.

Such estimates of expansion assumed the possibility of suitable accommodation. The present accommodation was very limited, and much of it entirely unsuitable. The Department was on a site incapable of free development. Consequently, as numbers of students have increased, staff and research workers have not kept up with them in numbers.

With a new building on a new site their programme of research could be planned without the frustration of knowing that a new piece of large equipment could not be housed, thereby leading to loss of able research workers and teachers. It was important to foster the reputation of Cambridge physics. This might contribute to the reversal of the trend from science, which was becoming noticeable in schools.

The Department of Physics produced a paper in January 1966 on *Research Space in the Present Cavendish Laboratory, Cambridge*. On an average, there were 21.8 square metres per head. Some 14.5 square metres of this could be described as research space, and the remaining 7.3 square metres as ancillary space. The U.G.C. recommended 11 square metres research space, and 3.9 square metres ancillary space. They felt that the extent to which they suffered from enforced inefficiency in use of space indicated that the U.G.C. figures needed generous revision.

Pippard drafted *General Notes on the Proposed New Cavendish Laboratory* in July 1965. He expressed the anxiety felt that full opportunity should be taken of utilising the advantages of an open site. He saw the problems of size, flexibility of research accommodation and services as particularly important.

In order to maintain contact between different groups, and between teaching and research, compactness was desirable. The usual way of achieving this was by multi-storey buildings, but these may not work well. Divided into standard rooms of modest proportions, they may prove highly restrictive. Heavy equipment cannot easily be accommodated on a higher storey, nor can higher floors be adequately protected from vibration.

Hence 'the ideal design of a research block probably confines experimental work to ground and first floors, with provision for a considerable amount of large equipment inside; but the possibility of attaching at almost any point a hut or other special building demanded by the state of research'.

It could not be stressed too strongly that 'a research laboratory is not a static piece of architecture, but must be capable of progressive remodelling inside and accretion outside without any sense of shame on the part of the occupiers at the abuse of an artistic conception'.

The general layout should consist of two parts, central and research. The central part should contain the administration, teaching laboratories, lecture theatres, libraries, workshops and common rooms. It would require 8970 square metres.

Research laboratories would require 12000 square metres, and academic staff, secretarial and seminar rooms 2400 square metres.

Research blocks should radiate from the central part. Research should, if possible, be confined to two floors.

Wide blocks rather than long corridors are desirable. The latter system is psychologically unsatisfactory, deterring chance meetings and fruitful gossiping. Cambridge experience suggested that a system of research rooms leading off from a central lofty, skylit hall provided a better atmosphere for research. It did not seem impossible to apply the idea on a large scale, especially if some experimental space was indirectly lit or skylit.

Only a few research workers were distressed by not being able to look out of a window, provided they can easily walk into an open space or wide hall. There was even some advantage in absence of windows, since plenty of wall space, for attaching apparatus to, is an essential need, and 'glass walls from floor to ceiling, such as are fashionable nowadays, are only a nuisance in a research laboratory'.

Pippard made models illustrating his ideas on the structure and arrangement of the laboratory rooms (see, for example, figure 43). The design of the New Cavendish owes a great deal to his conceptions, energetic exposition and persuasion.

It was not for nothing that Pippard's father was an engineer. The family interest in structures entered into the design of the New Cavendish, as that of John Clerk Maxwell had entered into the design of the Old Cavendish through his son.

A committee of the General Board of the University presently prepared a report on *The Long-Term Needs of Scientific Departments*. This was published in 1965. The ideas it adopted for physics were substantially the conceptions that had been expressed by Pippard and the Physics Department.

43   *One of Pippard's models of suggestions for the structure of the New Cavendish*

The complete document contained more than 30 000 words, and explored the problem thoroughly from many perspectives. The committee explained why they disagreed with the Board's acceptance of the conclusion in a report as late as 1963, that the redevelopment of the central sites in the city would meet long-term needs adequately.

Central development would involve high-density constructions, for which planning permission might be withheld. The clearance of old buildings and their replacement by new might cause intolerable delays. It left no margin for unforeseen developments, especially expansion in biological and medical sciences.

It posed 'the very serious, if not impossible, task of rehousing Physics on the New Museums Site except over an inordinately long period, involving a complicated series of decanting operations', and it did not take sufficient account of probable increases in students, research workers and staff.

They recommended the redevelopment of the old Addenbrooke's Hospital site, and the Downing site for the biological and medical departments.

For the physical sciences they recommended a large site in west Cambridge, 'initially for Physics and later for other physical sciences'.

The committee made an important recommendation of principle, that some existing barriers between departments should be removed with the formation of larger units on a Faculty basis. The traditional divisions between subjects had led to inflexibility, and hampered the deployment of limited resources with the greatest economy and efficiency.

The committee's ideas on regrouping and merging of sciences may have great significance for the Cavendish. It is possible that in the future the Cavendish will become the spearhead for a large part of Cambridge science, owing to the numerous and increasing connections of traditional physics with other traditional branches of science. Because of the severe congestion in the Old Cavendish, the New Cavendish was the first of the departments to be built on the new site. Its presence may, however, stimulate its development as the leader of a whole range of sciences grouped around and connected with physics.

The committee reported that the provision of modern accommodation for the Department of Physics was particularly urgent. The conversion of the former High-Tension Laboratory (completed in 1963) and the old Examination Schools (completed in 1965) were 'inadequate palliatives'.

They considered various peripheral sites, and especially Madingley Road. On the north side was one of 8 hectares, which already housed the Observatories, and the Department of Geodesy and Geophysics. On the south side the University possessed an area of 69 hectares. The Physics Department's radiophysics and 'thunderstorm field' occupied one small part.

There was ample space for adequate new physics laboratories, and also more science departments, which might be moved out later.

They thought that while research would be transferred to the new laboratory, undergraduate teaching could continue to be given in central buildings.

They envisaged the beginning of the building of the new laboratory to start in 1968–9.

They 'left open the question of the ideal long-term home for Applied Mathematics and Theoretical Physics'. They did not exclude the possibility of its being eventually 'incorporated in a physical sciences community in west Cambridge'.

They concluded that a new physics building on the west Cambridge site was an outstanding need.

Some people were diffident about moving to the new site, but Pippard was strongly in favour of it. Soon after the decision was made to move, the financial situation became much tighter, so the moves of other departments to the west Cambridge development were postponed. Though the New Cavendish went up by itself, the University Observatories, the Institute of Theoretical Astronomy and the Department of Geodesy and Geophysics were its neighbours on the other side of Madingley Road, so it was not isolated.

Architects and engineers were commissioned in January 1966 to report on the problems of the development of the Science site in west Cambridge, firstly with regard to the siting of the New Cavendish, and secondly with regard to the siting as a whole.

They analysed the physical attributes of the site, together with questions of land ownership and current planning proposals. They concluded that the best site was the area known as Vicar's Farm.

Then they sketched a pattern of growth to provide the best fit for apparent academic demands, which would at the same time leave the maximum degree of open-endedness, to make future adaptation to changing circumstances as painless as possible.

The authors were Robert Matthew, Johnson-Marshall and Partners; Pinto and Partners; L. J. Fowler; and James Nisbet and Partners. They delivered their report in September 1966, and were subsequently asked to design the New Cavendish. They commented on the excellence of the detailed and comprehensive brief that had been given to them, which indicated the thoroughness of the background work that had been put into it by the previous thinkers.

The chief purpose of the development plan was to provide a framework for the design of the New Cavendish.

The topography and landscape were discussed. The proposed Western Relief Road, which would be built sooner or later, though no one knew when, would cut the eastern part of the west Cambridge site in two. This covered about 125 hectares, and was 2 kilometres from the centre of the city. From it there was a fine view of the towers and pinnacles above a dense line of trees running parallel with the Backs. The well-loved footpath from Cambridge to Coton ran east–west across the site. These qualities would be destroyed by the intrusion of motor vehicles.

In examining the existing buildings they noticed the hut used by C. T. R. Wilson for measuring lightning potentials, and remarked that it might be of historic significance.

The development of an adequate system of road and path communication, which safeguarded cyclists and pedestrians, and preserved the environmental quality of the Backs, depended on the design of the Western Relief Road. This would be extremely important for the site, and for Cambridge.

By starting at the southern end of Vicar's Farm, phases could be built successively northwards, so that the traffic, noise, dirt and vibration retreated progressively away from the parts of the building already in use. This would provide a civilised environment within the site during the early period of development. Among the points for immediate action would be strategic tree planting and landscaping.

With regard to long-term development, it was thought that the total site could house at least 1000 people, besides the departments. The growth of pedestrian and cycle traffic would create a need for additional crossings over the River Cam. It was thought that the pedestrians would form one quarter, and the cyclists three quarters of this traffic.

J. R. Payne, the Secretary of the Faculty of Physics and Chemistry, acted as Secretary of the Building Committee, of which Pippard was Chairman.

The architects and engineers of the New Cavendish were Robert Matthew, Johnson-Marshall and Partners.

421

The cost of the New Cavendish construction was estimated at £3 805 000 at June 1966 prices. The construction would stretch over the period 1969–74. About five-sixths of the cost would be for internal, and one-sixth for external works. If prices increased at 5 per cent per annum, the cost would have risen to £4 645 000 by 1974.

The New Cavendish is a little more than $1\frac{1}{2}$ kilometres' walk from the Old. (Their relative positions are shown in figure 44). It is planned to have a usable area of about 16 000 square metres. It is divided into three main buildings:

1. The *Mott Building*, containing the solid-state research laboratories and offices, laboratories for advanced teaching and rooms for seminars. It has three storeys and 5800 square metres available.

2. The *Bragg Building*, containing the administration, library, stores, common rooms, lecture theatres and elementary teaching laboratories. It has two storeys and 5650 square metres.

3. The *Rutherford Building* will contain high-energy physics, radio-astronomy and other research laboratories and offices. There will be advanced teaching laboratories, seminar rooms and the main workshop. It will have two storeys and 4550 square metres. The site plan is shown in figure 45 and the floor plan in figure 46.

The Laboratory is entered from Madingley Road. There is plenty of parking space for cars and cycles, and walkers can travel by Adams Road and the Coton footpath.

The ground floor of the three-storey Mott building has half of the total floor area. Each floor in ascending order has a progressively smaller area. As the ground floor occupies half of the total floor area, and is on a substantial reinforced concrete slab, there are plenty of working places for heavy and vibrating machinery, and space for flexible arrangements of apparatus. The extended ground-floor slab is divided into sections in order to provide mechanical isolation.

The ziggurat design was proposed by the architect to meet Pippard's requirement that half the total floor space should be able to accommodate experiments sensitive to vibration. The ziggurat style of design prevents the sheer bulk of the building from becoming oppressive.

The low profile of the buildings meets the planning authorities' wish that it should accord with the flat East Anglian landscape.

In the Mott building there are 73 rooms on the ground or first level, 66 on the second level and 48 on the third and top level,

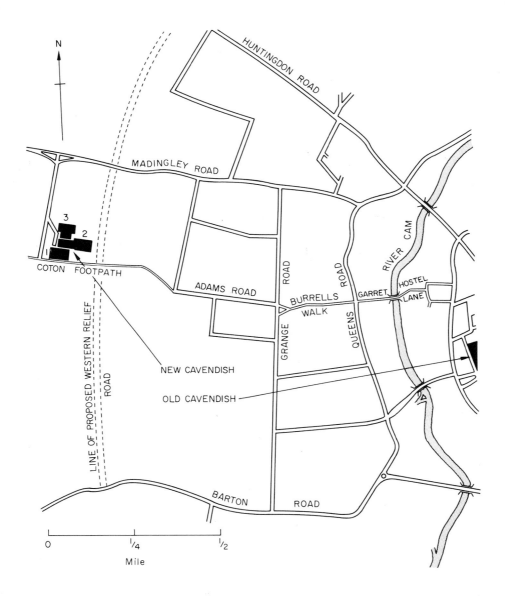

N

HUNTINGDON ROAD

MADINGLEY ROAD

3
2
1
COTON FOOTPATH

LINE OF PROPOSED WESTERN RELIEF ROAD

ADAMS ROAD

GRANGE ROAD

BURRELLS WALK

QUEENS ROAD

GARRET HOSTEL LANE

RIVER CAM

NEW CAVENDISH

OLD CAVENDISH

BARTON ROAD

0   1/4   1/2
Mile

44   *Relative positions of the New and Old Cavendish Laboratories*

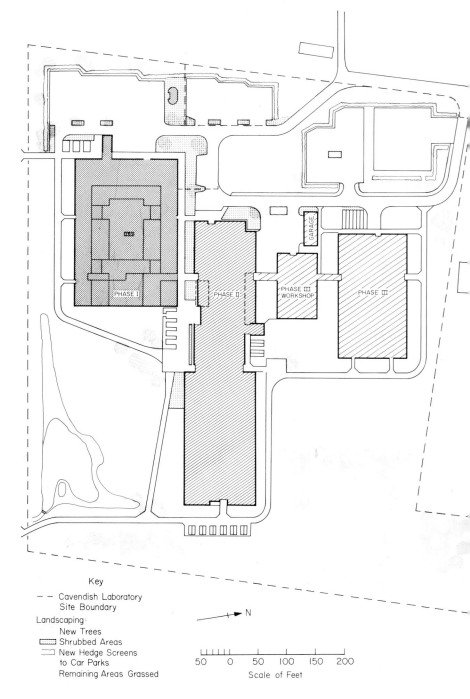

Key

— — Cavendish Laboratory
Site Boundary

Landscaping:
New Trees
Shrubbed Areas
New Hedge Screens
to Car Parks
Remaining Areas Grassed

N

50 0 50 100 150 200
Scale of Feet

45 *The site plan of the New Cavendish*

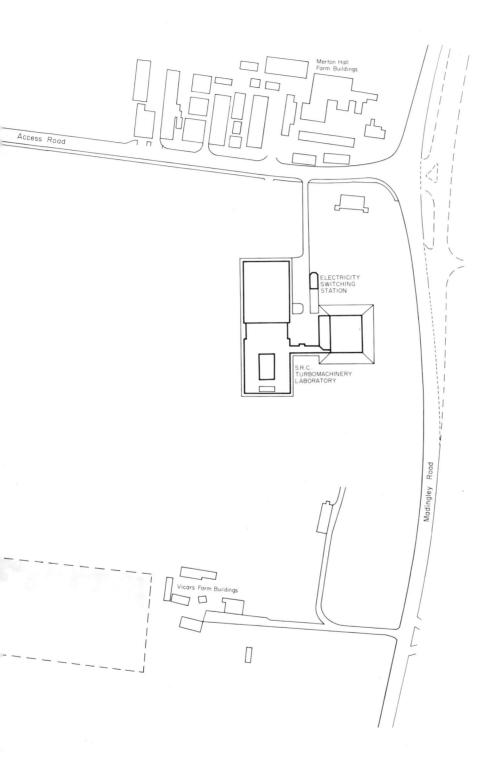

Access Road

Merton Hall
Farm Buildings

ELECTRICITY
SWITCHING
STATION

S.R.C.
TURBOMACHINERY
LABORATORY

Madingley Road

Vicars Farm Buildings

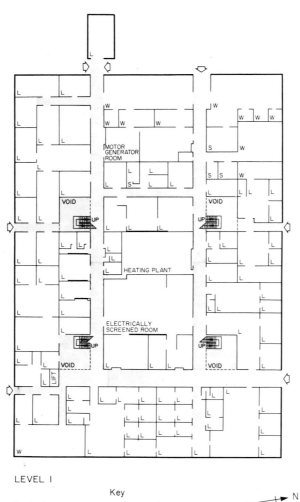

LEVEL I

Key

L   Laboratory type space   W   Workshop
    (laboratories, preparation   S   Stores
    rooms, dark rooms,   O   Offices
    special experiment rooms,
    teaching laboratories)

N

46   *Plans for the ground, first and second floors*
(first, second and third levels) in the Mott Building

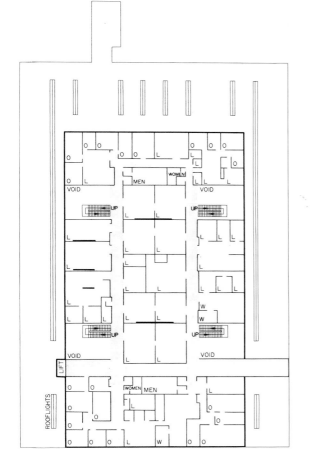

LEVEL 2

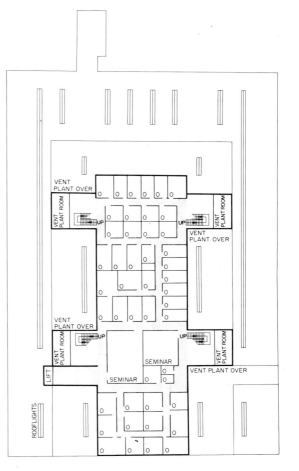

LEVEL 3

making a total of 187. In figure 46 various laboratories, preparation rooms, dark rooms and special teaching rooms are marked L; offices are marked O, workshops W and stores S. The uses of a number of rooms are marked by explanatory lettering.

From one point of view, the New Cavendish is a set of services, with a building to cover them. One important motive was sheer economy; it prompted economising on outside walls, and the construction of many inside rooms that can be used for appropriate purposes. The building consists of a light steel frame, upper-floor decks of precast slabs, very light timber roof decks and precast concrete external cladding with an 'aggregate' finish. The internal partitions are made of slabs of a compressed mixture of concrete and wood-shavings. Ease of adaptability has been a prime consideration. In 1874 they had not considered this problem, yet the old brick style of building is adaptable, more so than the recent Austin Wing. As usual, anything planned by Clerk Maxwell turns out to have unforeseen possibilities.

At all levels there are false ceilings, which conceal the steel structure and some of the services.

The building contains four vertical halls, around which the rooms are organised. They are designed so that workers can meet each other easily and casually in the course of their work, which had proved to be such an excellent and stimulating feature of the design of the Mond Laboratory.

The electricity supply system includes closed-circuit television in each of the halls at the first level, connected to a panel in the main preparation rooms adjoining the lecture theatres in the Bragg building. Lines from computer terminals run to a point in the Bragg building for connection to the University Computer Laboratory, or elsewhere.

A screened room is provided for apparatus that emits electrical interference and should not be used in the ordinary research rooms. It contains a mild-steel enclosure well serviced with electricity and cooling water.

The piped services include cooling water, gas and compressed air. The design contains protection against disasters, so that a leak of up to 4500 litres an hour can be made up from the mains supply. A leak exceeding this causes the circulation pumps to be switched off, and a klaxon to be sounded in the cooling tower.

Apparatus sensitive to the loss of cooling water has to be protected by fail-safe devices provided and installed by the user. The laboratory staff are warned that the cooling water quickly corrodes aluminium.

The sink waste fittings are made of polypropylene, and will stand a good deal of abuse; but a list is provided of forty-one chemicals that can cause rapid deterioration in the drains.

The compressed air is supplied to the work rooms at $1.4 \, \text{MN/m}^{-2}$.

In the laboratory built-in furniture is the exception rather than the rule.

A lake has been excavated in the grounds, in order to relieve the Cambridge drainage system in storms, when large quantities of water are drained off the buildings. The excess flood water can subsequently be discharged under control.

When the New Cavendish construction is sufficiently complete, scientists in the neighbouring institutes of Theoretical Astronomy and of Geodesy and Geophysics will be able to use their common rooms and cafeteria.

In the future, scientists here will have more contact with other scientists than with arts students. Many women students do not like dining halls; they prefer to make their own meals. The new Laboratory also presents social problems to the technical staff, many of whom like to go home and have a hot meal in the middle of the day. As most live on the other side of Cambridge they have to make a considerable journey when the streets are very busy. This is inconvenient for motorists; most of the students cycle.

Common rooms in which there are noisy activities are on the ground floor, while quiet ones are placed on the first floor.

By the spring of 1973 the Mott Building was sufficiently advanced for the solid-state physics research group to begin to occupy their new quarters, and it was expected that by the end of 1974 the whole department would be rehoused.

In the various ground-floor rooms equipment such as cryostats, vacuum pumps, electron microscopes, furnaces and ion accelerators are set up.

Electronics and teaching laboratories utilising lighter equipment, and staff offices are on the first floor. On the top are the theoreticians, in peace and quiet.

Careful thought has been given to the circulation of workers. The rooms occupied by a given research group are arranged in a three-dimensional cluster around one of the four halls, so that members of the group can easily call to each other, or meet near a blackboard to work out an idea arising from a chance encounter. The arrangement also allows a worker to pass by other groups, and have contact with them, without intruding. In this way, he can acquire some feeling of how others are thinking and going about their work.

The main routes through the halls converge onto a bridge connecting the Mott and Bragg Buildings. Later, another bridge will join the Rutherford to the Bragg Building.

The centre of the whole organisation will be in the Bragg Building, with the Head of the Department and the administrative staff. The library and main common rooms are open to all staff, technicians and students, who can meet in the appropriate places for lunch and social occasions.

The main workshop and stores are also near the centre. Further down the Bragg Building are two lecture theatres. There are practical laboratories for the first two undergraduate years. Only third-year lectures will be given in the new building; first- and second-year lectures will continue to be given in the lecture theatres in the city, used by all the science faculties. Practical work will occupy each student for one day a week, and besides working in the laboratories, he will read in the students' library, and relax in the common rooms.

In 1973 there were 350–400 first-year undergraduates, 200 second-year and more than 100 third-year. The research students numbered about 150, most of whom came from universities other than Cambridge, and from abroad. Altogether, with staff, and a fluctuating number of post-doctoral visitors, there are usually about 700 persons at work in the Laboratory on a normal day in term-time.

For the first time in living memory research workers will have adequate space for their experiments, and will not have recourse to cellars and corridors.

33

Pippard's View

Pippard (see figure 47) delivered his Inaugural Lecture as Cavendish Professor on 7 October 1971, the centenary almost to the day of James Clerk Maxwell's Introductory Lecture of 25 October 1871.

He observed that 'a century of untarnished greatness' lay behind him, and no man could sustain that tradition alone; the burden was on all who worked in the Cavendish. He believed that 'great traditions arise from the continuing ability to appreciate what is the right thing to do at any instant, and the fortitude to ensure that it is done'.

This implied that a tradition is not necessarily maintained by carrying on as before. He thought that the change now needed was that 'what we impart and learn should be controlled to a greater extent by what goes on in the world outside — inhabited, and to a great extent governed, by the non-scientist'.

In recent years physics had suffered a decline in prestige, one symptom of which was the loss of faith by employers in the ability of science to help them with their problems. This would not perhaps have occurred if physicists had paid more attention to the needs of society, and their duty to play a more positive role in it. He wondered whether physicists' ideals 'have been developed in the luxury of self-interest'.

The reaction against science, intellectual processes and liberal culture, though invalid, sprang understandably from detestation of industrial society. To this savage moral criticism were added the doubts of the economic value of science by the industrialists themselves.

In this situation it should not be specially meritorious to concentrate on research to the exclusion of other duties. The pursuit of knowledge for its own sake is rather easy and agreeable for those who

47 *Pippard*

are able to do it, if they give no thought as to whether anyone really wants the knowledge.

It would have been better for us all, even those who remained in academic life, if we had recognised that the maintenance and improvement of our present society, which owes so much to science for its creation and its sense of values, demand also from scientists that they play their full part in working within society to develop and humanise the ideals without which civilisation can become either sterile or self-destructive.

Hitherto, technically accomplished performers had been turned out, who only became educated citizens in any real sense by their own efforts, after they had absorbed their training in physics. Consequently the public tended to see the scientist as a brilliant technician, who must on no account be allowed to take over the controls.

The formal laws of physics had been worked out with extraordinary perfection. This tended to foster the illusion that all problems of nature were neatly, and even quickly, soluble. Most real problems, however, defy mathematical analysis; not that they cannot be mathematically described, but because the mathematical description is intractable. 'In practice, then, we resort to guessing and insight into how things are likely to behave far more than you would infer from looking at the syllabus of a physics course.'

He called for a serious effort to devise techniques by which people could be taught how to make reasonable guesses about nature. How wide from the fact such informed guesses could be, he was able to show by experiments. He asked his audience of 285 to guess how long a solid copper block 4.5 cm in diameter and length would take to cool down from room temperature to $-200°C$. Two members thought it would take 2 to 4 seconds, and one thought it would take more than half-an-hour; 170, or more than half, thought it would take 16 to 60 seconds. Actually, when the experiment was tried, it took 6 minutes 10 seconds. Thus the guess of an educated audience ranged over a factor of 600, and the most favoured answer was too low by a factor of about 20.

The experiment illustrated the great importance of qualitative factors in experimental physics. 'It is the intuitive feeling for what can and cannot happen that is the mark of the sound scientist.'

Was there any way in which this could be consciously developed by appropriate teaching? He showed by further experiments involving points that are very important in practice, but not easy to analyse, how this might be done. They were of the kind that arise in oscillatory

433

phenomena, similar in principle to those occurring in radio circuits, musical instruments, water waves and vibrations in bridges and structures caused by wind.

There is an enormous conceptual difficulty between laws and applications, so great as to cast doubt on traditional beliefs about the relations between different sciences and the nature and purpose of scientific theories.

The onset of instability can be worked out systematically, but in nearly every case of interest the mathematical difficulty of working out what will happen is prohibitive.

Even to predict what will happen in so elementary a system as an assembly of identical atoms cannot in practice be carried very far by abstract intellectual processes. When the behaviour of each atom in an assembly of, say, one hundred atoms is worked out by computer, types of behaviour are revealed that had not previously been guessed at when only paper calculations had been available.

The significant point is that the most important information gained from a very laborious and expensive numerical computation was not quantitative but qualitative; new forms of behaviour that had not previously been imagined were discovered.

Systems being changed slowly and continuously may suffer sudden discontinuities so great as to render impracticable any attempt to describe the behaviour as a modified version of what happened before.

It was for this kind of reason that it was no historical accident that chemistry developed independently of physics. Chemistry is not a mere branch of physics, and it is a delusion to believe that the application of physical methods to biological systems would enable living processes to be reduced to the rule of physical law.

Physics is not, in any useful sense, the fundamental science. It is applicable to the systems other sciences investigate only by selecting the aspects on which it concentrates. Its brilliant successes have been bought at the price of ignoring the infinite complexity of the real world.

The physicist, in contrast with most other intellectuals, has set himself a comparatively easy task. His success is as much a tribute to his subject's simplicity as to his own intelligence.

Pippard observed that it had been a long time since the fundamental physicists had discovered anything of profound interest to other scientists, apart from astrophysicists and cosmologists. In the 1920s they had developed a quantum mechanics that had been of profound significance to chemists, but no solid-state physicist, chemist or biologist expected to have his ideas altered by quarks.

What, then, are the true arguments for the pursuit of fundamental physics? It can be claimed that it has led to a cosmology that is demonstrably better than that of primitive societies and religions. But in so far as it has not led to the discovery of the fundamental laws of other sciences, it is of no other importance to mankind in general. In Pippard's view, anthropology has more to say about the universe in which we find ourselves, for it is more than an assembly of particles; it consists of an assembly of men, of thinking and feeling beings.

The justification of fundamental physics on the ground that it leads to valuable practical applications cannot withstand evaluation by even so blunt an instrument as cost–benefit analysis. It cannot be justified by commercial 'spin-off', or as a means of producing a stream of trained minds for other purposes.

Its justification on the ground that it is interesting may seem inadequate. Any study that inspires devotion and intellectual enthusiasm is, however, a major contribution to the general well-being. The disinterested use of intelligence distinguishes man from the beasts, and civilised from uncivilised societies. In these days, when there are shadows all around, this needs affirmation.

If Cavendish men are to keep faith with their great heritage, they should take a cool look at the claims of fundamental physicists.

The laws of physics are beautiful to contemplate and to work with, but this by itself does not provide a healthy diet. It is now necessary to tilt slightly more away from the academic contemplation of perfection, towards the difficult and less elegant problems of the real physical world.

The science graduate of the future will have to come closer to grips with the imperfections that make real-life problems so challenging, and resistant to tidy solutions.

Contemplation of the quite unforeseen effects of simple dynamical instability may contribute to a wiser insight into the problem of keeping a complex manufacturing process in stable operation, or into the interplay of economic forces that can throw the world monetary system into a catastrophic oscillation. It may throw light on the periodic rush of the lemmings to self-destruction, urged on, not by a death-wish, but by blind forces that permit no equilibration.

The message of physics may be that it can open a little window onto a reality far more multifarious than itself, and put into mankind's

435

hands a clue to the labyrinth where teachers can no longer lead, 'and often enough dare not even follow'.

Pippard was the son of A. J. S. Pippard who, after becoming a consulting engineer, occupied chairs in civil engineering at Cardiff, Bristol and Imperial College. Alfred Brian Pippard was born in 1920. He spent his early years in Bristol while his father was professor there, and, like Mott, was educated at Clifton College.

The professor of physical chemistry at that time was M. W. Travers, who had collaborated with William Ramsay in the liquefaction of the rare gases. Travers had an independent and original outlook, besides being a very skilful experimenter. He used to take Pippard to his laboratory and show him experiments involving low temperatures. So, in his boyhood, Pippard's interest in low-temperature phenomena was excited. He went up to Cambridge just before the outbreak of the Second World War, and graduated in 1941.

Like so many other young men of talent he was recruited into research on radar, which he pursued at Malvern until 1945. During this period he learned the physics of microwaves. It was evident that this technique could be used to explore experimentally and theoretically many phenomena in low-temperature physics and superconductivity. When he returned to Cambridge after the war he joined the Mond Laboratory in the Cavendish, and soon made notable contributions in this field. His experiments on the depth of penetration of currents in superconductors led to an important modification in the London theory of the electromagnetic properties of superconductors.

He investigated the anomalous skin-effect in superconductors, explaining it in terms of the mean free path of the conducting electrons being large compared with the skin-depth of penetration of high-frequency currents.

Then he measured the surface resistivity of single metal-crystals. His work on copper led to a plot of the Fermi surface of the metal, that is, the contour showing how the wavelength of the quantum waves governing the motion of the most energetic electrons depends on the direction in which they are moving. When this is combined with information derived by X-ray analysis of the lattice structure of the nuclei in the crystal, fundamental information can be derived about the geometrical pattern formed by the electrons.

In 1959 Pippard was awarded the Hughes Medal of the Royal Society for these researches, and in 1960 he was appointed John Humphrey Plummer Professor of Theoretical Physics.

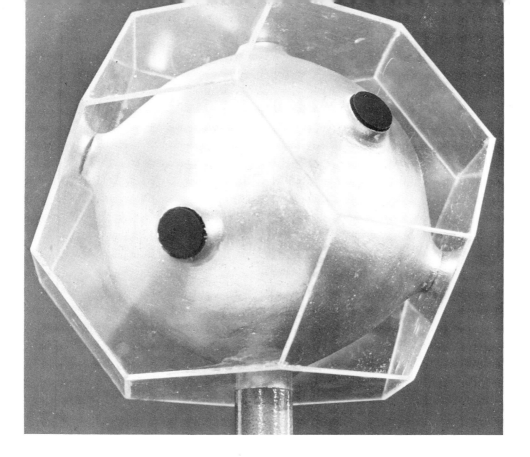

48   *Pippard's model of the Fermi surface of copper in its Brillouin zone*

Pippard was deeply influenced by Shoenberg, who was his supervisor in research for his doctorate. Shoenberg had devotedly investigated the de Haas–van Alphen effect. Consequently, when Onsager, during a visit to the Laboratory, realised that it could be turned into the most powerful single tool for studying Fermi surfaces, extensive practical knowledge of the effect was already available. This greatly facilitated the utilisation of the effect as a research tool. Without Shoenberg's work, this whole field would have progressed very much more slowly, and indeed might never have been properly started.

Pippard was appointed to the Cavendish chair in 1971. Rutherford and Bragg had been primarily experimental physicists, and Mott a theorist. Pippard, while an experimentalist in the Rutherford and Bragg tradition, is also a theorist. Like the first Cavendish Professors he combines both qualities.

Pippard's first view of physical research was formed in his youth, at the end of the Rutherford period. There were then perhaps twenty to forty people engaged in the front rank of fundamental physical research. Behind them was a supporting force about five times as large. In these conditions, a physicist with a good idea could spend much of his life working it out, with freedom from competitive anxiety. He could work pleasurably and happily.

This situation changed during the Second World War, when it became evident that the application of men and resources to physical problems could bring major military, political and economic results. If a physicist had a good idea he had to work it out very quickly, or watching competitors would pick it up, and work it out for him. This produced an almost hysterical pursuit of new results throughout the world, which destroyed the pleasure in research, and created a degree of instability and unhappiness, which has now spread to the whole of physics.

A striking example is the development of semiconductor physics since the war. It is a very difficult subject, requiring new ideas, new experiments, and particularly, new materials. The difficulty of obtaining the latter had formerly presented the crucial problem. Then an army of metallurgists was suddenly pushed in to make them, and physicists to measure them, so that very quickly the field was worked out. In this situation, many people have to change their line of work. When the mind is not allowed to dwell on a problem it becomes uneasy, and when large numbers of people have frequently to change the direction of their research, they, and the subject, become unstable and unhappy.

One answer to the situation is to choose a new subject where the former conditions hold—molecular biology, for example. Transference of the talented and imaginative to this subject tends to cause physics itself to decline. During the nineteenth century engineering had appealed to men of talent and imagination, but from the beginning of the twentieth, such men had been attracted to physics, leaving engineering in a decline. Pippard saw something of the same thing now happening to physics. It presented serious organisational problems in teaching as well as in research.

How and what should one teach the large body of physicists now required by a modern state? Britain needs about 100 000. Only a small fraction of these can engage in fundamental research—if they can find out how under present conditions—and enjoy the returns of pursuing it. How is physics to be taught to the rest, most of whom will not work in research or academic institutions? The Cavendish, like others, now has to think about these matters.

Pippard believed that it may be necessary to find a new use for the universities. Many of the large numbers of students now attending them consider that the universities do not provide what they need. The form that met the needs of the past may require remodelling to meet the needs of today.

Regarding the universities as the custodians of the intellectual side of the British people's culture, it is imperative for the future of that culture that they should be effective and inspiring, clearly contributing their essential part towards the pursuit and utilisation of science, for the happiness and benefit of the many categories of mankind.

Maxwell in his Introductory Lecture in 1871 had dwelt on the responsibility of Cavendish physicists to the general population, and Pippard returned to this theme on becoming Director of the Cavendish in 1971.

His view on the method of research, as consisting of a 'dialogue between experiment and theory', has had great influence on Mond and Cavendish physics. He gave a broad exposition of it after receiving the Dannie–Heinemann Prize in 1969. It might be broadly described as a Maxwellian combination of ingenuity and skill in experiment with a mastery of mathematical theory, in which each plays on the other alternately, until the problem is solved.

From the middle of the 1950s he devoted much attention to the description, as accurately as possible, of the motions of electrons in real metals, and to showing in detail how one metal differs from another. The data thus obtained enlarged understanding of the phenomena, and also stimulated the invention of new theoretical methods. These could be tested by experiment, and afterwards used in other areas of solid-state physics.

The Fermi surfaces for many elements have been determined, and show an extraordinary variety of shape. That of copper is a sphere drawn out at eight points (see figure 48), while those for graphite and arsenic look very odd, the former shaped like a bomb, and the latter like a plant seed.

Such extreme variation presents severe problems to the analyst. He may attack it from first principles, and in spite of its complication this approach is not unprofitable, for it helped to account for the magnetic properties of the element.

The second approach was to accept the Fermi surfaces as given, and see what could be deduced from them about the properties of the metals. This approach had also been very profitable, for some important properties were linked very directly to the geometry of the Fermi surface. For example, the de Haas–van Alphen effect, oscillatory magnetisation, was determined solely by the cross-sectional area of the Fermi surface, and not by its shape. Continuity, cyclotron resonance, and many other phenomena could be expressed very clearly in terms of the shapes of energy surfaces, without any formulae for them, and only the assistance of machines for differentiating and integrating hand-drawn curves.

This approach, by explicitly avoiding analytical expressions for the surfaces, was particularly valuable, because it concentrated attention on the way in which physical properties are related to shapes. It encouraged qualitative and intuitive thinking, and enabled the experimental physicist to carry out his traditional role, which involved much more than acquiring facts: he also had a strong motivation to find a coherent model of the phenomenon.

One of the experimenter's tasks is to provide rough explanations of phenomena for which no exact theory is possible, and for this he must acquire the right techniques of thinking about his problem.

Pippard felt strongly on this matter, considering it of great importance in educating physicists, and he returned to the topic in his Inaugural Lecture. There was a tendency to emphasise those techniques that provide an exact analysis of comfortably simple problems. This should be balanced by teaching students how to guess approximate answers when no exact answers can be found. This should be attempted where the guess can be tested only for performing an experiment, and not by doing a complicated calculation.

It was the method that he and many others had successfully used in thinking about electrons in metals in geometrical terms, without detailed mathematical analysis; he found it a useful aid to intuition.

For the experimental physicist particularly, pictorial methods had a power to stimulate creative thought where symbolisms were sterile.

Before the first Fermi surfaces were determined experimentally, numerous calculations had been attempted with very little success. Then a new generation of computers arrived, sophisticated enough to deal with the problem. The calculators were now able to check their results with the experimental data collected ten years before. After that had been accomplished, an experimenter was apt to feel embarrassed if his measurements contradicted theoretical predictions.

The success showed that the methods employed were probably adequate for investigating matters far deeper than Fermi surfaces. This liberated the theoretician and led him to collaborate with the experimenter on problems that previously seemed out of reach, such as the behaviour of alloys and the properties of metal surfaces in relation to low-energy electron diffraction.

It could also be hoped that a deeper understanding could be obtained of chemical bonds and reactions, of the interaction of electrons and lattice vibrations and thence into the condition of metals that become superconducting.

There was no question of theory guiding experiment or experiment guiding theory in any exclusive sense. Progress depended on a dialogue, which was an essential part of the advance of science.

Pippard held that the same relation was true between pure science and technology. Technology built on and made use of the results of pure science, and pure science learned its techniques and gathered ideas from the workshop of the technologist. He had himself started as a microwave engineer, who later applied this technique in pure research on the properties of superconductors and normal metals.

Two decades later it was a pleasure for him to see superconductors beginning to find large-scale application in industrial processes, and simultaneously the technical product in the form of superconducting magnets being utilised in pure research into the properties of metals.

Pippard believed that there was a parallel between the scientific dialogue he had described and the kind of dialogues required between nations, which should provide mutual reinforcement of the moral strength of each participant, without the exclusive dominance of one.

Farewell, and Hail!

The Old Cavendish has passed away, and the New Cavendish has arrived. It is a sign in the scientific field of the major change in Britain during recent decades.

Knowing the old Laboratory at its zenith and over a long period was a unique experience. There were other laboratories that were equally brilliant for a period under a director of genius, but there were none directed by such a succession of geniuses. This created a scientific authority that was strengthened still further by historical circumstances.

The spirit of the old Cavendish was expressed most powerfully within recent memory by Rutherford. He was the personification of greatness in science. His personality expressed in a simple clear-cut way, and to an intense degree, a combination of qualities made possible by the conditions of British science and history in his time. It was a splendid privilege to have known him, and the band of men he inspired. These were not merely gifted students who passed through his department for two or three years; they were very able scientists who stayed with him for an exceptionally long time before going to senior positions. They contributed to the exciting, pregnant atmosphere, in which it was confidently felt that any discovery was possible.

Taking leave of the old Cavendish revives glorious memories of the unparalleled succession of Cavendish Professors, of their research workers and their famous discoveries.

The foundation of the old Cavendish was based on long, many-sided and able discussions in the period between 1850 and 1870. A century later, the British situation was fundamentally different. New methods, means, policies and resources were required in a period

442

Front Row Dr J. R. Shakeshaft, Dr S. Kenderdine, J. Deakin, Dr J. G. Rushbrooke, Dr W. W. Neale, Dr B. D. Josephson, B. Elsmore, Prof. Sir Martin Ryle, Prof. A. Hewish, Prof. A. H. Cook, Prof. O. R. Frisch, Dr V. E. Cosslett, Prof. S. F. Edwards, Dr A. E. Kempton, Dr K. G. Budden, J. M. C. Scott.

Second Row M. C. Cross, R. P. Mount, Mrs L. Salih, S. M. Salih, A. Slingo, D. J. Munday, A. C. Steven, Dr M. B. Green, Dr G. G. Pooley.

Third Row Dr P. F. Scott, Dr J. E. Galletly, Dr R. E. Ansorge, A. B. Harris, D. J. Pittack, B. L. Fanaroff, G. Ioannidis, D. R. Ward, Dr A. C. S. Readhead, Mrs J. M. Riley, P. J. Duffett-Smith, R. Atkins, D. W. Bullett, S. J. Gurman, S. N. Henbest, Dr P. Gard, J. R. G. Armytage.

Fourth Row T. J. Pearson, P. J. Hargrave, D. W. J. Bly, Dr D. M. A. Wilson, P. J. Warner, D. M. Odell, Miss C. S. Harris, Dr T. A. Hall, Miss P. D. Peters, Dr A. Somlyo, J. E. Jennings, J. Robertson, R. C. Boysen, R. W. Barker, J. Schofield, Dr M. S. Longair, B. D. Turland.

Fifth Row Dr J. Dancz, A. G. Miller, Dr G. Wexler, K. R. Shaw, R. K. Cattell, A. H. M. Martin, K. J. E. Northover, R. T. Deam, I. Adsley, R. G. Palmer, W. O. Saxton, Dr R. Gerchberg, Dr J. Hertz, R. Raja, V. Urumov, I. P. Williamson, A. R. Gillespie.

Back Row Dr M. Saitoh, Dr J. E. Inglesfield, T. P. Fishlock, Dr M. J. Sik, S. F. Gull, M. J. Horner, M. J. Kelly, R. L. Johannes, M. W. Finnis, D. C. Threlfall, B. Knecht, P. Denton, Dr E. F. H. St G. Darlington, Dr E. Tosatti, A. J. B. Winter, E. A. C. Crouch, B. D. Hahn.

Front Row D. J. Stanley, G. L. Harding, A. T. Winter, Dr E. A. Davis, Dr W. Y. Liang, Dr L. M. Brown, Dr D. Tabor, Prof. A. B. Pippard, Dr A. Howie, Dr P. J. Brown, Dr J. E. Field, Dr M. M. Chaudhri, P. R. Ward, Dr U. Valdrè, Dr R. H. J. Hannink, Dr R. E. Winter.

Second Row B. K. Ambrose, Dr J. M. Wilson, Dr M. J. Murray, M. G. Bell, H. P. Hughes, Miss P. L. Gai, M. K. Hossain, I. C. Roselman, Dr B. J. Briscoe, Dr A. K. Pogosian, Dr A. M. Glazer, Dr J. A. Muir, Miss M. M. R. Costa, Miss J. Trotter, C. N. Guy, Dr N. Gane, J. R. Payne.

Third Row M. G. J. Gannon, W. Day, Miss J. A. Cross, L. G. P. Jones, C. G. Richards, O. Krivanek, C. A. Ferreira Lima, C. H. Hurst, J. T. Hagan, M. Dalipagic, H. M. Hauser, R. Kwadjo, Dr R. Turner, J. A. Williams, A. R. Beal, K. Ahmad.

Fourth Row D. Cherns, Dr J. J. Camus, R. T. Szczepura, G. N. Greaves, D. A. Gorham, K. N. G. Fuller, J. M. Lumley, T. M. Griffiths, D. C. B. Evans, P. J. Elliott, R. A. Abreu, A. M. Goldberg, P. G. Lurie, I. M. Hutchings, L. E. Haseler.

Back Row F. R. Allen, S. C. Harris, P. W. Forder, J. Davenport, N. Apsley, A. P. Troup, Dr J. C. Knights, A. C. Woodward.

when Britain was no longer the centre of an empire.

The responsible task of creating the new Cavendish, with the programme and facilities now needed for science in Britain, has rested with the governors of Cambridge University and the Cavendish Professors, in the period after the Second World War, and especially since the 1950s. The same kind of communal University effort has gone into the creation of the new Cavendish as into the old. The numerous, long, many-sided, contradictory discussions of the 1960s on the needs and future of Cambridge science, and of the Cavendish, parallel the similar activities in the 1860s.

The new Cavendish has been thought-out, almost hammered-out, to serve future British scientific needs. More attention will be paid to those aspects of physics of direct industrial significance and the training of relatively large numbers of well-qualified physicists who can contribute to research, teaching, administration and, in a significant new way, to general cultural and social ends. It is no longer sufficient just to make discoveries. Concern with the results of physics is now as important as physics itself. Physicists will be increasingly occupied with the integration of physics with the rest of science and human life. These and other concerns will be increasingly added to those that inspired the old Cavendish.

On 11 April 1973 there was an informal occasion in celebration of the moving into the completed parts of the new Laboratory. It was attended by Sir Nevill Mott and Lady Bragg, together with members of their families, including grandchildren. Staff, research workers and assistants walked through the beautiful rooms and green-carpeted corridors, inspecting apparatus being installed, and pausing to look through windows, observing the magnificent landscape.

The contrast with the old Cavendish could not be greater. Instead of grimy congested little workplaces and cellars there were light, shining rooms, surrounded by vistas of green fields, trees, and even a lake.

All presently assembled. Men and women stood or sat around, several accompanied by their infants. The Vice-Chancellor performed the naming ceremony for the Bragg and Mott buildings. Pippard suggested that the handsome lake, which had been made in the grounds for both practical and aesthetic reasons, should be named after J. R. Payne, the secretary of the building committee, who had borne most of the administrative burden.

With the new buildings, Cavendish physics starts on a second century of activity. The problems, historical situation and environment are different from those of the old Cavendish, but equally, and perhaps more, challenging.

Salutations to the new Cavendish, to Pippard, his colleagues and successors, and heartfelt wishes for an equally glorious future!

References

Maxwell

Introduction and chapters 1–4

Parentalia, or *Memoirs of the Family of the Wrens* compiled by his
son Christopher Wren and published by his grandson Stephen
Wren (1750).

Rudolf Dircks (ed.), *Sir Christopher Wren: A.D.1632–1723*, Bicen-
tenary Volume (1923).

John Ward, *The Lives of the Professors of Gresham College* (1740).

D. A. Winstanley, *Early Victorian Cambridge* (1940).

T. J. N. Hilken, *Engineering at Cambridge University, 1783–1965*
(1967).

J. G. Crowther, *Founders of British Science* (1960).

Seventh Duke of Devonshire, First Presidential Address. *Transactions of the Iron and Steel Institute*, 1 (1869).

Seventh Duke of Devonshire
Obit., *Journal of the Iron and Steel Institute*, 2 (1892).
Obit., *Proc. Roy. Soc.* 51 (1892).
Obit., *The Times* (22 December 1891).

Cambridge University Syndicate: *Report* (25 November 1868).
Amended Report (3 May 1870).
Professors' Report to Syndicate (3 December 1868).

Cambridge University Senate: *Decision to Establish the Cavendish
Professorship* (9 February 1871).

*Royal Commission on Scientific Instruction and the Advancement of Science:
Reports 1–8*, 3 vols. (1872–5).

A. W. Hofmann, *The Life and Work of Liebig* (1876).

Lewis Campbell and William Garnett, *The Life of James Clerk Maxwell* (1882).

R. T. Glazebrook, *James Clerk Maxwell and Modern Physics* (1896).

R. V. Jones, J. C. Maxwell. *Aberdeen University Review*, 33, nos. 100 and 101 (1949).

James Clerk Maxwell, A Commemoration Volume 1831–1931. Essays by Sir J. J. Thomson, Max Planck, Albert Einstein *et al.* (1931).

A History of the Cavendish Laboratory: 1871–1910 (1910).

W. D. Niven (ed.), *The Scientific Papers of James Clerk Maxwell*, 2 vols. (1890).

The New Physical Laboratory of the University of Cambridge. *Nature* (25 June 1874).

L. Rosenfeld. On the foundation of statistical thermodynamics. *Acta Physica Polonica*, 14 (1955).

A. Schuster, *Biographical Fragments* (1932).

J. J. Thomson, *Recollections and Reflections* (1936).

The Encyclopaedia Britannica: Ninth Edition, articles by J. Clerk Maxwell; reprinted in *Scientific Papers* (1890).

Norbert Wiener, *Cybernetics* (1948).

Cambridge University Reporter: 21 January 1873, June 1873, 24 March 1874, 1 June 1874, 22 March 1877, 26 May 1879.

Rayleigh

Chapters 5–6

Robert John Strutt, Fourth Baron Rayleigh, *John William Strutt, Third Baron Rayleigh* (1924).

A History of the Cavendish Laboratory: 1871–1910 (1910).

Lord Rayleigh, *Scientific Papers*, 6 vols., *1869–1919* (1899–1920).

J. J. Thomson, *Recollections and Reflections* (1936).

Letter from A. Schuster to Rayleigh, 8 January 1880. *Library, Cavendish Laboratory*.

Cambridge University Reporter: 26 January 1880, 13 April 1880, 30 September 1881, 15 October 1881.

J. J. Thomson

Chapters 7–11

Lord Rayleigh, *The Life of Sir J. J. Thomson, O. M.* (1942).

G. P. Thomson, *J. J. Thomson and the Cavendish Laboratory in his Day* (1964).

J. J. Thomson, *Recollections and Reflections* (1936).

J. J. Thomson, On contact electricity of insulators. *Proc. Roy. Soc.* (1874).

J. J. Thomson, *Applications of Dynamics to Physics and Chemistry* (1883).

A History of the Cavendish Laboratory: (1871–1910) (1910)

A. S. Eve, *Rutherford* (1939).

'Cavendish Laboratory Supplement'. *Nature* no. 2981, **118** (18 December 1926).

J. J. Thomson, *Conduction of Electricity through Gases* (1903).

J. J. Thomson, *The Corpuscular Theory of Matter* (1907).

Robert John Strutt, Fourth Baron Rayleigh, *John William Strutt, Third Baron Rayleigh* (1924).

A. Schuster, *The Progress of Physics* (1875–1908) (1932).

P. M. S. Blackett, Charles Thomson Rees Wilson 1869–1959. *Biogr. Mem. Fellows Roy. Soc.*, 6 (November 1960).

J. G. Crowther, *British Scientists of the Twentieth Century* (1952).

J. G. Crowther, *Scientific Types* (1968).

The Post-Prandial Proceedings of the Cavendish Society (5th edn 1920).

Cambridge University Reporter: 17 November 1884, 19 January 1885, 22 January 1889, 18 March 1889, 1 December 1890.

Rutherford

Chapters 12–20

A. S. Eve, *Rutherford* (1939).

N. Feather, *Rutherford* (1940).

M. L. E. Oliphant, *Rutherford: Recollections of the Cambridge Days* (1972).

C. D. Ellis, The Cavendish Chair. *Proc. Roy. Soc.*, **50** (1938).

J. B. Birks (ed.), *Rutherford at Manchester* (1962).

P. M. S. Blackett, Rutherford Memorial Lecture, 1954. *Physical Society Year Book* (1955).

P. M. S. Blackett, Rutherford Memorial Lecture, 1958. *Proc. Roy. Soc. A.*, **251** (1959).

P. M. S. Blackett. The old days of the Cavendish. *Revista di Nuovo Cimento, Serie I*, 1 (1969).

P. M. S. Blackett. The craft of experimental physics. *Cambridge Universities Studies*, ed. H. Wright (1933).

P. M. S. Blackett. Charles Thomson Rees Wilson. *Biogr. Mem. Roy. Soc.* 6 (1960).

P. I. Dee. Rutherford Memorial Lecture, 1965. *Proc. Roy. Soc. A.*, **298** (1967).

P. L. Kapitza. Recollections of Lord Rutherford. *Proc. Roy. Soc.*, **294** (1966).

Rutherford Centenary Celebrations. P. M. S. Blackett, P. P. O'Shea, W. B. Lewis, M. L. E. Oliphant, J. B. Adams, H. Massey, N. F. Mott and N. Feather. *Notes Rec. Roy. Soc.*, **271** (1972).

N. Feather, E. Rutherford. *Royal Society Reception Programme* (28 Oct. 1971).

N. Feather, Rutherford's Cavendish. *New Scientist*, 7, 598–600 (1960).

M. L. E. Oliphant and W. G. Penny, J. D. Cockcroft. *Biogr. Mem. Roy. Soc.*, **14** (1968).

Hommage à Lord Rutherford. *Pamphlet of World Federation of Scientific Workers* (1947).

J. D. Cockcroft. Some personal recollections of low energy nuclear physics. *Atom*, **84** (October 1963).

R. G. Stansfield. The Rutherford Centenary Celebrations, some personal recollections. *Quest*, **21** (Summer 1972).

T. E. Allibone. Rutherford, a century of nuclear energy. *Quest*, **20** (Spring 1972).

Sir John Hall, Trends in nuclear product improvement. *Atom*, **193** (November 1972).

C. E. Wynn-Williams, Electrical methods of counting. *Rep. Prog. Phys.*, 3, 239 (1937).

W. B. Lewis. The multi-electrode valve and its application in scientific instruments. *J. Scient. Instrum.*, **15**, No. 11 (1938).

T. E. Allibone. Sir John Cockcroft, O. M., an appreciation. *Br. J. Radiol.*, **40** (1967).

R. W. Clark, *Sir Edward Appleton, G.B.E., K.C.B., F.R.S.* (1972).

F. W. Aston, *Biogr. Mem. Roy. Soc.*, 5 (1945–8).

G. K. Batchelor and R. M. Davies (eds), *Surveys in Mechanics*, written in commemoration of the 70th birthday of Geoffrey Ingram Taylor (1956).

J. J. Thomson, *Recollections and Reflections* (1936).

G. P. Thomson, *J. J. Thomson and the Cavendish Laboratory in his Day* (1964).

J. G. Crowther, *British Scientists of the 20th Century* (1952).
J. G. Crowther, *Scientific Types* (1968).

Bragg

Chapters 21–25

W. L. Bragg. Physicists after the War, Royal Institution Lecture, 26
March 1942. *Proc. R. I.*, 32 (1944).

W. L. Bragg, *The History of X-ray Analysis* (1943).

Max Perutz. Sir Lawrence Bragg. *Nature*, 233 (1971).

Sir Lawrence Bragg, O.B.E., F.R.S.. The Austin Wing of the Cavendish
Laboratory. *Nature*, 158 (1946).

Sir Lawrence Bragg. Recent advances in the study of the crystalline
state, Presidential Address, Section A, British Association (1948).

Sir Lawrence Bragg. The Cavendish Laboratory. *J. Inst. Metals*, 75,
part 3 (1948).

Sir Lawrence Bragg. Research now in progress in the Cavendish
Laboratory. The Manchester Association of Engineers (9 January
1948).

W. L. Bragg and H. Lipson. A simple method of demonstrating diffraction grating effects. *J. Scient. Instrum.* (1943).

W. L. Bragg. A theory of the strength of metals. *Nature*, 149 (1942).

W. L. Bragg. A new type of X-ray microscope. *Nature*, 143 (1939).

W. L. Bragg. The structure of alloys and the strength of materials.
European Meeting of Joint Metallurgical Societies (1955).

Sir Lawrence Bragg, O.B.E., F.R.S.. Organisation and work of the
Cavendish Laboratory. *Nature*, 161 (1948).

Sir Lawrence Bragg. Half a century of X-ray analysis, The First Nobel
Guest Lecture (1965).

W. L. Bragg. First stages in the X-ray analysis of proteins. *Rep. Prog.
Phys.*, 28 (1965).

M. F. Perutz. The M.R.C. unit for molecular biology. *New Scientist*,
271 (25 January 1962).

J. D. Watson, *The Double Helix* (1968).

Sir Lawrence Bragg. How a secret life was discovered. *The Times* (16
May 1968).

M. F. Perutz. Bragg, protein crystallography and the Cavendish
Laboratory. *Acta crystallogr. A.*, 26 Part II (May 1970).

Sir Lawrence Bragg, O.B.E., F.R.S.. Giant molecules. *Nature*, 164
(1949).

Sir Lawrence Bragg, O.B.E., F.R.S.. The Cavendish Laboratory archives. *Nature*, 169 (1952).

W. L. Bragg. The X-ray analysis of proteins. *Proc. Phys. Soc. B.*, 65 (1962).

A discussion on the structure of proteins. *Proc. Roy. Soc. B.*, 141 (1953).

Sir Lawrence Bragg., O.B.E., F.R.S.. Budgets of the scientific departments of the University of Cambridge. *Nature*, 171 (1953).

W. L. Bragg. Department of Physics, Report for 1952–3. *Cambridge University Reporter*.

Derek J. Price. The equatorie of the planetics. *The Times Lit. Supp.* (29 February 1952).

J. A. Ratcliffe. Engineers, physicists and mathematicians. *Electronics and Power* (April–May 1971).

J. D. Bernal, L. Fankuchen and Max Perutz. An X-ray study of chymotrypsin and haemoglobin. *Nature*, 141 (1938).

D. Shoenberg. The Royal Society Mond Laboratory. *Nature*, 171 (1953).

Sir Lawrence Bragg. Notes of a speech at the farewell dinner at the Cavendish Laboratory, 1953. (Private communication).

M. F. Perutz and J. C. Kendrew, *Nobel Lectures: Chemistry 1942–62*, 647–70 (1964).

F. H. C. Crick, J. D. Watson and M. H. F. Wilkins, *Nobel Lectures: Physiology or Medicine 1942–62*, 749–821 (1964).

W. H. Taylor. The structure of the principal felspars. *Saertryk av Norsk Geologisk Tidsskrift*, 42, 2 (Feldspar vol.) Copenhagen (1962).

Configurations of the Space Group. *Minutes*, Cavendish Laboratory Library.

G. R. Crowe Obituary, *Nature*, 211 (1966).

J. J. Thomson, *Recollections and Reflections* (1936).

Lord Rayleigh, *The Life of Sir J. J. Thomson, O.M.* (1942).

J. G. Crowther, *Scientific Types* (1968).

Mott

Chapters 26–30

Sir Nevill Mott, F.R.S.: Award of Copley Medal; President's Anniversary Address, Royal Society, 30 November 1972. *Proc. Roy. Soc. A.*, 331, 285–286 (1972).

P. W. Anderson, Brian Josephson and macroscopic quantum inter-
    ference. *Proc. 12th Int. Conf. on low temperature physics* (1970).
D. Tabor, F. P. Bowden. *Biogr. Mem. Roy. Soc.*, 15 (1969).
N. F. Mott, *Atomic Structure and the Strength of Materials* (1956).
P. B. Hirsch. Seeing atomic defects in metals. Evening Discourse,
    Royal Institution, 24 November 1961.
G. I. Taylor. The mechanism of plastic deformation in crystals. *Proc.
    Roy. Soc. A.*, 145 (1934).
E. Orowan. Zur Kristallplastizität. *Z. Phys.*, 89 (1934).
M. Polanyi. Über eine Art Gitterstörung, die einen Kristall Plastisch
    machen könnte. *Z. Phys.* 89 (1934).
P. B. Hirsch. The impact of transmission electron microscopy in the
    science of materials. *Electron Microscopy in Materials Science*
    (1971).
Current Research in the Department of Physics (Pamphlet) Cavendish
    Laboratory (1970).
Cambridge University Reporter: 27 January 1954, 27 January 1958–9,
    1 October 1962, 23 November 1966, 26 February 1968, 10
    November 1969, 24 November 1969, 21 January 1970, 27 July
    1970.
J. P. Wilson (ed.), *Search and Research* (1971).
A. Hewish. The discovery of pulsars; from 'Pulsating Stars' a *Nature
    Reprint* (1968).
Martin Ryle. The 5-km radio telescope at Cambridge. *Nature*, 239
    (1972).
Cambridge radio telescope inaugurated. *Atom*, 194 (1972).
Mullard Radio Astronomy Observatory. (Leaflet) Cavendish Labora-
    tory.
S. Mitton. Newest probe of the radio universe. *New Scientist*, 56
    (1972).
Robin Page. Motorway threat to the new telescope. *The Daily Tele-
    graph*, 18 October 1972.
V. E. Cosslett. High voltage microscopy in Britain. *Fourth European
    Conference on Electron Microscopy* (1968).
J. Langrish *et al.*, Cambridge Scientific Instruments: Stereoscan elec-
    tron microscope. *Wealth from Knowledge* (1972).
The Cavendish Laboratory high voltage electron microscope. Press
    Notice, Ministry of Technology, 15 March 1967.
V. E. Cosslett, Outline of the Cavendish Laboratory high voltage
    electron project. Press Notice, *Ministry of Technology*.

Paul Jervis. Innovation on electron-optical instruments, two British case-histories. *University of Sussex Science Policy Unit.*

Otto Frisch. Tracks Galore (paper).

D. J. M. Davies, O. R. Frisch and G. S. B. Street. Sweepnik: A fast semi-automatic track-measuring machine. *Nuclear Instruments and Methods,* **82** (1970).

O. R. Frisch and A. J. Oxley. A semi-automatic analyser for bubble chamber photographs. *Nuclear Instruments and Methods,* **9** (1960).

A. H. Cook. Measurements of gravitation. British Association Annual Meeting, 6 September 1971.

A. H. Cook. The dynamical properties and internal structures of the Earth, the Moon and the Planets. *Proc. Roy. Soc. A.,* **239** (1972).

A. H. Cook. A new basis for measurement. *Physics Education,* **4** (1969).

Pippard

Chapters 31–33

A. B. Pippard. Reconciling physics with reality. Inaugural Lecture, 1971 *Cambridge* (1972).

A. B. Pippard. The electronic structure of metals — the dialogue between experiment and theory: Dannie-Heinemann Prize Lecture. *Jahrbuch der Akademie der Wissenschaften in Göttingen* (1969).

A. B. Pippard. The New Cavendish Laboratory. *The Times Higher Educational Supplement* (April 1973).

'Cavendish List' Cavendish Laboratory. (October 1972).

Report to the General Board of the Committee of the Board on the long-term needs of scientific departments. *Cambridge University Reporter* (8 December 1965).

J. R. Payne, Cavendish Laboratory: Notes on the building. Cavendish Laboratory (October 1972).

Approved Estimates of Expenditure: Physics: 1 August 1972–31 July 1973. Cambridge University Estimates, 1972–3.

Consolidated Accounts 1966–7; prepared for the U.G.C., 31 July 1967.

Cambridge University Reporter: 5 October 1880, 30 September 1881, 5 November 1937, 31 July 1946.

A. B. Pippard. Long-term planning for science in Cambridge. (paper) 9 April 1964.

A. D. I. Nicol. Memorandum on Space. (paper) January 1964.

A. B. Pippard. General notes on the proposed New Cavendish Laboratory. (paper) 12 July 1965.

Early account books of the Cavendish Laboratory: 1885, 1895, 1904, 1915, 1926, 1932.

University of Cambridge West Cambridge Site: Report on Development Plan: Robert Matthew, Johnson-Marshall and Partners, September 1966.

Professor Sir Nevill Mott. Notes on assessing the cost of a scientific department. (paper) June 1966.

'Speech by Professor A. B. Pippard in the Senate Discussion on the re-development of the New Museum Site, 4 December 1962' *Cambridge University Reporter*.

N. F. Mott. Departments of Physics, revised statement of Needs. (paper) 26 January 1965; revised 29 June 1965.

General

A. S. Eddington, *The Cavendish Laboratory* (1935).

Alexander Wood, *The Cavendish Laboratory* (1946).

Egon Larsen, *The Cavendish Laboratory* (1962).

G. P. Thomson, *J. J. Thomson and the Cavendish Laboratory* (1964).

A History of the Cavendish Laboratory: 1871–1910 (1910).

J. J. Thomson, *Recollections and Reflections* (1936).

M. L. E. Oliphant, *Rutherford: Recollections of the Cambridge Days*.

Romualdas Sviedrys, 'The rise of physical science at Victorian Cambridge'. *Historical Studies in the Physical Sciences*, Second Annual Volume (1970).

J. G. Crowther, *Fifty Years with Science* (1970).

The Cavendish Laboratory. Supplement, *Nature*, 118 (1926).

Museum of the Cavendish Laboratory: An Outline Guide to Exhibits. (pamphlet) (1954).

R. V. Jones. The Evolution of the Laboratory. Opening Address: Joint Meeting of the Institute of Physics and the R.I.B.A.

R. V. Jones. The Evolution of the Laboratory. *Science Journal* (January 1967).

Index

458

459

463